HURON COUNTY LIBRARY

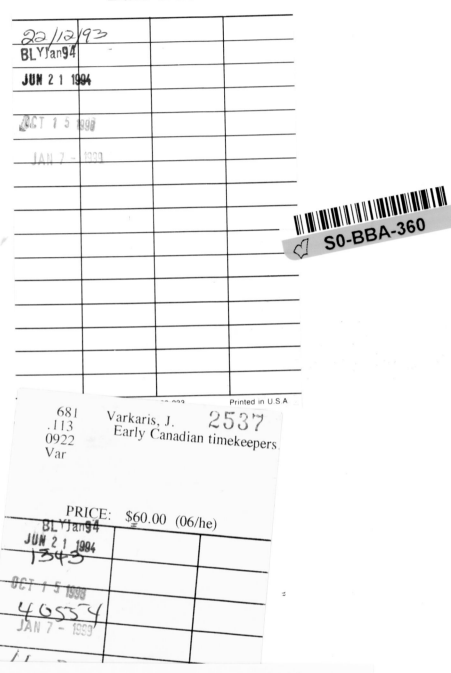

EARLY CANADIAN
TIMEKEEPERS

*Sir Sanford Fleming's watch has two dials for "Local" and "Cosmic" time,
dividing the world into twenty-four standardized time zones. The "Local" time
dial is shown here and the "Cosmic" time dial is pictured on page 182.
– Courtesy Smithsonian Institution, Washington, D.C.*

This fine brass dial bears the name of Jordan Post, first clockmaker of Upper Canada.

EARLY CANADIAN
TIMEKEEPERS

JANE VARKARIS & JAMES E. CONNELL

A BOSTON MILLS PRESS BOOK

Stoddart

Canadian Cataloguing in Publication Data

Varkaris, Jane, 1928-

EARLY CANADIAN TIMEKEEPERS

Includes bibliographical references and index.
ISBN 1-55046-073-0

1. Clock and watch makers – Canada.
I. Connell, James E.   II. Title.

TS543.C3V37   1993   681.1'13'0922
C93-093613-2

Design and Typography by Daniel Crack,
Kinetics Design & Illustration
Printed in Canada

First published in 1993 by
Stoddart Publishing Co. Limited
34 Lesmill Road
Toronto, Canada
M3B 2T6
(416) 445-3333

A BOSTON MILLS PRESS BOOK
The Boston Mills Press
132 Main Street
Erin, Ontario
N0B 1T0

The publisher gratefully acknowledges the
support of the Canada Council, Ontario
Ministry of Culture and Communications,
Ontario Arts Council and Ontario
Publishing Centre in the development of
writing and publishing in Canada.

THIS PAGE

Clock sold by Moses Barrett.
– Courtesy the Nova Scotia Museum, Halifax

PRECEDING PAGE

Dial detail of a tall case clock sold by
W. & G. Hutchinson, Saint John,
New Brunswick. (See page 40)
– Courtesy Barbara and Henry Dobson

TABLE OF CONTENTS

FOREWORD

*E*arly Canadian Timekeepers is the fourth and perhaps last volume of our series on the making and selling of clocks and watches in Canada. Jane and Costas Varkaris first ventured into the field with a little book entitled *Nathan Fellowes Dupuis, Professor and Clockmaker of Queen's University and his Family*. Shortly afterwards, they followed up with *The Pequegnat Story, the Family and the Clocks* which detailed the life and times of Arthur Pequegnat, Canada's most successful clockmaker. The third book, *The Canada and Hamilton Clock Companies,* was co-authored by Jane Varkaris and James E. Connell. It provided, for the first time, a study of these three little-known manufacturers. In this fourth book, Jane and Jim have compiled all the information they could find on a host of other makers and peddlers who have left their names on timepieces in this country.

These books have required a good deal of time and effort from each of us. For Jane, the driving force from the beginning, they represent the completion of twenty years of work. Research, of course, is never complete, and no doubt there is more information to be uncovered. To this end, we offer best wishes, sympathy, and support to anyone who may wish to take up the challenge. In the meantime, our task is finished, and we are happy to sit back and relax.

JANE VARKARIS   COSTAS VARKARIS   JIM CONNELL

This Hoadley "upside-down" wooden movement was used in a number of Moses Barrett clocks.
(See page 15)

ACKNOWLEDGMENTS

THIS BOOK touches upon the lives of more than 125 individuals who sold watches and clocks in Eastern Canada. A century or more has passed since most were active. Tracking down their products and activities has brought the authors into contact with an amazing number of descendants, collectors, dealers, museum staff, repairmen and others. Without exception these people have generously and enthusiastically contributed to the content and illustrations of the book. The authors are deeply grateful to all.

Special thanks must also be expressed to long-suffering family members and spouses who have aided, abetted, and on occasion criticized. Foremost among these are Marilyn Connell and Costas Varkaris.

Our list is long and if anyone has been omitted, we can only apologize.

W. Grant Allen, Stephen Alles, Adolph Amend Jr., Gordon W.D. Armstrong, Chris H. Bailey, Nellie Bell, Ray Benben, Clay Benson, Yves Bergeron, Peter Betts, Carl Beynon, Helen Boyce, Chris Brown, G.E. Burrows, W.H. Burton, Jack Busch, Stephen Chase, James Christian, Marjorie Clinkunbroomer, Donald Colquehoun, Robert Connell, Don Cook, David E. Cooper, John Crosby, G.C. Crowe, W.A. Cumming, William Daniels, Marvin De Boy, Henry & Barbara Dobson, Lilias Doucet, Julien Dubuc, Mike Dugan, Bernard Edwards, Robert Edwards, Donald E. Elgee, Kirk Fallin, Lucile Finsten, John R. Fletcher, Fred Foster, Kenneth Fram, Ward Francillon, Eugene Fuller, Wallace Furlong, Maurice Gagnon, Robert Galbraith, Terry Gallamore, Gord Gibbins, James W. Gibbs, Charles Gomberg, Valerie Gould, Fred Gottselig, William M. Graham, Douglas M. Grant, Peter Griffiths, Fred H. Haeflinger, W.J. Haggerty, Jim Halls, Irene & George Hartwick, John Haze, Alma Hess, John Higham, Al Hingley, Bryan Hollibone, Phyllis Horton, Martin Howard, Elizabeth Ingolfsrud, Paul Johnston, Elwood Jones, Joe Katra, Harry J. Kelsen, E. Forrest Keyes, Richard A. Kidd, Allen L. King, James Krause, Serge L'Archer, Fran La Flamme, Yvon La Flèche, Kathryn Laver, Paul Lavoie, M. Le Claire, Thomas J. Lees, S. Norman Lemieux, Herve Lenclud, Ira Leonard, Bess Lighthall, Fred Linker, Mildred Livingston, Glenn Lockwood, Stanley E. Lovell, Rebecca W. MacGregor, Kathryn MacIntosh, K.O. Macleod, Alan S. Marx, W.H. Maslin, Glenn E. Mastalio, Don Mathis, Michael Mazur, Grant McArthur, Elizabeth McArton, Hope McHale, John W. McMaster, Harold McNutt, Brenda Merriman, Catherine Milinkovich, Phillip P. Miller, Janet Muncy, Arnold Mooney, Charles Murray, George Neville, Rosemary Newman, Wm. Nixon, Sylvio Normande, Lela Olmsted, O. Osborne, Earl Pascoe, Mark & Susan Peer, Daniel Pelletier, Archie Pennie, Wayne Petti, Robert Phillip, Grady Pinner, J. Plewes, Audrey Poetschke, Harry Porter, Jonathan Pottel, Ralph & Patricia Price, R.M. Riguse, Fred Ringer, John Ruhland, John L. Richardson, Mary Rimmer, Kenneth Roberts, Scott Robson, Rebecca Rogers, Charles & Betty Rose, W.T.D. Ross, K. Rydall, Thomas Sandlands, Robert Sargent, William Savage, Lois Schroeder, Wm. G. Schwab, Craig H. Sebald, P. Seitl, Roy Sennett, Bruce Shoemaker, J.A. Sigurdson, V.S. Silmser, J.W. Sinclair, Jacob Smith, E.P. Spahr, Tony Sposato, Howard Stone, Allan Symons, Richard Taylor, Snowden Taylor, Frank Thornton, Mr. & Mrs. D. Thomas, Louise Throop, Loraine Tompson, William R. Topham, Nick Turnbull, Val Earle, Fred Varkaris, Jean Vaugan, George H. Werden, Margaret Welliver, Gail E. Wellner, James West, Robert White, Elda Williams, James Williams, Frank Willows, Brian Winter, Hugh Witham, Richard Withington, Aaron Wohl, Richard Wolton, Dixie Wood, Ronald Wright, William Yeager, Ken W. Yokom, Elmer Young, Thomas Zaffis.

We also acknowledge the financial assistance of the Canada Council and the Ontario Arts Council in researching the material for this book.

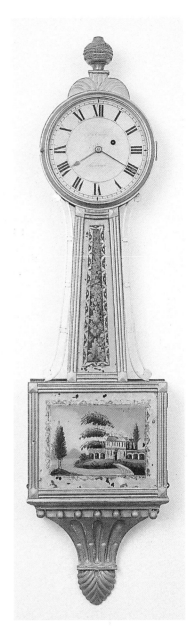

This fine wall clock by Martin Cheney of Montreal is constructed in the "banjo" style.
– Courtesy Stan Kirschner

CHAPTER 1
TIME MEASUREMENT IN CANADA

THIS VOLUME has been written with the objective of expanding the amount of published information on early horologers and their clocks and watches in Canada, from the earliest days to 1900. Many readers will already be familiar with a previous reference, *Canadian Clocks and Clockmakers* by G. Edmond Burrows. In his introductory remarks in that pioneer volume, Mr. Burrows expressed the hope that his book would inspire further research. The authors have taken him at his word and offer the present book to supplement existing knowledge, as well as to provide much new material. Since 1973, the date of publication of the Burrows book, a great deal of additional information has come to light. Canadian history has remained a popular subject and the authors are pleased to have discovered much that is new and relevant to this area of research. The clock manufacturing industry consisted of only a few companies, namely the Canada Clock Companies, the Hamilton Clock Company, and the Arthur Pequegnat Clock Company, and these have already been thoroughly documented in other publications. The thrust of this book is to report on the activities of that rather nebulous group of individuals who made or sold clocks and watches in earlier times.

SCOPE OF THE BOOK

Clockmaking as an industry can scarcely be described as having been a major factor in Canadian history. Compared with horological development in Britain, Europe, and the United States, the Canadian story is short and fragmented. Canada underwent a slow and difficult beginning with a small population, little wealth, and many physical difficulties. The market for luxuries, such as clocks, was always small.

Clocks were in daily use, however, from the earliest times, and were bought and sold throughout the years. In the preparation of this book, the authors soon realized that it would be necessary to define the scope of the work. Four categories of activity were identified.

In the first category are the many persons or organizations who simply purchased clocks and watches for direct resale in Canada.

The second category is clock importers, including many peddlers, who added identifying labels to their wares, suggesting that the seller was the principal or exclusive vendor of these marked items. Many American clocks, particularly those made by the Seth Thomas Company, were sold with these customized labels, which were printed for peddlers and agents. In addition, many fine English clocks were sold by early Montreal and Quebec City vendors. These clocks, both tall case and bracket, were characteristically British, but bore no sign or trademark indicating origin. However, they were marked with the seller's name.

The third category comprises a more innovative group of people, who imported movements and other clock parts, which were installed in cases of local manufacture. Movements of British, American, and occasionally Black Forest origin were used. There was a good economic incentive for this practice. Wooden cases were awkward and expensive to ship over long distances, especially when there was plenty of wood in Canada, as well as a skilled and willing body of craftsmen to make use of it. Many locally made cases were of the highest quality.

The fourth category comprises the small number of men who made complete clocks, "from scratch," entirely in Canada. These men often worked alone and encountered many difficulties, but their output was the most distinctive and intriguing of all the groups.

The authors established three criteria that had to be met before any individual could be included in this book:

(a) there must be "Canadian content"

(b) there must be a surviving clock or watch

(c) the surviving artifact must have been made before 1900.

Readers familiar with the rules imposed by the Canadian Radio-Television and Telecommunication Commission may wince at the mention of "Canadian content". However, if the reader will again refer to the four categories of commerce outlined above, he or she will see that the material of this book focuses on the second, third, and fourth categories. Little space is devoted to the first category, which deals with retailers who simply bought and sold clocks and watches. This effectively eliminates a large group of retail jewellers whose names appear in many city directories, and who may or may not have added their mark to commercial clocks and watches that moved through their shops. An extensive list of these retailers has been prepared by the late John E. Langdon in his book, *Clock and Watchmakers in Canada 1700–1900.* The location and approximate dates of operation for many retailers not documented in this volume can be found there. To be eligible for inclusion in this volume, a watch or clock must exist with a distinctive Canadian mark or label and often with additional local input.

The decision to include or exclude a vendor has at times led to lengthy discussions between the authors. Some compromise has been necessary in the case of certain individuals. As a result, the authors have, on occasion, stretched their own rules in order to include rather than exclude.

A few exceptional clocks have been featured in separate articles. These artifacts, such as the St. Sulpice Seminary clock, are of unique and historic importance.

Little has been said up to this point about watches. These were not manufactured commercially in Canada during the time frame of this book. Educational institutions, however, did exist to teach watchmaking, and watch cases were made commercially in Canada, so information has been included on these institutions and manufacturers.

EVOLUTIONARY FACTORS

To appreciate the developments in horology during this period of our history, some knowledge of the early settlements and nomenclature of the various regions in Canada at that time is necessary. Consequently, in the following paragraphs, the authors have summarized the factors relevant to the topic of this book.

The earliest major settlement was established by the French. In the *Revue d'ethnologie du Québec / 9,* a section on clocks written by Yves Bergeron reveals that, during the period of the French Regime, clocks were rare. It is known that Samuel de Champlain possessed a clock and there was a clock in 1694 in the Seminary of Quebec – the oldest building in New France that is still standing. It is probable that Montreal was the first Canadian city to have a public clock toward the end of the 17th century. However, it was expensive to import clocks into the country, as a heavy duty was levied on them. A few wealthy people may have had clocks in their homes, brought with them from their homeland. In the meantime, the majority had no access to clocks. They depended on the sundial, the hourglass, the town crier, and the angelus to regulate their days. Although the number of clocks in New France was never large, a few examples have survived. We have records of several clockmakers, but details of their activities are scanty. The existing clocks are typical of French technology of the time.

The fall of Quebec to the British in 1759 brought many changes. The provinces of Nova Scotia and New Brunswick were also opened for settlement, and there was a major influx of population from the British Isles, bringing clockmakers schooled in British techniques. For the next few decades, almost all clocks sold in Canada had British origins.

Because of the rapidly expanding population in Quebec and the land north of

Lake Erie, Lake Ontario, and the St. Lawrence River, most of which had been Indian territory, it became necessary to divide the area into districts, both for parliamentary representation and for military purposes. In 1788 the Governor-in-Chief, Lord Dorchester, created five new districts, Gaspé in Quebec and Luneburg, Mecklenburg, Nassau, and Hesse in what is now southern Ontario. The last four mentioned districts were renamed Eastern, Midland, Home, and Western in 1792. Thus began a complex series of divisions and changes of name until, by 1849, districts were completely replaced by counties. Whereas a detailed history of these changes is beyond the scope of this book[1] a review of the names and dates used in Canada in the 18th and 19th centuries may help the reader in understanding the terminology used when referring to clocks from different periods. From 1791 to 1841, the present Province of Quebec was referred to as the Province of Lower Canada. Present-day Ontario was called the Province of Upper Canada. In 1841 Upper Canada and Lower Canada were united into the Province of Canada. Lower Canada became Canada East and Upper Canada became Canada West. In 1867 Confederation took place and the Dominion of Canada was formed. At this time Canada East and Canada West became Quebec and Ontario respectively, and Nova Scotia and New Brunswick joined Ontario and Quebec to form the new nation, Canada.

Events in the United States produced the final significant influence on Canadian horology. After the American Revolution of 1776, several major waves of immigration to Canada took place from south of the border made up of Loyalists and, later, American Federalists. Included in these groups were a few clockmakers.

Then came the rather sudden development of a strong and innovative clockmaking industry in Connecticut. The activities of Eli Terry and others disrupted and eventually made obsolete all the old established traditional clockmakers. From that time on, clocks ceased to be a luxury and were sold in unprecedented numbers. American clocks became the principal clocks sold in Canada. As the market grew, a small Canadian clock manufacturing industry emerged, drawing its inspiration from Connecticut. In this book examples of clocks showing all these influences can be found.

CANADIANA CLOCKS

The term "Canadiana" clocks, used occasionally in this publication, is a rather loose designation and is probably not recognized by serious museologists. However, it is commonly used in the antique trade and by rank-and-file collectors. The clocks in question are simple, mass-produced clocks of American origin, fitted with customized or over-pasted labels stating the Canadian vendor's name and location. Labels were often pre-printed in the United States, but Canadian printings are known. The practice prevailed from approximately 1825 to 1875. The clocks are found with both wood and brass movements.

The specific wording on the labels varied greatly. As mentioned above, the Seth Thomas Company was particularly active in this field, supplying customized clocks for perhaps a dozen Canadian agents. The labels were quite precise, usually stating that the clocks had been made for the agent by Seth Thomas. In other instances – Moses Barrett is a good example – the Canadian agent claimed to have both made and sold the clock, although it was obviously the product of an American factory.

Other American makers were also active in supplying clocks for the Canadian trade. Chauncey Boardman supplied R.B. Field of Brockville and agents in the Maritimes. Peddlers such as Moses Barrett and the Young brothers in the Maritimes sold clocks that came from a variety of makers. One peddler, Hiram Hunt, sold clocks with American labels in Maine and similar items with Canadian labels in New Brunswick.

One of the consequences of the rapid increase in production in Connecticut after 1820 was a resulting increase in the number of clock peddlers. The selling environment became highly competitive. Canadian settlements, while smaller in size, offered a nearby market where competition was less intense, particularly since there was little local clockmaking activity. As a result, many American peddlers roamed Canada with clocks. Some clocks were imported complete and ready to run; in other instances, movements were brought in and cased locally. A number of clock sellers described in this book fall into this category, including the Twiss brothers, Horace and Charles Burr, R.B. Field, Riley and A.S. Whiting, Moses Barrett, Butler and Henderson, and Hiram Hunt. A number of other persons were active for short periods and left no permanent record behind them. The authors suspect that many of these people, such as H. Utley, Porter Kimball, R.W. and J.M. Patterson, George Steel, and the Young brothers, may have worked in Canada for longer or shorter periods and then returned to the United States. The clocks sold by many of these people fit the "Canadiana" designation.

In addition to the Canadiana clocks mentioned above, another group of clocks is often included under the term "Canadiana." These were imported from a variety of countries for the Canadian trade and often bear on the dial the name of the vendor, who was often a watch- and clockmaker. The Canadians who marked the clocks in this way usually had fair-sized businesses and therefore had some impact as horologers on their community. By no means all of the persons whose names are on clocks and watches of this group are mentioned in this book. The persons chosen sold many clocks and had an impact on Canadian horology at a local and often national level.

Canadiana clocks cannot be defined by any specific set of criteria, but since they exist in considerable numbers, they must be recognized as a loose-knit category.

MOVEMENT IDENTIFICATION

Wherever possible in this volume, the authors have attempted to identify the origin of imported movements. This task has been made easier by three important articles written by Dr. Snowden Taylor, research chairman of the National Association of Watch and Clock Collectors, and published in *The Bulletin* of the NAWCC as follows:

1. "Characteristics of Standard Terry type 30 Hour Wooden Movements as a Guide to Identification of Movement Makers."
   The Bulletin, Whole No. 208, Vol.XXII, No.5 Part 1 (October 1980).

2. "The Noble Jerome Patent 30 Hour Brass Weight Movement and Related Movements."
   The Bulletin, Whole No.221, Vol. XXIV, No.6 Part II (December 1982) and updates.

3. "Eight-Day Brass Weight Movements of the Noble Jerome Patent Era."
   The Bulletin, Whole No. 270, Vol. 33 No.1 (February 1991).

The reader will find frequent reference has been made to these articles, but for the sake of simplicity the full titles have not been repeatedly given.

FIG. 1

FIG. 1 The "Quebec" style is distinctive and readily recognizable.

In countries where clocks were produced in large numbers over a period of centuries, well-known national styles have evolved. It is more difficult to identify styles that might be unique to Canada. For the most part, movements were imported from other countries. Cases were frequently made in the Canadian provinces, but most craftsmen simply copied styles that they believed to be popular in Britain, Europe, and the United States. However, after working with the mass of photographs assembled in order to write this book, the authors feel that some unique style characteristics can be noted.

Most of these examples can be traced back to the province of Quebec. This is not surprising, since Quebec represented the largest block of population and one that predated the settlement of the other districts. In addition, the cultural forces were more diverse in Quebec, with its early French origins and strong subsequent influences from Britain and the United States. The unique nature of other Quebec domestic furniture is well recognized.

The development of a distinctive style of Quebec tall case clocks was first suggested in an article published several years ago by one of the authors in the NAWCC *Bulletin*.[2] This article discussed ten examples of the distinctive tapered-waist style, which was associated with cases produced only in the province of Quebec. In the intervening years, many other examples have been found, some of which are illustrated in this book. It is still fair to state that the style is uniquely Quebec, for while the clocks have migrated to many other areas, there are no examples known to the authors whose origins can be documented to any other source. Readers can see examples of this style in the sections concerning William Baxter, Thomas Cathro, Thomas Drysdale, William McMaster and others. Fig. 1 illustrates an unlabelled specimen with Quebec provenance.

Some uniqueness of design can also be seen in the clocks of Martin Cheney, sold in Montreal between 1809 and the 1830s. Cheney, of course, was American-born and his clocks generally follow popular American trends. However, many of his banjo and wall clocks are distinctive in appearance, with no American counterparts. The examples shown in Figs. 130 to 133 illustrate both unusual design and a family resemblance, suggesting that they came from the same shop. The basic "keyhole" shape is unique to Martin Cheney and several clocks of each style are known.

12

Cheney sold a number of other wall clocks. See Fig. 118. These clocks seem to have few, if any, exact counterparts south of the border. However, similar clocks were sold by another Montrealer, James Dwight (see Fig. 139), suggesting again that there was a common local source.

Some of the clocks sold by the Montreal firm of Savage and Son are also unique in design. The reader's attention is directed to the clocks illustrated in Figs. 187 to 191. Each of these clocks departs significantly from English tastes of the period. The movements in these clocks are typically English, but the details of the cases suggest a distinctive approach by local Montreal craftsmen.

The authors have had little success to date in identifying the names of specific cabinetmakers who made clock cases in Montreal. One exception is discussed in the section on Martin Cheney. Another exception is the clock shown in Fig. 135. This clock also has retained an original label stating that the case was made by J. Bte. Sancer. A number of other very similar, high-quality cases are known, probably also the work of Sancer. Another clock, illustrated in Fig. 102, bearing the name of William Baxter, Quebec, also contains a hand-written inscription, "Joseph Marcoux, 9 Septembre 1822 Quebec..." with a street address. Marcoux's name appears in two areas considered difficult to reach after the clock was made. This suggests that the signatures were made before the case was assembled. As there were a number of men of that name who were joiners and cabinetmakers during that period in Quebec City, Marcoux could well have been the maker. A third documented case-maker was Michel Rousseau, a cabinetmaker of the village of Lotbinière, one of whose clocks is pictured in this book.

A few unusual case styles can also be found in the output of three New Brunswick makers who are documented elsewhere in this book. They were William Plummer, Whitcomb Fairbanks, and James Melick. These men all sold clocks in Saint John. Interestingly enough, they were all related by marriage and were in business at approximately the same time. Almost a century and a half has passed since they were active. At this point, one can only speculate whether they used the services of a single cabinet shop or created these cases independently in a spirit of friendly competition.

Finally, some recognition must be given to the influence of Pennsylvania clockmakers on tall case styles of Upper Canada. A number of settlers of Germanic or Mennonite origin left Pennsylvania to settle in the Niagara Peninsula, Waterloo County, and York County. Their clocks display characteristics reminiscent of Pennsylvania styles of the early 19th century. In the section on Samuel Moyer, a fine example of such clocks is illustrated (Fig. 330) and the names of a few other makers are recorded. Other examples can be seen in the well-known reference *The Heritage of Upper Canadian Furniture* by Howard Pain (Figs. 754 and 1180). These clocks are not sufficiently numerous or unique to constitute a distinctive Canadian style. They do, however, illustrate how an outside influence can act as a stimulus. Unfortunately, in this instance the manufacture of hand-crafted clocks ceased before any really unique local style could emerge.

FIG. 2a

FIG. 2b

FIG. 3

CHAPTER 2

MARITIME PROVINCES AND NEWFOUNDLAND

JAMES AGNEW

James Agnew (d. 1850) was a Saint John, New Brunswick, clockmaker who ran his business from about 1834 to 1850. The Agnew clocks known to the authors are tall case clocks with serpentine, broken-arch cornices, brass finials and painted dials. Two Agnew clocks were reported in the New Brunswick Museum. One of the clocks, pictured in *Canadian Clocks and Clockmakers* by G. Edmond Burrows, was kept running. It carries the date of 1840. The other clock is pictured in this section.

The movements of these clocks have false plates and are probably of British manufacture.

Agnew's origin and the exact date of his arrival in New Brunswick are unknown. However, his name appears on an 1834 list of "Freemen of Saint John," where his occupation is given as a watchmaker. In July 1835 the *Courier* carried an advertisement for Agnew's watchmaker's and jeweller's shop located on Dock Street.

The great fire of 1837 destroyed Agnew's establishment, along with many others on the east side of Dock Street, forcing Agnew to relocate his shop. In the 28 May 1842 issue of the *Courier*, he advertised that his "watch, clock, silverware[1] and jewellery store had moved to the house previously occupied by Mr. Hendricks" and listed many items sold in his shop. Also, the land records show that Agnew purchased land in the Saint John area. By 1845 the business at the corner of King and Cross streets was doing so well that an advertisement appeared for a lad of fifteen or sixteen years of age to apprentice in the watch- and clockmaking business. The boy was to have "respectable references as to moral character, with a lad from the country preferred."

Advertisements for the business continued to appear in the newspapers until December 1850. Then a notice placed in the 7 December 1850 *Courier* reported that James Agnew, watch- and clockmaker, had been missing since 27 November. It was reported that Agnew had arrived at Eastport, Maine, aboard the *Admiral* on his return from New York. "At about 11 p.m. on a very dark night, he was preparing to board the *Maid of Erin* for completion of his journey to Saint John, but did not board the ship. It is believed that he fell from the wharf and was drowned."

By 24 December 1850, James's wife, Eliza, had settled her husband's estate and prepared to sell his land. She left for the United States and died on 10 January 1867 in Ronbury, U.S.A. Her will was probated in Saint John. The sum of $300 went to James and Eliza's daughter Letitia and was to "be free of interference and control of Letitia's husband." The remainder of the estate was divided among the six children: Letitia (b. 14 Sept. 1826 – d. 6 Feb. 1885), William B., Moore F. (b. 1835 – d. 30 Apr. 1894), James W. (b. 1838 – drowned 12 May 1867), Harriet, and Seymore. James, Eliza, and Letitia are buried at Fonthill Cemetery, Saint John, New Brunswick.

FIG. 2 *Clock sold by James Agnew.*
 – *Courtesy the New Brunswick Museum, Saint John*
FIG. 3 *Movement of a James Agnew clock.*
 – *Courtesy the New Brunswick Museum, Saint John*

MOSES BARRETT

Moses Barrett (b. between 1800 and 1810 – d. Dec. 1864) came possibly from New England to the Maritimes. He apparently spent a short time in Westmorland, New Brunswick, before settling in Yarmouth, Nova Scotia, and later in Amherst. His professions were those of "clockmaker" and "trader." In his clockmaking business, he often purchased wooden movements from Connecticut clockmaker Silas Hoadley and others, and it is said that he cased them during the winter months in his workshop. He also purchased entire clocks from Connecticut makers and added his own labels, which read, "manufactured at," followed by his location at the time, and a last line: "by Moses Barrett."

The only clock known to have a "Westmoreland" [*sic*] label was purchased from Silas Hoadley; the movement has been identified by Dr. Snowden Taylor as type 4.111 (now 4.1).[2] The case, with its glass insert at the top, was also identified as probably that of Silas Hoadley.

The label of the Westmoreland clock indicated, by the initials N.S. following that place name, that it had come from Nova Scotia. A search by the staff of the Nova Scotia Museum in Halifax found no mention of such a place in Nova Scotia. Scott Robson of the museum staff believes that the label has typographical errors in the name. The label should read, "Westmorland, N. B.," the county which "abuts the Nova Scotia border."

Many of the Barrett clock cases are ornately carved and are of pillar-and-scroll or column-and-splat style, with or without feet. The design found most frequently on the tablet depicts a scene with large mansions and trees. The clock illustrated in Fig. 4 has a pillar-and-splat case. The glass in the lower portion of the case is a recent replacement. The original finials are missing and have been replaced with finials copied, unfortunately, not from finials of a clock but from those of an antique bed.

Barrett purchased movements from several makers. One type of movement, made by Atkins and Downs, was designated 8.133 by Dr. Snowden Taylor. Clocks with this movement were sold while Barrett was living in Yarmouth. As clocks using this movement were made in 1831 and 1832, it follows that Barrett lived in Yarmouth during those years. Some of the clocks sold in the Amherst area have movements supplied by Silas Hoadley. These movements are the Hoadley "upside-down" movements. In this unorthodox movement "the great wheels start at the bottom of the plates, the escape wheel and anchor are at the bottom, and the winding arbors are above the centre arbor; also, the striking hammer is at the top."[3] This movement was probably devised to circumvent the Terry patents.

As most movements found in Barrett clocks are made of wood, it is questionable whether Barrett's clock-selling period extended much beyond 1840 when Connecticut makers began manufacturing more and more movements of brass.

Another feature of Barrett clocks that shows association with Hoadley is that some of the clock labels state that the clocks are "Franklin Clocks" and have pivots bushed with ivory, another Hoadley innovation.

Although most Barrett clocks utilize wooden movements, one interesting exception is illustrated in Figs. 7, 8, and 9. This clock is fitted with an early Joseph Ives–type 8-day strap-brass movement. A similar move-

FIG. 4

FIG. 5

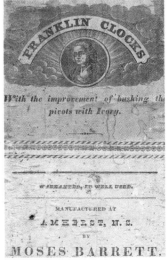

FIG. 6

FIG. 4 *Clock sold by Moses Barrett. – Courtesy the Nova Scotia Museum, Halifax*
FIG. 5 *Hoadley "upside-down" movement, used in a number of Barrett clocks.*
FIG. 6 *Label of the Barrett clock in Fig. 4.*

FIG. 7

FIG. 8

FIG. 7 *Fine 8-day triple decker clock by Joseph Ives & Birge & Ives and sold by Moses Barrett.*
FIG. 8 *Typical Joseph Ives weight-driven brass movement from clock in Fig. 7.*

ment can be seen in Fig. 43 of *The Contributions of Joseph Ives to Connecticut Clock Technology 1810–1862* (1988 edition) by Kenneth D. Roberts. The clock resembles triple-deckers sold by several Connecticut manufacturers.

In correspondence with the authors, Mr. Roberts has expressed the opinion that the Barrett clock was probably made by the firm of "Birge and Ives" or "John Birge." Both firms were short-lived, but can be dated to the 1831–1832 period. John Birge and Joseph Ives were connected with several enterprises during these years. Mr. Roberts has further commented that at this time Joseph Ives "became a roving consultant and made movements for himself and supervised those made by others."

These distinctive brass movements were in direct competition with the 8-day wooden movements of the period. They are contemporary with other clocks sold by Barrett and illustrate again that he sold the products of several makers. Clocks made by Joseph Ives and the various Birge firms are rare in Canada. The only other examples, from a somewhat later period, are illustrated in the sections describing the activities of R.W. Patterson and J.M. Patterson.

The fact that Moses Barrett met with great success on selling expeditions cannot be disputed. In a survey of clocks in Canada conducted in 1978, there were more clocks reported in private collections and museums with the name Moses Barrett on the labels than any other individual's name. Barrett clocks are pictured in *Canadian Clocks and Clockmakers* by G. Edmond Burrows and in *Antique Furniture by Nova Scotia Craftsmen* by G. MacLaren, and clocks are on display in museums under the jurisdiction of the Province of Nova Scotia.

Barrett's success must be attributed to the fact that he was a gregarious man who enjoyed "small talk." These characteristics were assets when he was travelling throughout the province peddling his clocks. The settlers welcomed him as a "bearer of news and gossip of the day." According to a report in the Archives of Nova Scotia, written by Harry Piers about 1936, he would "Leave a clock in the home and receive as part payment, his room and lodging." Also, according to Harry Piers, a rumour existed that Barrett was a lay preacher, but this has not been confirmed.

Additional information about Barrett's idiosyncrasies is provided in George MacLaren's book. He reports that Barrett was always accompanied by his cat, which he could not bear to leave behind, and that he preferred an uncut pumpkin pie for a serving and in fact resented anything smaller being offered to him. It has been suggested he might have been the inspiration for the itinerant peddler whom Thomas Haliburton made famous in his book, *The Clockmaker, or the Sayings and Doings of Samuel Slick.* Haliburton himself says in the book that he "first met Sam Slick when the latter was on the Eastern Circuit, somewhere between Colchester County and Fort Lawrence, near Amherst." In the year of publication of Haliburton's book, 1836, Barrett was indeed living at Amherst.

The *Nova Scotia Museum Reports of 1935–1936,* however, consider that Haliburton's Sam Slick is a composite whose character and sayings are built up from Haliburton's acquaintance with many peddlers while he was a circuit judge in Nova Scotia. Two chapters of Haliburton's book describe the clock-selling practices of the day.

It is not known exactly when Moses Barrett moved from Yarmouth to Amherst, Cumberland County. In his book *Silversmiths and Related Craftsmen of the Atlantic Provinces,* Donald MacKay suggested that Moses Barrett went to Amherst in 1830. However, the clocks with Atkins and Downs movements that were sold in Yarmouth must have been made in 1831 or

1832. Many clocks from the Yarmouth period do exist. Four of the six clocks pictured in the book by G.E. Burrows were sold while Barrett was living in Yarmouth.

He was certainly in the Amherst area by 1836, according to the land record documents in the Archives of Nova Scotia. While there, Barrett was involved in a number of land transactions, and it would appear that he lived in several locations in the Amherst area. According to G. MacLaren, Barrett "lived for a time two miles outside Amherst in a house later known as the Robert Barry House in which he had his home, workshop and barn under one roof." His place of residence during approximately the last ten years of his life was, according to the Museum Reports of about 1936, "near Embrees at Amherst Point, a village about four and one-half miles south-west of Amherst, Cumberland County."

FIG. 9

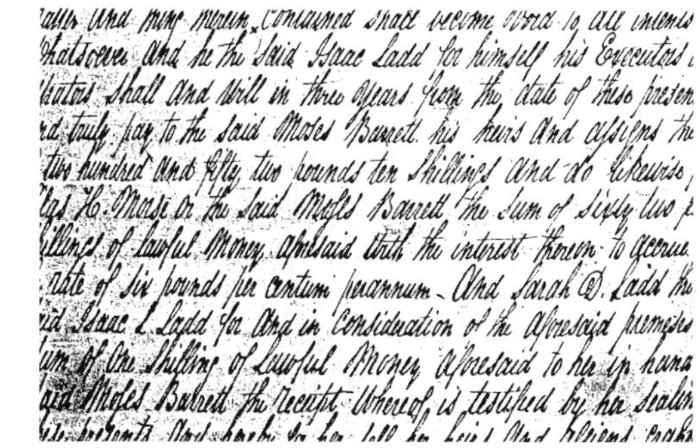

Moses Barrett was also involved in land transactions with Isaac Ladd and his wife, Sarah. The dates of these transactions range from 1835 to 1844 and the land involved was near Pugwash. Barrett continued to live at Amherst during these transactions. Ladd was a cabinetmaker and a partner of Moses Barrett from approximately 1835 to 1840. Therefore, clocks can be found with the names "Barrett and Ladd" on the clock label. Typical Hoadley "upside-down" movements were also used in clocks during the Barrett and Ladd partnership. One of the clocks in the Nova Scotia Museum in Halifax bears a label with the names Barrett and Ladd.

Moses Barrett died in December 1864. His will was probated on 10 January 1865. He was a bachelor and left the bulk of his estate to his brother Liceuda [*sic*].

FIG. 10

FIG. 12b

FIG. 11

FIG. 12a

FIG. 9 *Moses Barrett pasted his own label over the central part of the original label, claiming that he manufactured the clock.*

FIG. 10 *Excerpt of a deed involving Barrett and Ladd.*

FIG. 11 *This weight clock by Silas Hoadley is fitted with his famous upside-down movement and has a Barrett & Ladd, Amherst, label.*

FIG. 12 *A second clock with upside-down movement made by Hoadley. The label shows that the clock was sold during the Barrett-Ladd partnership. Note third winding hole for alarm.*

THOMAS BISBROWN

Thomas Bisbrown was a silversmith and watch- and clockmaker from Edinburgh, Scotland, who worked in North America from about 1784 until the early 1800s. His name was found on the dial of a tiger maple tall case clock, which was resold in 1990.

Three spellings of the name Bisbrown were used by him through the years: Brisbrown, Bisbrown and Bissbrown. However his silver mark, T.B. Brown, would suggest that his surname was Brown. Spelling the name a variety of ways was not unusual two hundred years ago. Also in his many advertisements, the same facts were given over the years and in different places, which would indicate that one man was involved.

Although it is not known when Thomas Bisbrown left Scotland, in November 1784 he was in Quebec City, where he signed a petition to the King. He was living opposite the Bishop's Palace on Mountain Street. He advertised in the *Quebec Gazette* on 15 December 1785 that he made all kinds of gold and silver work and repaired watches and clocks.

By September 1790 Bisbrown was living in Albany, New York, where he advertised his services in the *Albany Gazette*.[4] He guaranteed for a year all repairs on clocks and watches.

In July 1798 Thomas Bisbrown had arrived in Halifax, Nova Scotia, where he announced in the *Nova Scotia Royal Gazette* that he had commenced business on the Lower Side of the Parade in a house formerly occupied by Mr. Gordon, watchmaker. He had for sale "repeating, horizontal and plain watches, jewellery and silverware." He claimed that he had worked in the "First Cities of Britain." Advertisements appeared from August 1797 to July 1800 in the *Weekly Chronicle*.

FIG. 13

On 9 December 1798 a daughter, Anne Frances, was born to Thomas and Margaret Bisbrown. Nothing further is known about the child.

In July 1800 Bisbrown had an auction sale of all his stock of jewellery, watches, silver, and household furniture, as he intended to leave Halifax for Europe. He was still in Halifax in September of 1800 when he placed a notice in the newspaper that he would no longer be responsible for his wife Margaret's debts, as she had "left my house several weeks ago and behaved in a very improper manner to my great damage."

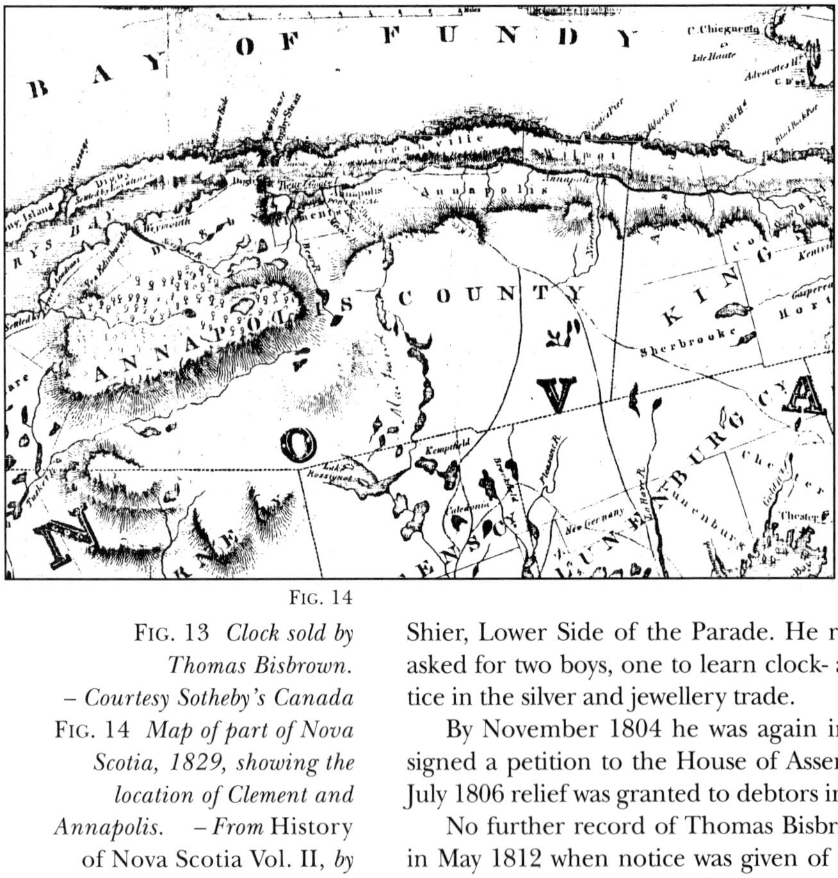

In spite of a Supreme Court ruling against him in March 1801 because of debt, he advertised in the *Nova Scotia Royal Gazette* that he was in business in the shop lately occupied by Mr. Martin

FIG. 14

FIG. 13 *Clock sold by Thomas Bisbrown. – Courtesy Sotheby's Canada*
FIG. 14 *Map of part of Nova Scotia, 1829, showing the location of Clement and Annapolis. – From* History of Nova Scotia Vol. II, *by T.C. Haliburton*

Shier, Lower Side of the Parade. He repaired clocks, watches, and jewellery and asked for two boys, one to learn clock- and watchmaking and one to be an apprentice in the silver and jewellery trade.

By November 1804 he was again in debt and was put in jail. From there he signed a petition to the House of Assembly about the conditions in the prison. In July 1806 relief was granted to debtors in the Halifax jail.

No further record of Thomas Bisbrown was found, but he was apparently alive in May 1812 when notice was given of the death at age forty-four of his wife, Margaret, who was buried with five other women in St. Paul's Cemetery.

ROMAN M. BUTLER AND
BUTLER HENDERSON & CO.

The names of Roman M. Butler, and Butler and Henderson are frequently seen on early shelf clocks with wooden movements that were sold in Nova Scotia around 1830. Roman M. Butler also sold clocks in New Brunswick. According to a clock survey conducted in 1978 by one of the authors, approximately 25 percent of clocks sold by Butler carry the name Roman M. Butler alone. Considerable information (which the authors will enlarge upon later in this section) exists on the firm of Butler and Henderson. Three Butlers and two Hendersons were associated with the firm rather than Roman Butler alone.

Although no records exist indicating their place of residence in the Maritime provinces, the clock labels show three locations as places of business, namely Clement, Nova Scotia; Annapolis, Nova Scotia; and Saint John, New Brunswick. The Annapolis clocks frequently display the names "Butler & Henderson" on the labels. In Clement, Nova Scotia, the name "Butler Henderson & Co." is found. Roman M. Butler alone sold clocks in Annapolis, Nova Scotia, and in Saint John, New Brunswick.

Historians at the Nova Scotia Museum believe that Butler and Henderson's place of business was along the Moose River. The choice of this region is puzzling, as from the mid-1820s to the early 1830s there was no industry in the area, with the exception of an iron foundry. One of their reasons for locating in the area might have been that they were obtaining their clock weights from the foundry, thus making the cargo they brought into port lighter.

Two styles of cases were used by Butler and Henderson. The first, a "short" transition style of clock, with and without feet, was used by both Butler Henderson & Co. and Roman Butler. The other style sold in both Clement and Annapolis was the pillar-and-scroll. Typical pillar-and-scroll clocks with labels are pictured in Figs. 17 and 21. The labels from both styles of clocks sold in Clement are identical, as are the labels from both styles of clocks sold in Annapolis.

The movements used by Roman M. Butler and Butler and Henderson were exclusively of the Terry-type, 30-hour, time-and-strike wooden variety. Close examination shows, however, that they purchased movements from several makers. Using the classification system of Dr. Snowden Taylor, one of these movements, illustrated in *Eli Terry and the Connecticut Shelf Clock* by Kenneth Roberts, is subtype 1.118, attributed to Eli and Samuel Terry or Samuel Terry (Bristol). The clock in Fig. 24 contains a subtype 8.35 movement whose maker has not been established. One clock bearing the label of Roman M. Butler, Saint John, New Brunswick, is shown in Fig. 25. The clock has a Taylor-type 8.132 movement attributed to Ephraim Downs. During this period in New England, clock movements had almost become a form of currency, often being bartered several times before being cased, so this disparity of makers is not surprising.

The authors have seen two Butler Saint John clocks and both have one distinctive feature. The cases are typical pillar-and-splat style except that all visible surfaces are mahogany. These clocks are usually found with pine splats and columns, painted black with gilt stencilling. The all-mahogany construction used by Butler is an interesting variation and the clocks themselves are quite attractive.

The exact date when these clocks were introduced to the Maritime

FIG. 15

FIG. 15 *Short transition style Butler Henderson & Co. clock sold in Clement, Nova Scotia. Label printed by Stubs & Son, St. Andrews, N.B.*

FIG. 16a

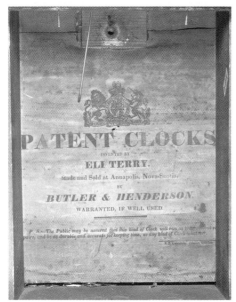

FIG. 16b

FIG. 16 *Butler & Henderson clock and label sold in Annapolis, Nova Scotia. Printer of the label, J.S. Cunnabelle, Halifax. – Courtesy Smithsonian Institution, Washington, D.C.*

market is subject to conjecture. An approximate date, however, can be deduced from the labels. The labels of most of the clocks in Clement, Nova Scotia, were printed in St. Andrews, New Brunswick, by Stubs and Son. Peter Stubs purchased the *St. Andrews Herald* newspaper and printing company in 1822 and the partnership existed until 1831 when Peter retired and his son continued on his own.

Whereas the majority of the clocks sold from Annapolis bore the names Butler and Henderson, Annapolis clocks sold by Roman M. Butler alone do exist. The labels of these clocks were printed in Halifax by F. & J. Holland (1830–1837) and by J.S. Cunnabelle, who was in business to 1836. Research done by Dr. H.G. Rowell of the Maritimes, who collected clocks before the Second World War, established that the "clock factory" at Annapolis Royal was an assembly plant. Clocks were brought into the area unassembled to avoid duty. However, stamps on the clock in Fig. 21 show that the duty has been paid.

The period in which the Butler and Henderson clocks appear to have been for sale was from 1828 to 1835, as indicated by the years when the printers were in business.

The names of Butler and Henderson are linked in correspondence concerning clocks as early as 1828. Clement, Nova Scotia, may have been their first place of business, as a bill of sale exists for one of their clocks and gives the date of 8 December 1830 on which the clock was sold to Mr. John Cameron of Pictou (see Fig. 19).

Butler and Henderson were originally clock dealers from New Hartford, Connecticut, who purchased clocks and movements from Samuel Terry. Kenneth Roberts also comments that they bought complete clocks from Terry prior to his move to Bristol in 1829. After 1829, movements were for the most part bought and cased in the Maritimes.

Correspondence of these men with Samuel Terry is on file in the Samuel Terry Collection, American Clock and Watch Museum in Bristol, Connecticut. Samuel Terry first was in business in Plymouth, and moved to Bristol in 1829, where he retired in 1835. An examination of accounts and letters written to Samuel Terry would indicate that there were at least five persons involved in the Butler and Henderson firm in New Hartford. There were three Butlers – William, the father, and two sons, Roman and Charles – and two Henderson brothers, James F. and W.T. (or W.F.), involved in the purchasing and selling of these clocks. A typical letter is shown in Fig. 26.

In the opinion of the authors, certain information found in the correspondence between Samuel Terry and the Butlers and Hendersons is of importance in understanding the relationship between maker and sellers. These are the main points:

1. In Samuel Terry's Day Book A an item appears dated 19 September 1828 – "Roman M. Butler and Henderson – 24 clocks at $9.00 per clock to be paid for in 6 months without interest."[5]

2. In the spring of 1830, 100 movements had been ordered by William Butler for his son's partner. W. Henderson was asked to accept responsibility for payment. Mr. Henderson assumed the responsibility and mentioned to Mr. Terry that "Butler and Henderson at this time are owing considerable money…and it is impossible for them to pay off the creditors as soon as due." Mr. Henderson referred to "my brother and his partner" in the letter and also suggested that the boxes of clocks be marked "Butler and Henderson as that is our name."

3. In April 1830 Jas. F. Henderson wrote Terry, asking if he would accept a lathe as part payment on the amount due for movements being made for Mr. Butler and himself. He also suggested in the letter that he might supply butter, beginning in the month of May. In a further letter, he urged that butter be used as part payment for these movements.

4. A document dated 27 May 1830 indicates that some payment was made to Samuel Terry by William Butler and James F. Henderson against a bill that had arrived in February.

5. William Butler ordered fifty movements, to be made by Samuel Terry for Butler's son in August 1830. In November, letters from W. Butler in New Hartford Center urged that the movements be delivered as soon as possible so that they could accompany other boxes being sent on a vessel leaving for Boston.

6. In December 1830 correspondence between J.F. Henderson and Samuel Terry referred to movements sent to Charles Butler, "my brother," suggesting that Charles Butler was, perhaps, a brother-in-law or a step-brother of J.F. Henderson. Charles Butler's name also appears in the Terry account books.

7. By January 1831 William Butler was urging that Terry send the order. He suggested that "with too much delay they may be out of business. Therefore, do not wait till spring."

8. In March 1831 James F. Henderson paid in lumber for movements sent to his "brother, Charles Butler." In October of that year, further movements were paid for with cheese and eleven doubloons worth $16 each.

9. A letter dated 5 March 1832 suggests that James F. Henderson was a regular supplier of lumber for Samuel Terry. Henderson answered a complaint voiced by Terry's son, who thought that a load of lumber was not sufficiently seasoned. Henderson was willing to replace the lumber "if the travelling was any how possible."

Existing evidence does not conclusively place the Roman M. Butler clocks earlier or later than the Butler and Henderson clocks. However, Roman M. Butler was working with Henderson in September 1828 – at the beginning of the clock-selling period. The partnership may have dissolved early in the 1830s, placing the Roman M. Butler Annapolis clocks and the New Brunswick clocks later than clocks sold in Clement, Nova Scotia.

FIG. 17

FIG. 18

FIG. 19

FIG. 17 *Butler, Henderson & Co. pillar-and-scroll clock sold in Clement, Nova Scotia. – Courtesy the Nova Scotia Museum, Halifax*
FIG. 18 *Movement from a clock identical to Fig. 17.*
FIG. 19 *Bill of sale for a Butler, Henderson & Co., Clement, Nova Scotia, clock.*

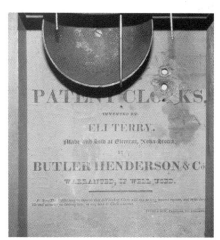

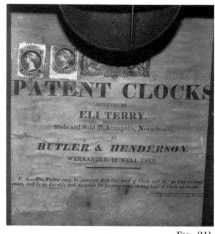

FIG. 20

FIG. 21b

FIG. 21a

FIG. 22

FIG. 23

FIG. 24a

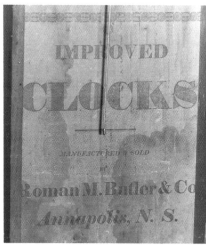

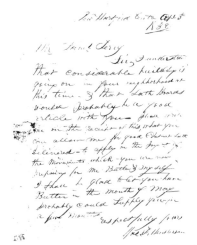

FIG. 24b

FIG. 25

FIG. 26

FIG. 20 *Label from a Clement, Nova Scotia, pillar-and-scroll clock, printed by Stubs & Son, St. Andrews, New Brunswick.*
FIG. 21 *Pillar-and-scroll clock and label sold in Annapolis. – Courtesy H.G. Rowell Collection, Dartmouth College, N.H.*
FIG. 22 *Short transition wood movement clock, by Butler & Henderson, Annapolis, Nova Scotia.* FIG. 23 *Butler & Henderson label from clock in Fig. 22.* FIG. 24 *Roman M. Butler & Co. clock and label, Annapolis, Nova Scotia.* FIG. 25 *Clock label by Roman M. Butler, Saint John, New Brunswick.* FIG. 26 *Letter to Samuel Terry from James F. Henderson. – Courtesy Samuel Terry Collection, American Clock & Watch Museum, Inc., Bristol, Connecticut*

FIG. 27

ROBERT H. COGSWELL

Robert Cogswell (d. about 1907) was a maker of chronometers, watches and nautical instruments in Halifax, Nova Scotia. He learned his trade in Glasgow and was established in business in Halifax from 1830 until 1902.

In 1864 Cogswell was associated with James Tweedle at 65 Upper Water Street, and in 1865 he purchased a business situated at 155 Barrington Street from William Crawford. He enlarged the premises and advertised that, as well as watch repairing, he corrected and reconditioned nautical equipment, adjusted compasses, and rated chronometers with a transit instrument on his premises. He advertised his goods as being "from the best makers and his work first class in every respect." He was responsible for the rating of chronometers of the Cunard Mail Steamers of the Bermuda and Newfoundland lines.

According to G.A. White in *Halifax and its Business*, published in 1876, Robert H. Cogswell, with his astronomical clock, was considered to be the standard authority on "true" timekeeping relating to Halifax, Boston, and Greenwich time. "He provided true time gratuitously to ships and shipping in the Halifax harbour and, from about 1857, was responsible for the noon day gun." For the latter service he was paid $100 annually by Ottawa after 1867.

He was in charge of railway time and also gave weather reports to the newspapers and included "thermometrical and barometrical" readings.

The address of his shop from 1873 to 1902 was 175 Barrington Street. In the early 1900s H.V. McLeod managed the business for a number of years for Robert's wife, Sarah.

ROBERT H. COGSWELL,
(SUCCESSOR TO WM. CRAWFORD)

CHRONOMETER, WATCH, AND

NAUTICAL INSTRUMENT MAKER,

OPTICIAN, &c.,

155 BARRINGTON STREET,

OPPOSITE PARADE,

HALIFAX, N. S.

KEEPS CONSTANTLY ON HAND AN ASSORTMENT OF

Chronometer Watches, Jewelery and Nautical Instruments,

CHRONOMETERS FOR HIRE AND FOR SALE.

Time taken by Transit Observations.

EVERY ARTICLE REPAIRED IN THIS ESTABLISHMENT GUARANTEED.

ESTABLISHED over 40 YEARS.

FIG. 28

FIG. 27 *R.H. Cogswell Shop in 1871. From* Rogers' Photographic Advertising Album. *– Courtesy the Nova Scotia Museum, Halifax*
FIG. 28 *Advertisement of R.H. Cogswell in* Rogers' Photographic Advertising Album.
  – Courtesy the Nova Scotia Museum, Halifax

FIG. 29

FIG. 29 *Clock sold by William Crawford in Halifax.*
– Courtesy the Nova Scotia Museum, Halifax

WILLIAM CRAWFORD

William Crawford was a watch- and chronometer-maker and silversmith in Liverpool and Halifax, Nova Scotia. In the Nova Scotia Museum in Halifax are a number of silver articles with his mark and a tall case clock with "Wm. Crawford, Halifax" on the dial. The hood of the clock is curved on top and access to the dial is through a hinged door in the hood. The dial is painted to give the impression that it is made of brass. The clock was made before 1842 when the firm became Wm. Crawford and Son.

William Crawford was born in Glasgow about 1795 and came to Liverpool, Nova Scotia, where in 1817 he advertised in the *Acadian Record* that he had for sale chronometers, quadrants, compasses, watches, and silver and that he also made repairs.

He had moved to Halifax by 1826 and was elected a member of the prestigious North British Society in 1827; he remained a member until 1847. Crawford erected an observatory on the upper floor of the Royal Acadian School. By 1832 the observatory was moved to Lockman Street. This observatory had most of the equipment necessary for Crawford to rate ships' chronometers. He signalled noon-time to the Halifax Citadel for the firing of the twelve-o'clock gun. Although not a particularly accurate method, it allowed ships' masters to rate their chronometers approximately. Crawford was also responsible for rating the chronometers of the Cunard Mail Steamers of the Bermuda and Newfoundland lines from the time of their commencement.

In 1832 Crawford applied to the House of Assembly for a reflecting telescope with sufficient power "to espy the satellites of Jupiter and their various occultations and eclipses and also a repeating circle." He would then be able to "regulate ship chronometers so perfectly as to indicate true longitude, with little chance of error, and thus more perfectly guide the property and lives of those who are engaged in foreign commerce." However, the sum of £100, which was considered appropriate, was made available through the Marine Insurance Association rather than directly to Crawford.

No instruments have been attributed to Crawford. Dr. Randall C. Brooks[6] considers it likely that he was assembling instruments from imported parts.

According to an advertisement in the *Halifax Journal* in 1831, his shop was at the corner of George and Barrington streets, but in April 1832 he "moved to the Lower Side of Grand Parade adjoining the Post Office."

A son, William, became a partner to his father in March 1842 and the firm was known as Wm. Crawford & Son. In 1858 the address of the business was 29 Barrington Street, and by 1863 it was 153–5 Barrington Street opposite the Grand Parade.

In 1865 Robert Cogswell purchased the business and announced that he was the successor to William Crawford.

R.H. EARLE AND THE EARLE CLOCK COMPANY

Reginald Heber Earle (b. 21 Sept. 1843 – d. 4 Oct. 1921), known as Heber, was a watchmaker, jeweller, and inventor who lived and worked for most of his life in St. John's, Newfoundland. Although some of his family earned their living from the sea, Heber – after a one-day experience on the water – decided to find an occupation on land. When he completed his education at the Church of England Academy in 1859, he was apprenticed to John Linberg, a watchmaker and jeweller. Heber showed exceptional skill and went abroad for further training. In 1865 he opened a store at 216 Water Street for the sale and repair of watches and jewellery.

Soon his interest was aroused by the new mode of transportation in Europe called the velocipede, a forerunner of the bicycle. The velocipede had a large wheel in front and one or two wheels in back. According to Earle, the existing design had drawbacks, which included fixed handlebars and a wheel that could not be turned from a straight position. Also, the tires were not air-filled. When Earle was forced to close his store in 1868 due to the expansion of the building owner's business, he went to London, England, to purchase better watchmaker's tools. At the same time he purchased a velocipede and had it shipped to Newfoundland. When it arrived he and a carriage-maker, Sam Carnell, proceeded to improve upon the design of the velocipede and to lighten the wheels. On 26 May 1869 Earle used his improved version to cycle from St. John's to beyond a place called Topsail and back. The event made Newfoundland history.[7]

In 1868 Heber Earle purchased a house on Military Road, where he stored his equipment. In July 1869 he and his bride, Eliza Haddson, were living in this house, where he also had his shop. Shortly after Easter in 1878 three of his seven children, two sons and a daughter, died of diphtheria.[8]

Between 1869 and 1871 Earle repaired watches in his home and began experimenting with the possibility of making luminescent paint so that his watch and clock dials would glow in the dark. It took him almost six years to perfect his paint. By 1877 Earle had moved his watch- and clock-repairing business back to his original Water Street shop, where he continued to work at his profession. Unfortunately, his trip to New York to sell his luminescent paint was postponed because of deaths in his family. By the time he got to New York in 1879, he found that similar paint had been patented in the meantime and luminescent paint was being successfully sold to the industry.

Earle returned to his business, but in 1892 the great fire in St. John's destroyed his shop and all his equipment. He found employment elsewhere as a watchmaker and continued to invent. Among his inventions was a table with a net surface for the use of fishermen to spread their fish for curing, a device to send signals under water for use in submarines, and an "Earle's Masthead Emergency Light," which the navy considered using.

In addition to his other talents, Earle was a gifted musician and his services as an organist were in demand in the many churches of St. John's. His book, *The Organist Friend,* was used by church organists for over one hundred years.

Early in 1912 Earle left Newfoundland for Saint John, New Brunswick, where he began the Earle Clock Company. The clocks he sold were ornate "Gingerbreads."

FIG. 30a

FIG. 30b

FIG. 31a

FIG. 30 *The "Puritan" clock and label manufactured by The E. Ingraham Co. and sold by R.H. Earle.*
FIG. 31a *The "Detroit" clock label. See Fig. 31b.*

25

FIG. 32

FIG. 31b *The "Detroit" clock made by*
The E. Ingraham Clock Co.
for the Earle Clock Co.
Fig. 32 *Watch sold by Benjamin Etter.*
– Courtesy the Nova Scotia Museum, Halifax

The E. Ingraham Company was one clock supplier used by Earle. The name of the Earle Clock Company on the label indicates that Earle had ordered a "lot" of clocks from the E. Ingraham Company. Two styles of cases are shown in Figs. 30 and 31. On the Ingraham movement in the Detroit model are the patent dates of 8 October 1878 and 11 November 1879. The plates of the movement are made of brass-plated steel.

On the label of the Puritan model, three patent dates are given. On 13 April 1897 patent number 580,747 was given to W.H. Wright for a clock bell. On 23 November 1897, W.H. Wright received patent 594,309 for a base for clocks. Both patents were assigned to the E. Ingraham Company of Bristol, Connecticut. On a third date, 14 September 1897, William H. Wright was assigned patent number 589,886 for clock gearing.

The Sessions Clock Company of Connecticut also supplied movements for cases thought to have been provided by the Earle Clock Company. Although the patent for the cases was applied for, it was never granted. The Earle Clock Company was in business for part of a year only. Heber Earle returned to Newfoundland and frequently travelled. He died on 4 October 1921 and is buried in the St. John's Cemetery.

Information about Earle's ancestry reveals that he was the youngest of the seven sons of Henry Earle, who immigrated to Newfoundland from Reading, Berkshire, England, in 1832. Henry was the Admiralty tailor. He lived and worked in Haymarket Square, where he catered to the tailoring needs of bishops and professional people, as well as military and naval officers. After the fire of 1846, he relocated at 212 Duckworth Street. Heber's mother was a talented musician and gave music lessons in St. John's.

PETER ETTER AND BENJAMIN ETTER

Peter Etter (b. about 1755 – d. May 1798) and his brother Benjamin Etter (b. 1763 – d. 18 Sept. 1827) were both watchmakers, jewellers, and silversmiths in the Maritimes. They were sons of Peter Etter Sr. (b. 1715 – d. 24 Mar. 1794), who emigrated from Berne, Switzerland, in 1737. In 1752 he settled in Braintree (Quincy), Massachusetts, where his sons were born. After the American Revolution, the father and sons[9] settled in Halifax, Nova Scotia, as United Empire Loyalists.

Peter Etter arrived in Halifax in 1776 with the British army, under the command of Lord Howe. In the Maritimes he enlisted in the Royal American Fencible Regiment and served for a time at Fort Cumberland, New Brunswick. After leaving the army about 1778, Peter went to Halifax as a watchmaker and it is believed that he worked for John Finney, a watch- and clockmaker from Charleston, Maryland. In July 1780 he announced that he was "commencing watchmaking in all its branches next door to Mr. Welsh's near the Market where the celebrated time piece of the late Mr. Finney stands." Peter's brother, Benjamin, was working for him, probably as an apprentice.

Peter became the proprietor of "Peter Etter's Jewellery Store" and advertised in the *Nova Scotia Royal Gazette and Weekly Chronicle* in May 1781 as a watchmaker and clockmaker, "next door to Mr. William Welsh's near Market House." In the same newspaper, in the issues of December 1782 and January 1783, he announced that he had moved "to the House on Hollis Street Formerly occupied by Mr. Sparrow…had for sale watches, jewellery, silver spoons etc."

In 1786 Peter married Letitia Patton, the daughter of a wealthy landowner, Mark Patton of Westmorland County in New Brunswick. Mark

Patton was one of the rebels who supported Jonathan Eddy in the siege of Fort Cumberland. He was also the father-in-law of Col. John Allen, leader of the insurgents.

Shortly after the birth of their son Peter, on 12 February 1787, the family left Halifax for New Brunswick and opened a shop at Fort Cumberland (Beausejour) in Westmorland County. According to Huia Ryder, in *Antique Furniture by New Brunswick Craftsmen*, it was at this location that Peter Etter made the "working parts" of a tall case clock built for William Chapman, a New Brunswick craftsman. The clock had only one hand.

Peter's wife died and in 1791 he married Sarah H. Wethered, daughter of Samuel Wethered, adjutant for Fort Cumberland.

In May 1798 the ship on which Peter was travelling from Boston to Cumberland sank, and he was drowned. On 19 June 1798, the *Nova Scotia Royal Gazette* announced that Sarah Etter and William Allan were administering the estate of Peter Etter of Westmorland.

Benjamin was a younger brother of Peter Etter; he apprenticed with Peter and worked with him in his shops from 1780 to 1785. As well as being a watchmaker, Benjamin was a working silversmith. Examples of silver made by him and by his son Benjamin B. are in the Nova Scotia Museum.

When Peter Etter left Halifax in 1787, Benjamin took over his shop. In an advertisement in the *Halifax Weekly Chronicle* on 14 January 1794, he announced that he was a watchmaker located on George Street. In 1796, as a watchmaker, jeweller, and silversmith, he also "sold nautical instruments and military accoutrements." In November 1798 he had moved again "to lower side of the Grand Parade adjoining Mr. Richardson." One year later, he was in business in Richardson's store. At this time he took as a partner James Tidmarsh and the business was known as Etter and Tidmarsh. The partnership was dissolved on 26 August 1803.

In March 1806 Benjamin Etter was in business on the corner of George and Barrington Streets with his son-in-law, Thomas Hosterman, and in May 1813 he turned over his business to his son Benjamin B. and Thomas Hosterman. This partnership lasted until 1 December 1815. In 1813 Benjamin Etter was granted by the Crown the south half of Lot 2, Letter E in Calendar's Division of Halifax.

Benjamin Etter trained a number of apprentices, among whom were: Daniel Bessonett, his first wife's younger brother; John Etter, his son; Benjamin Etter, his son; Samuel Black, brother of his son-in-law.

The Nova Scotia Museum has in its care a watch with "Benjn Etter, Halifax N.S." engraved on two parts of the watch. On one part is the number 4569. On the other part is the number 2569. William M. Graham, horologer, comments on the watch: "The photo shows an English watch with verge escapement which certainly could have been made in the first quarter of the nineteenth century as suggested by the death of Benjamin Etter in 1827. The photo shows the dust cap which fits over the movement and can be locked in place by sliding the curved steel spring to engage slots in two pins on the movement. The discrepancy of 2,000 in the numbers is hard to explain. Having the last two digits the same is sufficient to identify the parts in the maker's shop when a batch of watches is in process.

"The movement, or at least the 'ebauche' (rough or unfinished movement) was probably made in Prescott, Lancashire. Thousands of local watch repair men, all over Britain, finished these ebauches, carved out the ornate balance cock and engraved their names. Hence the use of the

FIG. 33

FIG. 34

FIG. 33 *Clock sold by Whitcomb Fairbanks.* – *Courtesy New Brunswick Museum, Saint John*
FIG. 34 *Fairbanks Clock in the York-Sunbury Museum, Fredericton, New Brunswick*

27

FIG. 35a

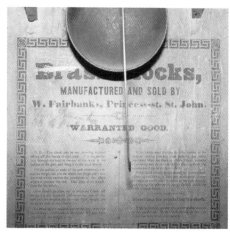

FIG. 35b

FIG. 36

FIG. 35 *Chauncey Jerome movement and label from a Fairbanks OG clock.*
FIG. 36 *Label of a Fairbanks OG clock.*

word 'watchmaker' for a watch repair man which persists to this day. These verge watches have a very modest value unless made by a famous watchmaker."

In addition to being a prominent businessman in Halifax as a watchmaker, jeweller, and noted silversmith, Benjamin Etter had other enterprises. In 1783 he advertised "to be sold by Benjamin Etter – opposite Barrack Gate – linens, woolens, sugar and candies." His wealth, however, was accumulated in part through his ventures as a privateer. According to Donald C. Mackay, Etter was part owner of an armed brig, the *Earl of Dublin* and an armed schooner, the *General Bowyer.* In one profitable cruise, the schooners *Peggy* and *Nancy* and a Spanish ship were captured and gold bullion was taken from another Spanish ship – "all considered his Majesty's enemies."

On 18 September 1827, Benjamin Etter died at the age of 84. He had nineteen children by three wives. He is buried at St. Paul's Cemetery, Halifax.

WHITCOMB FAIRBANKS

Whitcomb Fairbanks (b. 12 Jan. 1810 – d. 9 Sept. 1877) was a manufacturer and distributor in Saint John, New Brunswick. Among the items sold in his establishment were clocks, and such clocks carried the following inscription on their clock labels – "manufactured and sold by" W. Fairbanks.

Although it is probable that many of the clock cases were made in Saint John because Fairbanks was engaged in the manufacture of furniture and related products, the movements were imported from the United States. A Fairbanks clock in the New Brunswick Museum contains a clock movement made by Chauncey Jerome.

Fairbanks also used Seth Thomas movements for his clocks and such a clock may be seen at the York-Sunbury Museum, Fredericton.

Many of the Fairbanks clocks illustrated in this book and elsewhere are of a distinctive design, as can be seen in Figs. 33, 34, and 37. These clocks appear to have been designed by Fairbanks and, while he was no doubt influenced by American styling of the time, he was not content to make exact copies. In addition, he did sell OG clocks that do not have any unique characteristics and it is possible that he imported cases as well in some instances.

Approximate dates when the clocks were sold by Fairbanks can be established by examining the labels. Clocks sold by him before 1848 carry the inscription "W. Fairbanks". The movement and label of such a clock are shown in Fig. 35. The clock is an OG with a wooden dial. Clocks manufactured and sold after 1848 have the inscription "Fairbanks & Co." In this second period, Israel Hawes was his partner. The printers of Fairbanks labels included H.P. Sanction and F.W. Clear.

Recently, another "W. Fairbanks" OG clock was brought to the attention of the authors. There is a resemblance between its label and the one illustrated in Fig. 35, but they are not identical. No street name is given on the label. The clock is fitted with a movement, designated type 5.412 by Snowden Taylor, which was manufactured prior to 1847 by a number of American firms, including Forestville Manufacturing Company and S.B. Smith and C. Goodrich Company.

Whitcomb Fairbanks was a native of Sterling, Massachusetts, where he began to manufacture furniture. He moved to Bangor, Maine, and learned the picture-framing business. It was here that he met Israel Hawes of Brooksville, Maine, a carpenter by trade.

28

In 1844 Fairbanks moved to Saint John, New Brunswick, and established himself as a dealer in clocks and picture frames. His shop was on Princess Street in his early years in Saint John. Hawes joined him in 1845 as one of his employees, then in 1848 became a partner. Together, they started a planing shop.

The business flourished, and in order to accommodate the necessary machinery for moulding, planing, and general jobbing work, they leased the tannery of C.J. Melick on Union Street. They also maintained a warehouse and retail facility at 23 Germain Street. Hawes managed the mill and Fairbanks looked after the retail store on Germain Street, where he sold looking-glasses, picture frames, clocks, and silver. There are silver articles with the Fairbanks silver mark in the New Brunswick Museum at Saint John and a dessert spoon in the York-Sunbury Museum.

In spite of his diverse activities, Fairbanks considered himself to be primarily a clockmaker and identified himself as such to the census takers.

In 1866 a fire destroyed the Union Street plant, and it was rebuilt, restocked, and put into full production. In 1869 the plant burned for a second time. In 1870 they erected a plant on City Road. This large, well-equipped factory was advertised as "manufacturing sashes, doors and blinds." The mill, located near a branch of the recently built railroad, also did "planing, sawing and turning."

According to a book written by J. Hannay and W.K. Reynolds, *Saint John and its Business, 1875*, Fairbanks was "scathed and scarred, but not dismayed" by the setbacks he had suffered. Unfortunately, his trouble was not over: in 1877 his Germain Street furniture warehouse, filled with new stock, was destroyed by flames.

Whitcomb Fairbanks's personal life story reveals connections, by marriage, to other clockmen of the day. Whitcomb married Elizabeth Jane Plummer (b. 25 Mar. 1823), daughter of Thomas Plummer and his first wife Elizabeth (d. 1828). The marriage took place in St. Andrews Church on 31 August 1848. Thus, as reported later, he also became related to clockmaker William L. Plummer. His partner, I.B. Hawes, was also a son-in-law of Thomas and Mary Plummer.

The family of Whitcomb Fairbanks included two daughters, Louise, who married Joseph Tyler and lived in Massachusetts, and Lotty, wife of Charles Flewelling of Saint John. Mary F., who was seventeen years old at the time of the 1871 census, may have been a third daughter who did not survive past 1877.

The family belonged to the Congregational Church and lived in the parish of Portland, then north of the city but, since 1889, part of Saint John.

In 1877 Whitcomb Fairbanks died, leaving no will. He had a personal estate of $20,000 and owned real estate valued at $2,000.

FIG. 37

FIG. 37 *Distinctive 30-hour case style used by Fairbanks & Co. A similar case was used by another maker in Saint John (Wm. L. Plummer), but was not used elsewhere.*

FIG. 38 *Fairbanks & Co. label from clock in Fig. 37 with engraving of the shop.*

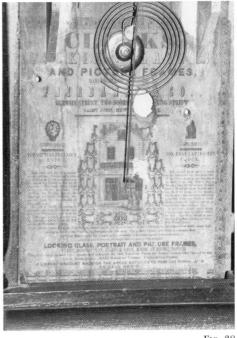

FIG. 38

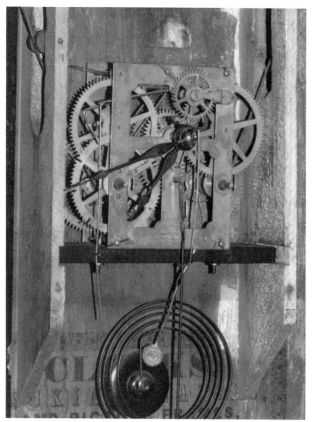

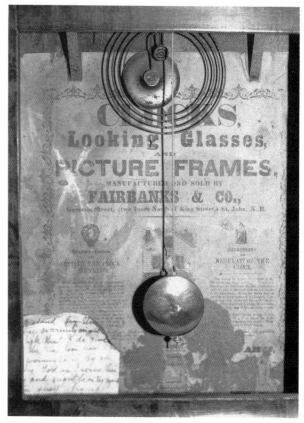

FIG. 39

FIG. 40

138 FAIRBANKS & CO.

FAIRBANKS & CO.,

23, Germain Street, St. John, N. B.

PLAIN & ORNAMENTAL GILDERS, &c,

Keep constantly on hand a well selected assortment of

PICTURES LOOKING GLASSES,

Clock Materials, &c.

OLD FRAMES RE-GILT; CLOCKS CAREFULLY REPAIRED.

SAWING

PLANING MILL,

No, 153, Union Street, Opposite the Hay Scales,

All Kinds of

Job Planing, Sawing, Tounging, Grooving, Morticing, Trimming and Jig Sawing, done with dispatch.

Every description of Wood Mouldings stuck to order and on hand for sale. Wood Turning executed in a superior manner. Mahogany, and Black Walnut, Planks, Boards, Joist and Veneers of all kinds.

FAIRBANKS & Co.

FIG. 41

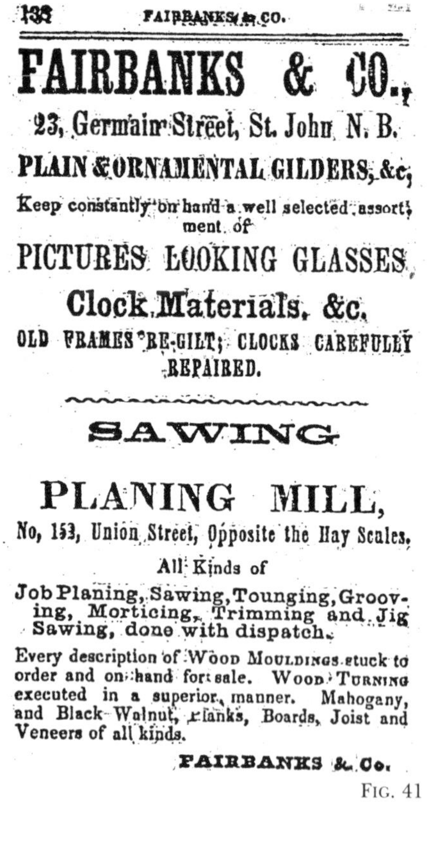

W.S. FLETCHER

William S. Fletcher was a watch- and clockmaker who had a shop in Pictou, Nova Scotia, in the 1800s. The earliest reference to his business is in the 4 December 1839 issue of the *Mechanic and Farmer* of Pictou, where he offered for sale an assortment of watches, jewellery, silver, etc., and advertised "watches and jewellery repaired at shortest notice."

While in Pictou, William Fletcher imported movements for tall case clocks and possibly entire clocks from Great Britain and sold them in Pictou, Nova Scotia. A clock in a private collection has the name "Fletcher, Pictou" on the dial. In the arch of the dial is a picture of a lady in pink in a pastoral scene. Another tall case clock, with no name on the dial but known to have been purchased in Pictou, may also have been sold by William Fletcher.

Sometime before January 1852, William S. Fletcher moved to Charlottetown, Prince Edward Island, where he placed an advertisement in the *Weekly Advertiser* announcing that he "sells and repairs jewellery, clocks and watches...."

A William Fletcher appears in the 1866–67 Nova Scotia directory as a watchmaker in Wallace, Nova Scotia. The name did not appear in the 1868 issue. It is not known if this was the same person who lived in Pictou and Charlottetown.

FIG. 39 *Waterbury movement in the Fairbanks clock shown in Fig. 37. (Taylor type 2.411)*
FIG. 40 *Another label from a Fairbanks & Co. OG clock.*
FIG. 41 *Advertisement from the Saint John City Directory of 1857.*

CHARLES GEDDES

Charles Geddes (b. 25 Jan. 1750 – d. 24 Nov. 1807) was a clock- and watchmaker who was in business in Halifax, Nova Scotia, from 1783 until his death in 1807.

Several clocks that bear his name are known. One such clock is in the Henry Ford Museum at Dearborn, Michigan, and is illustrated in Fig.43. It is a bracket clock and was described by Donald C. Mackay as "the most beautiful North American clock" in his book, *Silversmiths and Related Craftsmen of the Atlantic Provinces*. The clock has an ornate brass face and is inscribed "Charles Geddes New York" and so obviously dates from his years in that city. The styling of the clock is typically British, suggesting that Geddes's name may have been engraved on an imported product. It is interesting to note that the clock is virtually identical to two others, illustrated in this book in Figs. 228 and 138. The latter clocks were sold by François Valin in Quebec City, and François Dumoulin in Montreal. All three men were in business at approximately the same time and it is quite possible that they all obtained clocks from a common source. Another of Geddes's clocks, a tall case clock sold in Nova Scotia, is in a privately owned collection in Canada.

An advertisement in 1795 in the *Royal Gazette* offered a reward for the return of a silver watch bearing the name Charles Geddes. The advertisement appears to prove that there were watches that bore his name, although none has been located.

Charles Geddes was born in Edinburgh, Scotland, the third child of James Geddes (b. about 1713 – d. 18 Oct. 1755) and Lilias Grey. James was a watch- and clockmaker of Edinburgh. A clock made by him was reported in *A Handbook and Directory of Old Scottish Clockmakers* by Wm. J. Hay.

Although his father died when he was only five years old, Charles followed his father's profession and became a watch- and clockmaker and silversmith, probably apprenticing with his uncle in the firm of Johnson and Geddes "of the Golden Ball, Paten Street, Leicester Square, London." In 1773 Charles Geddes moved to Boston, Massachusetts. An advertisement in the *Boston News-Letter* of 14 October 1773 announced "Charles Geddes, clock- and watchmaker, & Finisher, from London, at his shop below the Sign of Admiral Vernon, in King Street, Boston, makes, mends, cleans, repairs and finishes all sorts of clocks and watches, in the best and neatest manner, and upon the most reasonable Terms." In Boston he joined the Royal North British Volunteers and followed the British Army to New York in November 1776. There he established his business, first located on Queen Street and later at 21 Hanover Square.

As a Loyalist, Geddes left New York in 1783 and was given a grant of land in Shelburne County, Nova Scotia.[10] However, he settled in Halifax, opening a shop on Hollis Street. He advertised his arrival in the local newspaper, the *Nova Scotia Gazette and Weekly Chronicle*, on 22 July 1783. The advertisement read in part as follows:

> Charles Geddes, Watchmaker from London lately from New York, has opened a shop in Hollis Street, late Mr. Sparrow's where he intends carrying on the watchmaking business – and as he flatters himself he has acquired a thorough knowledge of his Profession, he hopes for encouragement from the Gentlemen of the Navy and Army and the respectable Inhabitants of this town and Province....

FIG. 42

FIG. 43

FIG. 42 *Charles Geddes.*
FIG. 43 *This fine bracket clock is inscribed "Charles Geddes New York."*
– Courtesy the Henry Ford Museum, Dearborn, Michigan

FIG. 44

FIG. 44 *Clock sold by John Geddie.*
– Courtesy the Nova Scotia
Museum, Halifax

He did not stay long at this location because on 19 August of that year he announced that he had opened a shop on "Granville Street next door to Mr. M'Cra's." Again in May 1784 he announced his "removal to the house fronting the Parade, formerly occupied by Mr. Howe, Printer" and he remained there until his death. He continued to advertise in the newspapers for twenty years.

In addition to his active business life, Charles Geddes contributed much of his time to his church. Apparently a very religious man, he signed "a solemn covenant with God" in Edinburgh when he was twenty years old and renewed the covenant at least yearly until his death. This covenant is reproduced by John Grier Stevens in his book, *The Descendants of John Grier with Histories of Allied Families.* Moreover, Charles Geddes was a vestryman and church warden in St. Paul's Church in Halifax for many years.

The Masonic lodge was also an important part of his life and it is partially through the records of the lodges in London, Boston, and New York that the life of Geddes can be followed. In Halifax in 1784, he became a member of lodge No. 211 and rose in the lodge to be the master in 1787. He was treasurer from 1790 to 1797. In 1791 he was one of the twelve past masters or grand stewards with white rods.

He continued to be associated with the army in Halifax and was for many years an officer in the Halifax militia, rising from a commissioned second lieutenant in January 1788 to breveted major in November 1804.

As was customary for prominent businessmen in their communities, Charles Geddes added his name to the various addresses and proposals that were printed in the newspapers. On 21 February 1786, he signed an agreement to support and obey laws regulating trade. However, on 27 June 1792, he signed a memorial to the House of Assembly protesting the duty of 2½ percent on articles of British manufacture. In 1802 he signed an address to Lt. Col. Edwards of the Royal Surrey Regiment and to officers of the Royal Fusiliers on their departure from Halifax.

Geddes was also a member of various Halifax societies. On his arrival, he joined the prestigious North American Society. He became vice president in 1789 and president in 1806. According to the annals of the society from 1768 to 1893, "the position of President…was at this time a most distinguished one…demanded wealth, polish and acquaintance socially with the world…so that really the number who could accept the office was limited to but a few…."

Geddes also belonged to a literary group that met weekly at the Pontiac Inn. The members read papers on social and scientific subjects, after which the evening continued in "conversation, song and toasts." The Duke of Kent joined them when possible.

Although his business was successful, and Geddes died a wealthy man with an estate of around £100,000, he accumulated his wealth during the Napoleonic Wars. He purchased condemned prizes and cargoes, captured from the French by the British fleet. These treasures were sold through the Admiralty Court in Halifax.

When Geddes arrived in Halifax, he was preceded by John Geddes, who had arrived around 1755 and was thought to be Charles's uncle. Charles's brother Robert, a cabinetmaker, arrived around 1755, and Charles followed later. Both Robert and Charles eventually owned property on the Musquodoboit River, and they belonged to the same societies, but they belonged to different churches; Charles was Anglican and John and Robert were Presbyterian.

Charles Geddes married Ann Bleigh (b. 1742 – d. 7 April 1836) on 2 December 1773. Ann was from Boston, Massachusetts. Their son James (b. 2 Nov. 1778 – d. 5 Aug. 1834) was born in New York. He became a surgeon and moved in 1807 with the military to Kingston, Upper Canada. He married Sarah Gamble (b. 6 Apr. 1788 – d. 18 Dec. 1857) and they had eleven children, born on St. Joseph Island and at Kingston. James died in Kingston in 1834.

On 8 August 1799 a daughter of Charles and Ann, Mary, married Lieutenant Hulton Rowe of the 7th Royal Fusiliers.

Charles died on 24 November 1807. In 1836 his wife was buried beside him in St. Paul's Cemetery, Halifax.

JOHN GEDDIE

John Geddie (b. about 1778 – d. 27 April 1843) was a Scottish clockmaker who came to Pictou, Nova Scotia, in 1817. John, the son of a cooper, was born in Banffshire on the Moray Firth, Scotland. He married Mary Menzies (b. about 1791 – d. Apr. 1851). Both John and Mary came from backgrounds that supported evangelical missionary work of the Church of Scotland.

Geddie apprenticed in Scotland as a clock- and watchmaker and in his own shop made tall case clocks. The Napoleonic war, however, drove the Scottish economy into a depression, and consequently, the people had no money to buy luxury items such as John's "beautiful clocks." The depression left its scars on John by forcing him to go into debt and, eventually, to lose his business. In 1817 he decided to immigrate with his family to Pictou, Nova Scotia, where many of the early settlers were Scottish. They brought with them their musical instrument, an organ, now on display in the McCullough House Museum, Nova Scotia.

Upon arriving in Pictou, Geddie and his family found rather difficult conditions. The area was recovering from an infestation of mice, which took place in 1815, and from the following "year without a summer" – in 1816. The "year of the mice" brought a complete catastrophe, which created great hardship in Nova Scotia. The "year without a summer" was equally catastrophic. Snow and ice persisted into July and crops could not be planted; as a result there was a great shortage of food.

In spite of these conditions, Geddie was not discouraged. Having settled his family, he started making clocks similar to those made in England and Scotland. The people of Pictou readily accepted the clocks, and John worked steadily. His business must have been financially rewarding because, in addition to supporting his family, he was able to send money to Scotland to pay all his debts.

Clocks with the name "John Geddie, Pictou" on the dials are treasured by museums and collectors. One such clock may be seen at the Nova Scotia Museum in Halifax. This broken-arch clock was made between 1820 and 1830. The case is mahogany with maple string inlays. The secondary wood is pine. Geddie's cases were made mainly of mahogany, but at least one case was made of local (Nova Scotia) pine. The case of the clock in the museum has square bracket feet and stands 82¾ inches high and is 18¼ inches wide. The movement of the clock is English in style but may have been made by Geddie.

According to the *Dictionary of Canadian Biography*, at least sixteen Geddie clocks are known. The cases vary from finely matched mahogany to roughly finished local cases. Dials are decorated with scenes of ships, hunting scenes, portraits of ladies, and landscapes. One clock is also dec-

FIG. 45a

FIG. 45b

FIG. 45 *Geddie clock in the First Presbyterian Church, Pictou, Nova Scotia*

FIG. 46a *Clock sold by John Geddie, Pictou, Nova Scotia*

orated inside the case. Its weights and pendulum are painted with bright flowers. Another Geddie clock in Nova Scotia is dated 1824.

There is a definite difference of opinion as to whether some of the movements of the Geddie clocks were actually made by him or only imported by him, and his name painted on the dials. Two Nova Scotia Provincial Museum reports on this issue differ in their views. One states that "these tasteful and skillfully made clock works" were made by Geddie and he, as well, painted the decorations on the dials. The other report, which is based on the opinion of F.J. Tobin, a clock repairer of Pictou who examined a clock, concluded that Geddie was only a cabinetmaker. F. Baird, a historian, also suspects that the movements in Geddie's clocks were imported from Great Britain.

Certain collectors, however, who have considerable knowledge of British and American clocks are convinced that the movements were made by Geddie. There is some evidence to support this view. For example, two persons related to John Geddie believed that "the works" were made by him. The Rev. John Geddie, a great-nephew of John Geddie, stated that John "made the works all by himself." This statement appears in the writings of Rev. George Patterson, a famous Pictou historian whose books include *A History of Pictou* and *Missionary Life Among the Cannibals: Being the Life of the Rev. John Geddie, D.D.* Mrs. George Musgrave of Halifax, another of John Geddie's descendants, also believes that "John made the whole clock including the works, cabinet and painting of the dial."

One cannot rule out the evidence of the oral traditions handed down in the family. The authors are inclined to accept these statements, particularly in view of the existence of clockmaking equipment in the list of personal effects left behind by Geddie[11] at the time of his death. The presence of "one clock and watch wheel engine," as well as lathes, clock barrel cutter, vises and clock tools is significant.

It can be argued, too, that some of Geddie's prosperity can be attributed to the greater profits to be reaped from making complete clocks rather than merely casing and selling them.

In addition to his clockmaking business, Geddie sold watches, which he imported from Britain. A watch paper exists bearing the date 1836. John lived and had his shop in the same building. In 1836 a notice appeared in the newspaper, *The Bee,* announcing his "removal" from his residence "nearly opposite to a new house belonging to A. Patterson, Esq." He thanked the public for their patronage and hoped to "enjoy a continuation of their favors." A further notice appeared in a June 1838 newspaper announcing the arrival of watches "to be disposed of at a moderate profit."

Rev. George Patterson wrote about John Geddie's character in a book written primarily about John's son. Geddie is described as "An Israelite, in deed in whom was no guile…. He was a modest, unobtrusive man – not a leader, but always present to help in every good work. He feared God, and was gentle and loving in disposition."

John Geddie's social life revolved around the activities of the First Presbyterian Church, Prince Street, Pictou. He was elected Elder, an office that he held until his death. In addition, he was a teacher in the Sabbath School. Geddie also served as overseer of the poor. When the town council refused to pay for some purchases he had made as overseer, he was arrested in November 1835. The people of Pictou were incensed at this injustice and wrote angry letters, which appeared in *The Bee,* demanding his release. He was given his freedom and the bills were paid.

John Geddie died of "paralysis" on 27 April 1843 during a trip to West River. In 1847 an advertisement to dispose of his clocks stated, "For sale 6 eight day clocks out of the estate of the late John Geddie...apply to Mr. James Simpson, clock- and watchmaker."

At the time of his death he was owed £275 and his possessions were worth £252. He is buried in Laurel Hill Cemetery, Pictou.

According to his obituary in *The Guardian*, on 12 May 1843, he was an "efficient member of the Society for prayer...he was a person of genuine and ardent piety, exemplary as a Christian...felt a lively interest in all that belongs to the prosperity of the Kingdom of Christ and was held in much esteem and respect by all who knew him."

John and Mary Geddie had four children: Jane, born in Scotland (d. about 1840), married Rev. Dr. Wm. Fraser of Pictou in 1834. Mary Ann (b. 1812 – d. 27 Apr. 1862) married James Johnson. A third daughter married Mr. Henderson of California. John Jr. (b. 10 Apr. 1815 – d. 14 Dec. 1872) married Charlotte L. MacDonald (b. 1822 – d. 1 Jan. 1916). As a boy, John helped his father build clocks. He became a minister of the Presbyterian Church and the first foreign missionary appointed by any colonial church in the British Empire. Detailed accounts of his life appear in *Missionary Life Among the Cannibals* by Rev. George Patterson and *Misi Gete* by R.S. Miller.

FIG. 46b

FIG. 46b *Dial detail of the clock in Fig. 46a.*

~ GEDDIE ESTATE INVENTORY ~

We the subscribers having been appointed to appraise and value the goods and chattels of the late John Geddie watchmaker do value and appraise the same as contained in the following inventory:

Contents of drawer No. 9	£2.0.0
Contents of drawer No. 8	1.0.0
Contents of drawer No. 7	0.7.6
Contents of drawer No. 6 (watch glasses)	10.7.0
Contents of drawer No. 4	0.1.6
Contents of drawer No. 3	0.3.0
Contents of drawer No. 2	1.3.6
Contents of drawer No. 1	0.1.0
Box of drawers containing above	0.5.0
One small box tools	0.3.0
One box watch keys	0.4.0
Main springs	2.8.1½
A lot of watch hands	0.12.0
15 chains for inside of watches	1.7.0
One clock barrel cutter	0.7.6
One box Brass and Copper	0.12.0
One shop lamp	0.1.3
Two small table vices [*sic*]	0.15.0
Four magnets	0.6.0
Ten dozen watch Dials	5.0.0
Two boxes Spectacle Eyes	0.5.0
Two doz. watch keys	0.4.0
One glass Case	0.5.0
Two Eight day clocks	16.0.0
	£43.18.4½

Page 2 [of inventory]
Amount brought forward £43.18.4½

One Eight day small clock without case	5.0.0
Four fine clock cases	8.0.0
Two large vices [*sic*]	1.15.0
One hand lathe	1.0.0
One small hand Lathe	1.2.6
One large Drawer	1.0.0
A lot of clock tools	3.10.0
One clock & watch wheels engine	10.0.0
20 clock work	50.0.0
28 watches	86.0.0
One shop stove	3.0.0
One cooking stove	4.10.0
One cow	4.0.0
Wearing apparel under £10.0.0	0.0.0
One chest of Drawers	5.0.0
One Birch Table	1.5.0
1/2 dozen chairs	0.13.0
One cup-board	0.7.6
1/2 Dozen silver Tablespoons	6.0.0
1/2 Doz. silver Teaspoons	1.0.0
One silver ladle	3.0.0
A set of books	2.0.0
Carpet	0.10.0
Bedding etc.	8.0.0
Kitchen utensils	0.15.0
A lot of old silver	2.8.9
	£252.17.1½

Pictou 5th May 1843 [signed]   R. Dawson
Wm. S. Hidet [?]
James Allan
Appraisors

In 1843, the exchange was roughly $4 to one English pound.

GORDON GILCHRIST

Gordon Gilchrist (b. about 1760 – d. 1846) was a Scot who came to St. Andrews, New Brunswick, and worked as a carpenter and builder from about 1819 to 1835. Little is known about the man. However, he was the first cabinetmaker on record in St. Andrews. He crafted many fine pieces of furniture in the Regency style. One such specimen of his craft is an elaborately carved pulpit in Greenock Church in St. Andrews, which he made without nails. John Warren Moore, a well-known cabinetmaker, was apprenticed to Gordon Gilchrist of St. Andrews and was very active in St. Stephen in the 1840s.

A case for a tall case clock known to have been made by Gilchrist in 1822 is in the New Brunswick Museum. It has a pediment of carved thistles. An illustration of the clock may be seen in *Antique Furniture by New Brunswick Craftsmen* by Huia G. Ryder.

FIG. 47 *Clock sold by William Gossip. – Courtesy the Nova Scotia Museum, Halifax*

WILLIAM C. GOSSIP

William Gossip came from Plymouth, England, to Halifax, Nova Scotia, in 1822. He was a master smith and was employed by the Ordinance Department in Halifax and worked in "His Majesty's" lumberyard. Following the death in 1829 of James Deckmann, the keeper of the Halifax Town Clock, William Gossip added these extra responsibilities to his own duties and in return was given lodging in the building under the clock and an additional annual salary of £26. He was responsible for keeping the clock wound and for cleaning the dials. In order to accomplish the latter task, he was required to venture out onto a 5-foot-wide ledge, with no guard rail, at the base of the clock dials.

In the 1820s or 1830s, Gossip sold a bracket clock made of mahogany with brass inlays. The brass face of the clock is signed "William Gossip, Halifax, Nova Scotia." The clock is now in the possession of the Nova Scotia Museum, and according to the *Museum Report of 1925*, there is also in existence a tall case clock of mahogany inlaid with brass attributed to William C. Gossip.

Little is known of Gossip's personal life. The only William Gossip listed in the Census of Halifax in 1838 was a printer with a household of eleven members. It is possible that William Gossip, clockkeeper, was part of that household of which William Gossip, the printer (and perhaps father) was the head.

It is known and recorded by Dorothy Evans in an article called "The Old Town Clock" in the *Atlantic Advocate*, January 1962, that T. Gossip, whose relationship to William Gossip is unknown, was paid to help replace the coppered cupola in 1826.

THE HALIFAX TOWN CLOCK

The Town Clock[12] in Halifax, Nova Scotia, has been an important landmark in the city for almost two centuries. It stands on the eastern face of Citadel Hill, facing Halifax Harbour. In 1803 the location was close to the military barracks, and the military, who were responsible for its care and repairs, referred to it as the "garrison clock." However, to the people of Halifax, including Governor Wentworth, it was the "Town Clock."

Founded by Sir Edward Cornwallis of England in 1749, Halifax has always been associated with the military, and the star-shaped fort called the Citadel was immediately built to counter the French threat at Louisburg, Nova Scotia. Now, Halifax Citadel National Historic Park, with the Town Clock in full view, is the most visited National Park in Canada.

When the Duke of Kent, father of Queen Victoria, was in Halifax in 1800, he saw the need for a clock and had the plans drawn up by the Royal Engineers for its construction.

The clock was obtained from the firm of Benjamin Vulliamy, British clockmakers, at a cost of £339. The mechanism, which weighs about 1,000 pounds, is driven by three 125-pound weights, one weight each for time, hour-strike, and quarter-hour-strike. The shaft into which the weights descend is 45 feet deep. A 12-foot pendulum, suspended from the clock, also descends into the shaft.

The clock's three bells strike twice on the quarter hour, four times on

Fig. 48

Fig. 49a

Fig. 48 *The Halifax Town Clock.*
Fig. 49a *A view of the movement of the Halifax Town Clock.*
  – Courtesy Environment Canada - Parks, Halifax Citadel National Historic Park

FIG. 49b

FIG. 50a

FIG. 49b *Another view of the movement of the Halifax Town Clock. – Courtesy Environment Canada - Parks, Halifax Citadel National Historic Park*

FIG. 50a *A clock sold by H. Hunt & Co.*

the half hour, six times at the quarter to the hour, and eight times on the hour, after which the hour is struck. The clock is wound every week by cranking each of the three weights to the top of the tower. An impulse is delivered to the pendulum at two-second intervals.

An article by Dorothy Evans in the *Atlantic Advocate* in 1962 is quoted in the treatise by Barbara Schmeisser, *Town Clock, 1803–1860, a Structural and Narrative Study*.[13] The article states that, "The frame was made of heavy hand-wrought iron exactly and securely fitted together and supporting a strange assortment of brass cog wheels and pulleys on which were wound ropes that were fastened on the weights...one wheel marked off from one to sixty...enabled the clockkeeper to set the time without actually seeing the hands...."

Originally the clock had three dials 7 feet in diameter, and a smaller dial that faced the nearby Citadel. In 1836 a long poem, "To the Town Clock," which mentions the dials, was written by the Hon. Joseph Howe:

...A double face, some foolishly believe,
Of gross deception is a certain sign;
But thy four faces may their fears relieve,
For who can boast so frank a life as thine...

The original dials were black with white Roman numerals. In later years this colour scheme was reversed. Most recently the dials have been coloured dark blue with gold painted hands and Arabic numerals. The dial facing the Citadel is 6 feet in diameter and the other three are now 8 feet.

The entire structure that houses the clock cost £450. The architect is thought to have been William Hughes, a master builder at His Majesty's Dockyard. According to Alan Gowans, architectural historian, this basic classic design was not unusual for the period, so no one bothered to record the architect. The design was a three-tiered, irregular octagon.

There are few details available about the construction, which was probably started in 1802. On 20 October 1803 part of the clock was mounted in the tower. It was not until 1805 with the arrival of Alexander Troup, a sergeant in His Majesty's 98th Regiment and a watchmaker by trade, that the movement was put into working order (see p. 56). The clock tower measures about 46 feet from the bottom of the colonnade to the top of the cupola, giving the structure an overall height of about 60 feet. The part of the tower that houses the clock is approximately 16 feet in diameter. A detailed description and illustrations of the structural aspects of the building and changes made to 1860 can be found in the book by Barb Schmeisser.

The north and south sides of the building below the clock tower are 30 feet long and the east and west sides are 51 feet in length. The height of this part of the structure is approximately 39 feet, allowing for about 1,500 square feet of living space under the clock. Although the first clockkeeper, James Deckmann, lived elsewhere, many clockkeepers and their families did occupy the living space below the clock.

The second clockkeeper, master smith William Gossip, who assumed this duty in addition to his own after the death of Deckmann in 1829, did live in the building, receiving in addition to his lodging the annual sum of £26. At that time a repair to the clock costing £49 was made. William Gossip continued to be in charge of the clock into the 1850s.

By 1857 the city of Halifax assumed the cost of paying the clock-keeper, and in the mid-1860s the person filling this position was elected yearly. According to A.M. Payne in an article in the *Halifax Herald* of 31 July 1896, William Crawford and John McCulloch were two of the successors to William Gossip. It has, however, been many years since the building was occupied by the clockkeepers.

The clock and clock tower have weathered many disasters, perhaps protected by the ghost that is said to haunt it. The 1917 Halifax explosion, the greatest man-made explosion prior to Hiroshima, killed 1,700 people and left many others injured and homeless, but did little damage to the clock. The clock also escaped with little harm in 1945 when the Royal Canadian Navy magazines on the Dartmouth side of the Halifax Harbour blew up, causing heavy property damage; and when, on V.E. day in the same year, there were riots in Halifax.

The clock has been restored and repaired many times in its 190-year history. J. Cyril Tanner reported in an article in the *Southender* that major replacement parts for the old Town Clock were obtained from the tower clock that was salvaged from the Navy's victualling depot, which was torn down in the late 1940s. The clock's most recent restoration was carried out in 1992.

FIG. 50b

H. HUNT

Hiram Hunt, an American, sold clocks in St. Stephen, New Brunswick, some time between 1841 and 1845. H. Hunt & Co. sold OG clocks to both the wholesale and retail trade, as indicated by the labels affixed to the clocks shown in Figs. 50 and 51.

The movements of the clocks were made by H. Welton Co., who began making movements in 1837 in Terryville, Connecticut. As a result of probable litigation with Chauncey Jerome regarding infringement on the Jerome patent, Heman Welton asked Eli Terry Jr. to design a movement that would not infringe on the patent. In order to accomplish this task, the count wheel was moved to the back of the movement.

Hiram and Heman Welton took over the business of Eli Terry Jr. in 1837 (Eli Terry Jr. died on 21 May 1841) and proceeded to manufacture "Welton" movements. In 1845 they were forced into bankruptcy and the factory later was used to make locks.

In *The Book of American Clocks*, Brooks Palmer asserts that Hiram Hunt made both tall case and shelf clocks in Bangor, Maine, and that he was associated with the clock business from 1806 to 1866.

The clocks sold in St. Stephen, New Brunswick, carry labels printed by "S.S. Smith, Printer, 13 West Market Place, Bangor." On the labels the name of the town St. Stephen is misspelled, "St. Stephens."

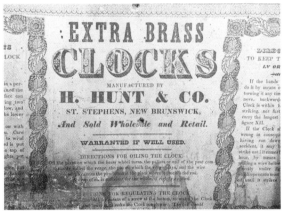

FIG. 51

FIG. 52

FIG. 50b *A clock sold by H. Hunt & Co.*
FIG. 51 *Label of clock in Fig. 50a.*
FIG. 52 *Welton movement found in the clock in Fig. 50a.*

THE HUTCHINSON FAMILY

FIG. 53   FIG. 54a

FIG. 54b

FIG. 53 *Elegant tall case clock sold by W. &*
G. Hutchinson, Saint John, New Brunswick.
See page 3 for dial detail picture.
– Courtesy Barbara and Henry Dobson
FIG. 54 *George Hutchinson clock in the New*
Brunswick Museum, Saint John.

During the 19th century four generations of the Hutchinson family served the people of Saint John, New Brunswick, as jewellers, gold- and silversmiths, and watch- and clockmakers. Also, because of the proximity of the sea, they were experts in chronometers and nautical instruments.

William (b. before 1794 – d. before 1863), the founder of the company, was born in Londonderry, Ireland, where he learned the watchmaking trade as his father, William, had before him. In 1819, while sailing to Philadelphia to take over his brother's business in the New World, William and his wife were shipwrecked near Liverpool, Nova Scotia. His wife, Sarah (d. about 1822), was unable to travel further by sea, so the family decided to settle in Saint John, New Brunswick. There William set up shop on the "Old Coffee House Corner" on Prince William Street. In Philadelphia, Thomas Hutchinson, thought to be William's ailing brother, died in 1820.

Shortly after the birth of William's son William[14], in 1821, his first wife died, and in 1823 he married Sally Nisbet, daughter of Thomas Nisbet, a cabinetmaker who made cases for Hutchinson clocks (see p. 48).

In 1820 William's brother George (b. about 1774 – d. 19 Nov. 1877) and his wife, Mary (b. about 1794 – d. about 1884), came to Saint John and formed a partnership with William. During their time in business, the Hutchinsons imported clock movements from Great Britain, had them cased by resident cabinetmakers, and sold the clocks. The York-Sunbury Museum in Fredericton, New Brunswick, has on display a clock made during the brothers' partnership.

In 1834 William's brother George left the partnership and William continued the business. In May 1842 he advertised in the *Courier,* "Watch and Clockmaking in all its branches, jewellery made and repaired, silverware, Nautical instruments touched, repaired and warranted." Smythe, a silversmith, and Larson, a goldsmith, were employed as foremen.

George established a business on Dock Street in Saint John. For about twenty years he advertised in the newspapers as a watch- and clockmaker. A wall clock made by George about 1840 can be seen in the New Brunswick Museum. The case is mahogany and it is 71 inches in height. A picture and description of the clock can be found in *Canadian Clocks and Clockmakers* by G. Edmond Burrows. A tall case clock with the name of George Hutchinson on the dial is also on display in the New Brunswick Museum.

After George left the partnership, he managed a shop alone until he was joined in 1846 by his nephew George. George Hutchinson Jr. (b. 1818 – d. 6 July 1891) learned the trade from his father William and his uncle George. George Jr. married Margaret Wallace (b. 1825 – d. about 1900.)

When William retired in 1856, George and his uncle George purchased William's shop and carried on the business as G. and G. Hutchinson on Prince William Street, Old Coffee House Corner.

The elder George Hutchinson retired in 1860, and George Jr. continued the business. The business was moved several times, but by 1875 it was back on the Old Coffee House Corner, Prince William Street. In this period, the firm expanded the jewellery manufacturing branch, hiring goldsmiths and silversmiths.

The addresses of the business from 1857 for the next thirty years were as follows:

1857: G. and G. Hutchinson, 19 Prince William Street
1863-64: G. Hutchinson, 3 North Side Market Square
1865-66: G. Hutchinson, 70 Prince William Street
1874: G. Hutchinson, 38 Prince William Street
1880: Hutchinson & Co., 49 Charlotte Street (Market Building)
1884: G. Hutchinson, 74 Prince William Street.

In 1874 George Jr. was joined by his son Daniel L. (b. 1855), who was learning the trade. On 20 June 1877 the Hutchinson family lost the shop and all the contents of the store and safe in the great fire of Saint John, which "laid two-thirds of the city to ashes in nine hours."

George Jr. and his son re-established the business and formed a company specializing in wholesale and retail jewellery. The shop was located in the Market Building where it remained in operation until 1883.

Shortly after William's arrival in New Brunswick, he was asked to take responsibility for the winding and repairing of the clocks of the city, a responsibility that he eventually passed to his sons. Most prominent of the city clocks was the one in the tower of Trinity Church (erected 1791). In 1812 it was fitted with a clock made by Barraud of Cornhill, a maker in London, England. The cost was £221.19s sterling. The clock had three dials, with a fourth added in 1857. The Hutchinsons cared for the clock until the church was destroyed in the great fire of 1877. The tower with the clock was the first to fall.

In 1870 an observatory on Fort Howe Hill was completed and the "time ball," described in the *Daily Telegraph* of 1 March 1870, was installed.[15] It was dropped at one o'clock every weekday. George Hutchinson was appointed director of the observatory and the time ball at a salary of $500 a year. Because the time ball was not easily seen from the harbour, it was moved to the roof of the customs house on the waterfront in 1873. However, the fire destroyed the observatory and a temporary ball was mounted on the roof of the Anchor Line warehouse until the new customs building was completed in 1880. A permanent time ball was installed on the north tower. George Hutchinson continued as director of the new observatory. His son Daniel was employed at the observatory and his second son, Norman, worked with his father in the meteorological department at the building. In the 1880s the equipment was updated to ensure a more accurate recording time. After George's death around 1891, Daniel became director at the Saint John Observatory.

During the period that the Hutchinson family was in business in Saint John, other Hutchinsons were watchmakers and jewellers in other towns. No relationship between the families has been established. A business in St. Andrews, New Brunswick, was operated by a William Hutchinson, commencing about 1835. It has been speculated that William from Londonderry, father of William, George, and Thomas, followed his sons to America and established this business, which continued for about twenty years. Sometime during the existence of the St. Andrews business, a George Hutchinson took over. In 1845 George Stickney, watchmaker, advertised that he was "located in the shop in St. Andrews of the late George Hutchinson, Water Street."

John Hutchinson was a jeweller and goldsmith in Saint John prior to 1833. It is not known to what Hutchinson family John belonged. In the early 1830s he had studied dentistry in New York and returned to Saint John in January 1833 to establish a practice. He advertised as a jeweller and dentist, stating that he had "performed many experiments in human teeth, his mode of plugging the filing [*sic*] are on the most improved plan...will devote a portion of his time to tuning piano fortes also." His business was on Germain Street, close to Trinity Church.

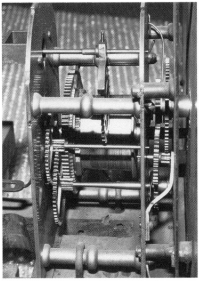

FIG. 55

184 G. & G. HUTCHINSON.

G. & G. HUTCHINSON,

HAVE FOR SALE AND TO HIRE,

CHRONOMETERS

OF THE MOST APPROVED MAKERS.

Chronometers Accurately Rated

By a Fixed Transit Instrument and Astronomical Clocks.

ALL NECESSARY REPAIRS DONE WITH PROMPTNESS AND SKILL.

WATCHES & JEWELLRY

Of all descriptions for Sale, and Repaired.

BAROMETERS,

SEXTANTS, QUADRANTS,

SPY GLASSES,

COMPASSES, &c., &c.

CONSTANTLY ON HAND, AND REPAIRED AT THEIR

Nautical Instrument Establishment,

19, Prince Wm.-street

Corner Market Square, Saint John, N. B.

N. B.—All Chronometers left at this Establishment are insured against loss or damage by Fire.

FIG. 56

HUTCHINSON,

ESTABLISHED 1819.

IMPORTER AND REPAIRER OF

Chronometers, Watches, Clocks,
Nautical Instruments, &c.

Spectacles and Eyeglasses in all styles correctly adjusted to sight.

SOLID GOLD JEWELLERY, with or without Gem Settings, manufactured to order by experienced workmen in this establishment.

CHRONOMETERS on hire and accurately rated by comparison with Astronomical Clocks corrected by Solar and Siderial Observations with transit instrument.

GEO. HUTCHINSON, Jr.,

No. 2 IMPERIAL BUILDINGS,

PRINCE WILLIAM STREET, ST. JOHN, N. B.

FIG. 57

FIG. 55 *Movement of the Hutchinson clock in Fig. 54. – Courtesy the New Brunswick Museum, Saint John*

FIG. 56 *Advertisement from Saint John City Directory, 1857.*

FIG. 57 *Advertisement from the Saint John City Directory, 1874.*

JOHN JURY AND FAMILY

John Jury and his descendants maintained a clock and watch business for over ninety years in Charlottetown, Prince Edward Island. At least one tall case clock bears his name. John Jury's position as an important horologer is assured because he appears to have been the earliest clock- and watch-maker in Prince Edward Island.

He was in business by 1813 when he advertised in the *Weekly Recorder* of 27 September 1813 for an apprentice, a boy between twelve and fourteen years of age to learn the business of clock and watch repairing. "He must be a good steady boy with Education and well-recommended. Terms not less than five years.... I cannot expect any premium in this new country."

In a further notice that appeared in the *P.E.I. Register* of 10 July 1824, Jury advertised that he had "imports to sell and he did repairing. He devotes his time entirely to his business." However, the 29 October issue indicated other pursuits. He had "three rooms and a cellar for let.... Also having erected a kiln for drying hops, malt and grain, the same will be done upon the most accommodating terms. For sale by retail, a quantity of hops." A further advertisement in June 1825 announced the arrival from London of a large assortment of jewellery items and watch materials. At that time he also had to let a blacksmiths' forge and a large room.

In spite of the fact that in 1830 he opened a school of music in St. Paul's, the Anglican parish church in Charlottetown, the Jury watchmaking and jewellery business continued. In 1839 John Jury placed a notice in the *Royal Gazette* in which he expressed his concern about the inaccuracy of the time given to the inhabitants of Charlottetown, and the Island in general, because the noon gun time

> is always too fast, not only by the true mean time, but by the apparent time also. The gun is in error 9 – 12 minutes by the Sun time and 15 – 18 minutes from true time as it suits them.... I wish to inform you that Captain D. and me took the Sun with the Quadrant at 12 o'clock on Thursday the 6th of December 1838 and we proved my time to be correct, the Sun being too fast on that day 8 minutes and 47 seconds. It don't want an Astrologer to find out that the gun is too fast by the Sun for any man that has an almanak [*sic*] can see by the Sun's rising and setting, that the gun time is too fast. But what am I thinking about? Keeping time with the Sun! It is impossible for all the Watchmakers, Doctors or Soldiers in the world to keep time with the Sun, as the Sun is always going too fast or too slow the whole year round. The Third of November last, the Sun was too fast by 16 minutes and 16 seconds...the rightest time it is in the whole year is on the 15th of June; then it is right within one second.
>
> A Captain at sea must first take the Sun and allow the equation and bring his chronometers into true mean time; and by keeping that time, I have known a chronometer to be regulated so as to keep time for a whole year and not vary 2 seconds; and a patent lever, with maintaining power may be made to keep time for a whole year without varying very little; also a good eight-day clock...may be made to go the whole year without altering.... So I hope you see it is impossible for you to regulate your clocks, watches and timepieces by the Sun.... I Have known people in this place to alter their watches every night by the gun – more fools they! For my part, I have no trouble with my timepiece, but wind it up every Monday night, as I understand keeping the true time I can assure you that you will never have the right time in this town until you keep true mean time. [signed] J.J.
>
> Postscript: Tues. January 1st, Sun slow three minutes and fifty seconds.

The business was eventually managed by John A. Jury, John Jury's son, who trained under his father and, according to John E. Langdon, "then spent two years in one of the principal towns of the United States and in one of the first shops for business." In 1869 the shop was on Kent Street, Charlottetown, moving in 1870–1871 to Queen's Square, North, Charlottetown.

The third generation of Jurys was represented by George G. (b. about 1853 – d. between 1904 and 1908.) The business had again moved and was on Grafton Street in the 1880s. George incorporated the business and Jury and Company continued to prosper until the death of George G. Jury. His sons, Horace, Percy, and Russell, did not continue the business.

Another son of John A. Jury, Thomas W.D. (b. 1863), was listed as a watchmaker in the census of 1881. No further information was found about Thomas Jury.

RICHARD UPHAM MARSTERS

Richard U. Marsters[16] (b. 31 Oct. 1787 – d. 11 Feb. 1845) was a chronometer maker in Nova Scotia. According to the newspapers of the time, he was considered to be the first person to make "practical chronometers in North America."

In 1801, at the age of fourteen, Richard Marsters was apprenticed to a watchmaker and silversmith, David Page Sr. of his home town, Onslow, which was situated about 60 miles from Halifax. After completing his training (usually seven years as an apprentice and one or two years as a journeyman), Marsters remained with Page until 1817. He then went into business for himself in Halifax, a busy port and naval base.

An advertisement dated 31 March 1817 appeared in the *Acadian Reporter*, giving his place of business as "the House of Mr. George Innis, merchant, No. 2 Sackville Street opposite Richard Tremain and Co." In October 1819 he moved to the house on Sackville Street belonging to Richard Tremain and Co. Marsters also took an interest in nautical instruments; besides timekeepers and jewellery, he also adjusted and repaired compasses and quadrants. His ability as a craftsman was indicated by advertisements in which he advised the public that "any kind of watch wheels will be made and gilded, agreeably upon short notice...."

Apparently, Marsters detected deficiencies in the chronometers that passed through his shop. He was determined to make chronometers that, along with other improvements, would compensate for extremes of temperature. In order to learn more about the manufacture of chronometers, Marsters went to England, where he remained for a year. Because several features of his own chronometer resemble those of chronometers made by Thomas Earnshaw (1749–1829) of England, it is suspected that Marsters may have worked with him during his time in England.

On his return to Halifax, he offered chronometers for sale that he had constructed to "withstand the greatest vicissitudes of climate to which chronometers are exposed." He also advertised that he would make to order 1-day to 8-day chronometers. Several testimonials attested to the accuracy of chronometers made by R.U. Marsters, including a letter from D.W. Watson, Lieutenant of the Royal Navy, who wrote of the excellent service chronometer number 20 had given.

When he first started to manufacture chronometers, Richard Marsters realized that he had little money to further his research. He knew that much time would elapse before the excellence of his improved chronometers was recognized and appreciated by the ship owners. Therefore, on 16 February 1826 he petitioned the House of Assembly in Nova Scotia for funds to "perfect many of his inventions and improvements with respect to adaptation and temperament of the machinery." He was granted £98 to "purchase a Transit Instrument to

FIG. 58

FIG. 58 *Views of the Marsters chronometer. – Courtesy W.R. Topham*

43

further the important views of said Marsters in relation to his contemplated improvements in chronometers."

In February 1828 he advertised that he had "set up a temporary observatory by which he is able to rate all time-pieces with great exactness." Nautical instruments and chronometers were checked here. By 1831 he had chronometers for sale.

In late 1831 or early 1832, Marsters left Halifax for New York. His interest in chronometers continued, and in 1832 the *Halifax Journal* announced to the people of his former town that Marsters had exhibited a marine chronometer at the annual fair in New York. Newspapers in both Nova Scotia and New York claimed that "it is the first machine of the kind manufactured in the United States." The newspapers also stated that "Mr. Marsters not only made watches and chronometers in the Province (of Nova Scotia) but that he had erected the first Transit Instrument for the benefit of ship owners and mariners in the North American Colonies in Halifax."

By 1834 Marsters had returned to Nova Scotia, and the census of that year lists him as a chronometer maker in Falmouth, Nova Scotia. Although he was in Windsor, Nova Scotia, in 1838, Falmouth remained his home. Only one of the chronometers made by Richard U. Marsters is known to exist – number 765. This chronometer is on display at the Marine Museum of the Atlantic in Halifax, Nova Scotia.

David Cooper[17] of Boulder, Colorado, has examined the pictures of the Marsters chronometer and was able to make several observations. Like the American-made Little & Elmer chronometer,[18] Marsters "mounted the dial with feet leaving depressions in the dial.... The dial is inscribed with the name of the maker with his number, rather than the maker of the ebauche, as is the case with contract pieces. Most interestingly, both the Little & Elmer and the Marsters chronometers are not really two-day chronometers, the Marsters being only one day six hours. Further, the indicator dial is an upside-down configuration with the up and down being at the bottom of the indicator dial.... Like Little & Elmer, it would appear that Marsters also made his own tub and box.

"Some items are particular to Marsters. The seconds bit is unusually close to the center wheel (hour, and minute-hand post) which would indicate that the clock has a rather small gear train clustered around the center wheel and balance. This small gear train would mean that he was a watchmaker first and a chronometer maker second."

Mr. Mercer, an authority on chronometers, suggested to David Cooper that the Marsters chronometer might have an entirely new escapement invented by Marsters. An examination of the actual chronometer would be needed to confirm this supposition.

"The outer edge of the dial is rounded rather than being flat.... It would also appear that the plate jewels are held in with three screws instead of the usual two. Some other features particular to Marsters are the very poor finish of the plates which are fastened with pins, highly unusual; the unusual hair-spring stud; and the ratchet wheel of only two teeth. There is also a rather odd cutout in the back plate for what would appear to be a detent bridge of some sort which tends to support Mr. Mercer's suggestion concerning an original escapement."

Mr. Cooper believes that Marsters bought the completed frame (available at the time). He also finds that the turning on the columns and the fusee stop spring screw is quite fine compared to the rather heavy-handed finish of the bridges. "Whether Marsters made the barrel and fusee and chain would be open to question. The fusee is definitely of an unusual configuration. Fusee engines would have been a great rarity in (North) America at the time, since they were expensive and England was less than anxious to have such trades carried on (outside Britain). It would seem that he (Marsters) made enough chronometers to justify the cost of the engine."

44

It seems strange that more of the Marsters chronometers have not survived. Mr. Cooper thinks that some fatal flaw must have rendered them unrepairable after a certain period of time. He suggests that one possible problem was the driving of a large balance with a small gear train and a heavy main spring. This would cause great wear on the gears, eventually making the instrument useless and unrepairable. The other suggestion Mr. Cooper made was that, because of the short length of time between windings of this one-and-a-quarter-day instrument, it was risky to have on board. A delayed winding would make it useless. The chronometers may have been replaced by instruments that went for a longer period of time between windings.

In addition to his interest in chronometers, Marsters was engaged in a number of other pursuits. In 1817 he claimed that he had invented a water or propelling wheel for use on steamboats and described its effectiveness in the *Acadian Recorder*. He also made teaspoons and medals. In 1819 he produced a silver medal for the Provincial Agricultural Society. It is now in the Historical Society Museum in Wolfville. In 1820 he made a gold medal, now in the Nova Scotia Museum, for the Halifax Regiment of Militia.

From the year 1829 until he left for New York, Marsters was in the boarding house business and he advised the public that he was "filling up a Boarding House, formerly known as Millars Hotel."

Unlike many horologers Richard U. Marsters did not come to the land that became Canada from elsewhere; he was a native of Onslow, Nova Scotia. His grandfather, Jonathan,[19] with his two brothers, Abraham and Moses, all sons of Abraham and Deborah (Knowlton) Marsters and their families left Manchester and Salem, Massachusetts, in their own vessel and arrived in Falmouth in 1760 and occupied land vacated by the Acadians. Jonathan and his wife, Mary, brought with them eleven children.[20]

One of Jonathan's sons, Nathaniel (b. 1758 – d. Sept. 1843), father of Richard U., moved to Onslow in 1784, where he became a merchant. Nathaniel was also a Member of Parliament from 1806 to 1818, a magistrate for thirty years, a registrar of deeds and probate and a coroner from 1820 to 1843. Nathaniel married Sarah Upham (d. 29 Nov. 1789). Their son Richard was born on 31 October 1787 and his brother Robert K. was born on 22 November 1789. After Nathaniel married Lynda Lyns, three more children were born.

In 1819 Richard Marsters married Ann McKay, widow of Hector McKay, a merchant from Dornoch, Scotland. It is believed that there were two daughters, Ruth and Sarah Ann. In the 1830s Marsters's wife and children moved to the United States and in 1838 a notice appeared stating that he would no longer be responsible for her debts.

Richard U. Marsters was related to the Brown family, in which the father, sons, and grandsons were watchmakers and silversmiths for a period of almost one hundred years. Richard's aunt married William Brown (b. about 1750 – d. about 1795) from Tipperary, Ireland, who originally trained for the priesthood. After his death Deborah and her five sons went to live with her father, Jonathan. Both she and her father were killed in a fire in her father's house in 1804. Her sons escaped. Deborah's son William Brown had descendants who were watchmakers and silversmiths.

Richard U. Marsters died on 11 February 1845. His will, dated 9 December 1844, gave his occupation as "chronometer maker." His estate comprised property worth £521, a bequest from his father's estate of £1,250, and bedding worth £2. His entire estate was left to his daughter Ruth, but the executors were unable to trace her whereabouts.

His watchmaker's tools and other articles were appraised at £15.7.9.

JOHN McCULLOCH

John McCulloch (b. 1821 – d. 29 Nov. 1875) was a gold- and silversmith and a watch- and clockmaker in Halifax, Nova Scotia, from 1844 to his death. He came from Glasgow, Scotland, in 1837 and apprenticed to Peter Norbeck, silversmith. On 14 August 1844 he advertised that he was commencing business on Granville Street near George Street. An advertisement of 1854 placed by McCulloch, "watch and clockmaker and jeweller," gave the address of his shop as 36 Granville Street.

FIG. 59

FIG. 59 *McCulloch clock.*
– Courtesy the Nova Scotia
Museum, Halifax

He was known primarily for a large clock with gilt figures that he built about 1850. It was made in the shape of the monument to Sir Walter Scott in Edinburgh, Scotland. In 1851 it was exhibited in England in the Great London Exhibition in the Crystal Palace. In 1852 the clock was raffled in London and was won by William Smith of Bathurst, New Brunswick. The clock won honourable mention in the Nova Scotia Industrial Exhibition in Halifax in 1854. It was eventually donated to the Public Archives of Nova Scotia and is now in the possession of the Nova Scotia Museum.

In addition to the clock, John McCulloch is known for the beautiful jewellery that he fashioned from gold and silver. In 1866 he won an award for gold jewellery at the Exhibition in Paris, France. Also, first prize for jewellery in native gold was awarded to him in the Nova Scotia Industrial Exhibition of 1868.

His talent is manifest in the medals that he made. In 1862 the *Colonial Standard* described medals made for the Halifax Mayflower Rifles and the Militia Co. of Engineers, Dartmouth.

As his business expanded he advertised in 1863 for youths to apprentice as silversmiths and watchmakers. One of McCulloch's apprentices, Thomas C. Johnson (1853–1923), began business as a watchmaker and jeweller in Halifax in 1874. Johnson's sons, Charles and Albert, were taken into the business in 1891 and descendants continued the business until 1974.

Also in 1863, McCulloch opened a branch shop in Liverpool, Nova Scotia, with a partner, Andrew W. Carten, who managed the Liverpool shop. His address in Halifax in 1868 was 83 Granville Street.

In addition to his business life, he was very active in Halifax civic affairs. He was an alderman for many years and, while on the city council, was considered to be one of its most diligent and useful members. He was, however, defeated twice as a candidate for the office of mayor. He was a school commissioner and one of the principal movers in establishing a Halifax public garden, which was opened in August 1867.

According to the book by Donald Mackay,[21] John McCulloch was related by marriage to several watchmakers and silversmiths in Nova Scotia. He was married in June 1858 to Mary Jane, daughter of Robert and Ann Kerr. There were no children.

When his health failed, he lived with his wife's niece, to whom he gave a number of pieces of gold jewellery that he had fashioned. He died on 25 November 1875 at fifty-four years of age.

46

JAMES G. MELICK

James Godfrey Melick (b. 24 May 1802 – d. 8 May 1885) was first listed as a watchmaker at Saint John, New Brunswick, in 1824. One clock bearing his name is in the New Brunswick Museum. As can be seen in Fig. 60, this is a rather curious little clock. It has the shape of a tall case clock, but is less than half size. Its unusual lines suggest the work of a local craftsman. The time-only movement has the general appearance of those used in weight-driven banjo and wall clocks of the period. The authors have not examined the clock at first hand, so it is difficult to speculate on the movement's origins. The clock has a brass dial, which by the middle to late 1820s would have been somewhat old-fashioned. It would seem probable that the clock dates from the early period of Melick's business career.

James G. Melick was a jeweller and watchmaker in Saint John for nearly forty years. In 1828 his first place of business was on the "North side of Market Square near Mr. Baldwin, Surgeon." By 1835 he had moved to the east side of Market Square. In July 1845 he announced that he was moving his shop to Prince William Street "nearly opposite to L.H. De Veber and Son." However, a fire in Saint John demolished his store a few days after he had opened it. After the fire he was missing a "drawer containing watchmaker's tools," for which he advertised in August 1845.

In 1846 his shop was at 21 Prince William Street. On the second floor of the shop was a daguerreotype studio, and Melick took the owner Robert Toules as his partner. Although the partnership lasted only eight months, Melick ran the studio for five years. During this time he set up his watchmaking shop at 54 Dock Street. The city directories list his shop at 21 Prince William Street in the 1850s and at 54 Dock Street in the early 1860s. He retired in 1864 and purchased a residence in Hampton, New Brunswick, where he resided until his death. His son, Frederick C., was a watchmaker and jeweller. He was in business from the mid-1860s until some years into the 20th century.

James's ancestors were German, and after their arrival in America the families changed the spelling of the family name so that a number of variations exist, namely Moelick, Mellick and Melick.

James's great-grandfather, Wilhem Moelick, lived in Bendorf, near Coblenz, in Germany. In 1735 Wilhem and family came to Philadelphia on a ship called the *Mercury*, landing on 29 May. One of his sons, Gottfried (b. 1724 – d. 11 Sept. 1760), changed his name to Godfrey in America. Godfrey settled in Greenwich, New Jersey, in 1747, and on 20 May 1748 he married Margaret Folkenburger (b. Feb. 1733 – d. 26 Dec. 1799). They farmed 365 acres near the village of Bedminster, known as the "Lesser Crossroads." Godfrey and Margaret had ten children, two of whom (William and John) became United Empire Loyalists. William (b. 1753 – d. 27 April 1808), a currier, came from New Jersey to settle in Saint John, New Brunswick. His daughter, Mary Melick (b. 26 Jan. 1806 – d. Nov. 1884), married Thomas Plummer, and his granddaughter, Mary Crown, married William Plummer, the cabinetmaker who dealt in weight clocks in Saint John. Mary C. Plummer's sister, Sophia (d. 1886), married Israel Hawes, the partner of W. Fairbanks.

John Melick (b. 1762 – d. 6 May 1856), a cordwainer, was born in Warren County, New Jersey. His great-grandson, Andrew D. Mellick Jr., wrote a book in 1889 entitled *The Story of an Old Farm*, which told about the family life and times and gave a detailed genealogy of the family.

In 1783 John came to Saint John, New Brunswick, and in July 1793 he married Mary (b. 31 Jan. 1777 – d. 19 Apr. 1857), daughter of Joseph

FIG. 60

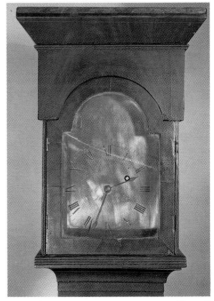

FIG. 61

FIG. 60 *Clock inscribed James G. Melick. – Courtesy the New Brunswick Museum, Saint John*
FIG. 61 *Close-up of the Melick clock. – Courtesy the New Brunswick Museum, Saint John*

FIG. 62

FIG. 62 *Movement of the Melick clock.*
– Courtesy the New Brunswick Museum,
Saint John

Beck of Philadelphia. There were nine children.

Their fifth child, James Godfrey, was the Saint John clock- and watchmaker. James's first wife was Debora (b. 1805 – d. 2 Oct. 1827), daughter of Daniel and Mary Smith. Their marriage took place on 28 October 1826. Debora and their son Daniel James both died in October 1827. On 15 February 1835 James married for a second time, to Caroline Fairweather (b. 17 Aug. 1807 – d. 11 Feb. 1888) of Millstream, New Brunswick. James and Caroline had seven children: George Godfrey (b. 19 Oct. 1836), Mary Eliza (b. 27 Aug. 1838 – d. 5 Oct. 1858), James William (b. 19 Mar. 1840 – d. 3 May 1889), Frederick C. (b. 17 July 1843 – d. 1907), watchmaker and jeweller in Saint John, Deborah J. (b. 26 June 1845), Catherine O. (b. 15 Nov. 1847), Andrew W. (b. 28 July 1850 – d. about 1900.)

James Melick and his family belonged to the Church of England.

THOMAS NISBET

Thomas Nisbet is considered by historians to be one of the most important cabinetmakers in New Brunswick in the 19th century. Among the many pieces of furniture he made are cases for clocks sold by his son-in-law, William Hutchinson. One such clock may be seen at the York-Sunbury Museum in Fredericton, New Brunswick.

Nisbet was born in 1776 in Dunse, Scotland, where he learned the trade of cabinetmaking from his father. On 28 March 1803 he married Margaret Graham in Glasgow and worked there for some time. Thomas and Margaret had three sons and four daughters, one of whom, Sally, married William Hutchinson in 1823.

In 1813 the family immigrated to Saint John, where as an accomplished craftsman Thomas easily obtained his freeman's papers, thus gaining the right to vote, and established his business on Prince William Street in Saint John.

During 1815 and 1816 he advertised in a number of New Brunswick newspapers. One advertisement announced that he had received a quantity of mahogany from Jamaica. This is the first documentation of a Maritime cabinetmaker importing lumber for use in furniture. By late 1816 Nisbet's fame had spread to Fredericton, where he had arranged for orders to be left at the home of Mr. Mark Needham.

Nisbet was one of two local Georgian cabinetmakers to place a paper label, advertising himself as a cabinetmaker and upholsterer, on his furniture. About forty pieces of his labelled work are known. However, it has been established that he made many other pieces. Documentation exists proving that a great deal of his work was purchased by the province to furnish Government House, the residence of the governor of the colony. This furniture was marked "G.H.N.B." Although a fire in 1825 destroyed Government House, some pieces of furniture were saved from the fire. New furniture was ordered from Nisbet for the rebuilt Government House and he apparently had trouble collecting the full £940 owed to him for this furniture. It is known that in 1842 a mahogany double counting-house desk for the mayor's office cost £11, so the order of £940 for furniture for Government House represented a large number of pieces.

It was believed by Harry Piers and Donald Mackay that Nisbet labelled only furniture made for customers outside Saint John. Apparently, Nisbet thought that his extensive newspaper advertisements promoted enough business in the Saint John area. Many labelled pieces have been found in Fredericton and St. Andrews.

Nisbet's son Thomas, born in 1810 in Scotland, apprenticed with his father and went into business with him in the early 1830s. The label was then changed to "Thos. Nisbet and Son, Cabinetmakers and Upholsterers." When Thomas Nisbet Jr. died on 7 February 1845, his brother Robert, born in 1820, went into the family business. After the death of his father, Robert continued the business until the fall of 1855, when everything was sold in an auction.

Thomas Nisbet Sr. was engaged in many enterprises outside his business. He co-founded the Saint John Mechanics' Whale Fishing Company and was president during its existence, from 1836 to 1850. He was president of the Saint John Hotel Company in 1844–45 and served as an executive member of the Saint John Mechanics' Institute. Also, by July 1827 he was given the rank of Captain in the Saint John City Militia.

Announcing his death, the 28 December 1850 edition of the *New Brunswick Courier* reported:

> Died this morning, Mr. Thomas Nisbet...who had been long a highly respectable inhabitant of this city where, by his upright conduct and fair dealings, he gained the respect of a large circle of friends.

WILLIAM T. PARSONS

The clock in the old House of Assembly, Colonial Building, in St. John's, Newfoundland, which was described in a letter from the Newfoundland Historical Society as a "stubby, round clock," displays the name of W.T. Parsons and gives the date of 1855. The Colonial Building was the seat of government from 1850 to 1960. It is now a provincial historical site and houses the Provincial Archives of Newfoundland and Labrador. William Thomas Parsons (b. 1813 – d. 4 Oct. 1902), born, as were his parents, in Newfoundland, was a watchmaker and jeweller in St. John's before 1846. On 9 June 1846 fire destroyed his premises, stock, and tools. He claimed £55 for some of the damage.

FIG. 63

FIG. 63 *The Parsons clock.*

He then re-established his business at 158 Water Street, St. John's, where by 1870 he was joined by his son, William T. Parsons Jr.

On 30 October 1873, William T. Parsons and his family left Newfoundland for Canada.[22] According to the city directories, W.T. Parsons opened a business in Ottawa, first on Sparks Street and then on Sussex Street. An advertisement in the 1873–74 Ottawa Directory stated that "always on hand was a large stock of watches, clocks and jewellery – hair work[23] of beautiful patterns and unique designs, a specialty."

After a stay of several years in Ottawa, Parsons and his family moved to Toronto, where he established his business first at 328½ Yonge Street, then across the street at 219 Yonge Street, and finally at 266 Simcoe Street.

By 1880 W.T. Parsons's son George (b. 1859 – d. 19 June 1914), a druggist, his wife, Margaret, and his family were living in Dundalk, Ontario. In the mid-1880s William T. and his wife, Jane H. (b. 1821 – d. 20 Apr. 1899), moved to Dundalk and opened a watch repair and jewellery shop. He remained in business until about 1897 when he and his wife, along with their son George and his family, returned to Toronto.

William Thomas Parsons died at the age of eighty-nine on 4 October 1902, following a fall on College Street, Toronto.

WILLIAM H. PATERSON

William H. Paterson, born in 1828 in St. Andrews, New Brunswick, became a clock- and watchmaker and jeweller in Saint John, New Brunswick. He established his shop in the early 1850s, and in 1857, his address was 36½ King Street. In 1861 he was listed as "a Freeman" in the city of Saint John.

After two or three more moves on King Street, he remained at 78 King Street from 1868 to 1877. After the fire in 1877 in Saint John, he re-established his business at 6 Bell Tower Avenue and began working with William Purchase.

In the 1850s William H. Paterson sold weight clocks imported from the United States. One such clock is illustrated in *Canadian Clocks and Clockmakers* by G.E. Burrows. On the movement of the clock is the name of its maker, Chauncey Jerome, New Haven, Connecticut, placing the date of manufacture of the clock somewhere between 1845 and 1853. The overpasted label is printed with the information "William H. Paterson, King Street, St. John, N.B." and advertises clocks, watches, and jewellery made and repaired. Below his name is a picture of the Saint John harbour.

Although no proof has been found, it is probable that William H. Paterson was the son of D. Paterson of St. Andrews who was a watchmaker on Water Street in 1810. W.H. Paterson was married twice, first to Mary (b. 1833) and then to Sophia (b. about 1832). William H. Paterson's sons, John B (b. 1856), William (b. 1858), and Charles S. (b. 1860), all apprenticed with their father. Charles later became an accountant. William began working with his father, William H., in 1876 and continued his father's business into the 20th century. His four other children were Hubert (b. 1864), Ira (b. 1866), Mary Elia (b. 1868), and Elizabeth (b. 1871).

A brother of William H. Paterson, Alexander Y. (b. 1834 – d. 2 Sept. 1907), spent some time in the United States in the 1850s, then began business as a watchmaker and jeweller in St. George, a town near St. Andrews. His son, A.Y. Paterson Jr. (b. 1861), had a watch repair and jewellery shop on Adelaide Street in Portland (now part of Saint John). The business continued on Main Street, Saint John, into this century.

The Paterson families were Scottish and belonged to the Wesleyan Methodist Church.

Fig. 64 A clock and its label, sold by William L. Plummer. – Courtesy the New Brunswick Museum, Saint John

FIG. 64a

FIG. 64b

WILLIAM L. PLUMMER AND PLUMMER & MITCHELL

William Lawrence Plummer (b. in U.S.A. about 1824 – d. after 1889) was a clockmaker and cabinetmaker in Saint John, New Brunswick. He came to New Brunswick in 1845, and his name was included in the list of "Freemen" of the city of Saint John in 1851. While he lived in St. John, Plummer sold clocks labelled with his name.

Clocks also exist that were sold during the period when Plummer was in partnership with a man named Mitchell, and the names "Plummer & Mitchell" appear on the labels of these clocks. Both labels advertise the fact that these firms made looking-glasses and picture frames as well, Plummer in his shop on the corner of Germain Street and Market Street, and Plummer and Mitchell at the shop on "Germain Street, four doors North of King Street at the sign of the mammoth clock."[24] It has not been established whether Plummer sold clocks with Mitchell before or after he worked alone.

The clock pictured in Fig. 65 is in original condition. The half column clock in Fig. 66 has a replaced dial and tablet. The label pictured in Fig. 67 was used in the three known Plummer and Mitchell clocks; it has an engraving of a boat flying an American flag. On the lower part of the label used by Plummer, and also Plummer & Mitchell, is the following statement:

> Every person in want of a good Time Keeper should inquire for Clocks made at this Establishment, the reputation of which stands high, not only in this Vicinity, but throughout the provinces. Their country Friends, who reside at a distance from any Clock Maker, would find it much to their advantage to purchase a good, warranted article instead of an inferior one – always liable to more or less expense.

On 5 November 1850, by special licence, William L. Plummer married Mary C. (b. about 1833 – d. before 1889), whose name also happened to be Plummer, in the Congregational Church in Saint John. Through his marriage, William Plummer became related to several other clockmen of New Brunswick. Mary C. was the daughter of Thomas Plummer and Mary Melick (d. 1884). William Plummer thus became related by marriage to James G. Melick, clockmaker, who was a cousin to Plummer's mother-in-law, Mary Melick Plummer. He also became related by marriage to W. Fairbanks, who was the son-in-law of Thomas and Mary Plummer. Fairbanks was one of the witnesses of the Plummer-Plummer marriage.

By 1864 William and his family had moved from Saint John to Fredericton, New Brunswick, where William ran a liquor and oyster saloon on Wilmot Alley. In the early 1870s William moved back to Saint John and, with his son Arthur, lived for a short time with his mother-in-law, Mary Plummer, and her step-sister Elizabeth. Although details of the whereabouts of William after this are sketchy, it is known that he moved to New York, where he was living in 1889. His date of death is not known.

FIG. 65 *Plummer & Mitchell OG clock. Movement is Taylor type 2(10)3, maker unknown.*
FIG. 66 *Plummer half-column clock. Note similarity to Fig. 37. Movement made by Morse & Co.*
FIG. 67 *Label of Plummer & Mitchell clocks.*

FIG. 65

FIG. 66

FIG. 67

Fig. 68 *Plummer & Mitchell*
clock with unusual styling. Only
one example is known. The 30-hour weight
movement is made by Morse & Co.
Movements by this maker are relatively rare.

JOHN E. SANCTON

John E. Sancton (b. 1833) was a watchmaker in Bridgetown, Nova Scotia. His name is engraved on a sterling silver pocket watch that is in the possession of the Nova Scotia Museum. On the interior of the watch are the words "made expressly for John E. Sancton by L. Remond Locle" and the number 10596.

According to William M. Graham, "the silver-cased watch was made 'for the trade' in Locle, Switzerland, a town well known for its vast production of watches in the nineteenth and twentieth centuries. Many small firms (and a few famous ones) assembled and finished watches in Locle. In earlier times, many parts were made by farmers as a cottage industry during the winter months. Baillie's *Watchmakers and Clockmakers of the World* lists three Remonds, two in Geneva and one in Paris, but none of these seem to apply."

John E. Sancton was the son of Thomas Sancton, merchant. The family lived in New Brunswick, where John and two of his sisters were born. They then moved to Nova Scotia, where they lived between 1856 and 1859. John married Rowena (b. 1838), and there were at least three children, Augusta (b. 1859), William H. (b. 1861), and Minna (b. 1863).

John Sancton was in business from at least 1864. By 1900 his son was in business with him, and the name of the firm was changed to John E. Sancton & Son.

FIG. 69

FIG. 70

JUSTIN SPAHNN

Justin Spahnn, watchmaker and silversmith, was born in 1803 at Chaux-des-Fonds, Switzerland. He began his apprenticeship there and completed it when he arrived at Philadelphia in 1819. He then worked as a journeyman silversmith.

By 1823 he had moved to Fredericton, New Brunswick, and advertised his newly opened shop in Mr. J. Macpherson's house where he "repaired and cleaned watches, clocks, jewellery, etc." By the mid-1830s he had purchased a lot for a shop on Front Street (Queen Street), Fredericton, and was making silver for his own shop and for other customers, such as James Agnew of Saint John.

After the death of his first wife, Catherine, in March 1836, he married Elizabeth Macpherson and had at least one daughter, Elizabeth, who was married in 1851.

During the 1840s Justin Spahnn had at least two apprentices. One was his nephew Andrew Macpherson who, with Spahnn's widow, carried on the business from 1856 to 1861. The other apprentice was James White (see p. 60), one of the few watch- and clockmakers born and trained in New Brunswick.

Justin Spahnn died in Fredericton on 4 October 1856.

FIG. 69 *The Sancton watch.*
  – *Courtesy the Nova Scotia Museum, Halifax*
FIG. 70 *Clock with 30-hour Terry-type movement, sold by George Steel, Horton, Nova Scotia.*
FIG. 71a *A typical George Steel clock.*
This clock contains a Terry-type movement, although the label makes no reference to Terry.

FIG. 71a

53

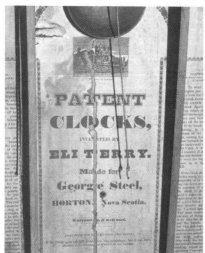

FIG. 71b

FIG. 72a

FIG. 72b

GEORGE STEEL

George Steel sold clocks from about 1830 to 1837 in Horton, Nova Scotia, near present-day Wolfville. The date 1831 was found in one of his clocks.

When searching for their ancestors, the Steel(e) families of (or from) Nova Scotia were unable to uncover any information on George Steel, the clock peddler. The only document that may be relevant is a marriage record showing that a George Steel married Mary Ann Lingree on 27 January 1833, by licence. No occupation was given for this man. The authors have found no additional information.

Steel, however, seems to have had considerable success in selling clocks. One clock is on display in the Haliburton House, a Nova Scotia Provincial Museum. Another is in the collection of the Museum of Canadiana, Aylesford, Nova Scotia. G. Edmond Burrows, in *Canadian Clocks and Clockmakers*, illustrates four "Steel" clocks, including the two mentioned above. The authors are aware of over a dozen Steel clocks, a relatively high number by one vendor to have survived.

The clocks are all from the transition period of New England wood movement clockmaking. Most of the examples contain Terry-type movements and some of the labels are marked, "Patent Clocks Invented by Eli Terry and made for George Steel." Note the engraving of the Tower of London on the label in Fig. 72. Other labels simply declare "Improved Clocks Manufactured for George Steel."

Movements by two American makers have been identified. In the numbering system devised by Dr. Snowden Taylor, one movement is Subtype 9.223, by Chauncey Boardman or Boardman and Wells. The second style of movement is Subtype 8.221, by Langdon and Jones.

Two clocks reported to the authors appear to have been fitted originally with Torrington-type "east-west" 30-hour wood movements. As luck would have it, both clocks were found without movements, but the case fittings, absence of pulleys, and a shadowy outline left by the movement confirm that Torrington movements were once present. These are the only clocks known to the authors to have been sold in Canada with these distinctive movements. It is interesting to note that both were accompanied by the labels acknowledging Eli Terry's patent. George Steel was obviously not much concerned with this conflict between makers.

Close examination of the surviving clocks reveals a notable variation in the shape of the splats. The four clocks illustrated in *Canadian Clocks and Clockmakers*, for instance, employ three splat forms, none of which is common. One short transition clock is known to have carved columns and splat.

It would appear that George Steel obtained clocks from a variety of sources. In all likelihood most of these clocks were assembled entirely in the United States. However, the existence of varied splats and the strange combination of a Torrington movement with a "Terry Patent" label suggest that he might have assembled some of the clocks himself.

FIG. 71b *The label of the George Steel clock in Fig. 71a.*
FIG. 72 *A Terry-type movement from a George Steel clock. The label in this clock acknowledges Terry's patent.*
FIG. 73 *(Next page) This Steel clock, which has undergone restoration, was fitted with a Torrington movement, but the label refers to Terry's patent.*
FIG. 74 *(Next page) A Trenaman clock pictured in an article by Catherine McLeod,* Canadian Collector, *Sept.–Oct. 1973.*

JOHN TOBIN

John Tobin advertised that he manufactured brass and wooden clock movements at Beaver River, Yarmouth County, Nova Scotia. According to George MacLaren's *Antique Furniture by Nova Scotian Craftsmen,* an advertisement appeared in an issue of the *Eastern Chronicle* in 1847 asking for business.

In the clock survey conducted by Jane Varkaris in 1978, two OG clocks were reported that bore the name of John Tobin, Clare, Digby County, Nova Scotia.

JOHN TRENAMAN

John Trenaman (b. 1792 – d. 18 Dec. 1868), a native of Devonport, England, was a watch- and clockmaker who was established in Charlottetown, Prince Edward Island, before 1820. His first place of business was on Queen's Square, Charlottetown. By September 1839 he had moved to Grafton Street, where he remained until he died.

Five tall case clocks are known to have the name Trenaman on the dial. One dial reads "J. Trenaman, Charlotte, P.E. Island." Two others are marked "J. Trenaman, Charlotte Town."[25] Another clock, pictured in Fig. 74, reads "Trenaman, P.E. Island." On the back of the pendulum, a name and date, "W. Gregor 1842," have been scratched in script. This clock appears to be of English origin and was in the possession of Trenaman's great-great-nephew. Two OGs with the Trenaman name on the dials are also known to exist in Prince Edward Island. These clocks have not been viewed by the authors and the maker is unknown.

In addition to the clocks that keep alive the memory of John Trenaman, there are watch papers that bear his name. A white watch paper in the Harry Birnbaum collection of watch papers has printed on it: "Trenaman, Charlottetown P.E. Island, Queen's Sq."

J. Trenaman's name appeared occasionally in the various newspapers of Charlottetown. However, he did not advertise for customers. Several other businesses gave Trenaman's location to identify their own. He was also mentioned in the late 1820s as one of two persons appointed as "Fence Viewers" in Charlottetown. Another notice that appeared in the *Royal Gazette* in 1846 announced that John Trenaman, clockmaker, had purchased land from Elizabeth Burnett.

Twice, John Trenaman placed advertisements in newspapers. In June 1832 he asked that the person to whom he had lent a volume of Durham's sermons return it. In June 1834 he advertised that he was selling a large house on King Street and a store on the same lot fronting on Dorchester Street. Another house he owned was also for sale.

John Trenaman's name was also in the newspaper when he served as a juror on 7 May 1840 at the Foster inquest and on 23 September 1843 at the MacPherson inquest. Trenaman, as foreman of the jury in Mrs. Foster's inquest felt obliged to write a letter to the *Colonial Herald* regarding the fact that she had died of alcoholism. He was concerned that "so many nests of iniquity and intemperance should be tolerated. Also, we would like to record our utter abhorance [*sic*] of individuals who persist in dealing out to the unfortunate drunkard, that draught which, alas! so often hurries him unprepared into the presence of his Maker."

John Trenaman married Mary Collins (b. 1790 – d. 5 Dec. 1852) in England. They came to Prince Edward Island with their eldest child, John (b. 1812 – d. 16 July 1881). Their other children, Louisa (b. 1818 – d.

FIG. 73a

FIG. 73b

FIG. 74

30 Mar. 1844), Amelia (b. 1822 – d. 30 Jan. 1865), Thomas (b. 1825 – d. 10 July 1852), James (b. 1830 – d. 25 Dec. 1852), and Mary Ann (b. 1823 – d. 1903), were born on the Island. After the death of his wife in 1852, he married by licence Mrs. Catherine Mason Griffiths (b. 1812 – d. 23 Nov. 1895), widow of Captain Griffiths, on 24 March 1855. Catherine was born in Ireland.

Other records of the Prince Edward Island Museum and Heritage Foundation show that in 1833 John Trenaman was one of the first trustees of the Methodist Church in Charlottetown. He and Mary and their family were all Methodist, but Catherine was a Bible Christian.

John, Mary, and three of their children are buried in St. Paul's Anglican Church Elm Avenue Cemetery in Charlottetown. The information on the tombstone of John Trenaman is illegible because it was apparently pushed into the ground upside down. Catherine Trenaman is buried in the People's Cemetery.

A grandson of John Trenaman, Thomas Trenaman Stumbles (b. 1853 – d. 1884), was the son of William Washington Stumbles (b. 1823 – d. 1900) and Mary Ann Trenaman. Thomas was listed as a watchmaker and jeweller in Summerside in 1880.

ALEXANDER TROUP AND FAMILY

Alexander Troup Sr. (b. 1776 – d. 30 Dec. 1856) was born in Aberdeen, Scotland. While in Halifax, he left His Majesty's 98th Regiment and, being a watchmaker by trade, completed the installation of the Town Clock (see p. 38). From 1805 until his death, Alexander was a watch- and clockmaker and silversmith on Argyle Street in Halifax.

From a survey of clocks in Canada conducted by one of the authors in 1978, ten tall case clocks with the name of Alexander Troup on the dials are known to exist in private collections and museums in Nova Scotia and Ontario. A picture of such a clock is shown in Figs. 75a and 75b. Another "Troup" clock is pictured in *English–Canadian Furniture of the Georgian Period* by D. B. Webster. Available information about several clocks reveals that they have a serpentine broken-arch cornice and are made of mahogany veneering over pine. Several known clocks are decorated with maple string inlays and block inserts. It is not known if Alexander Troup made any of his movements, but the movement of at least one clock was British. The cases were made in Nova Scotia and it is possible that they were made by James Thompson, cabinetmaker, whom Troup chose as one of the executors of his will.

In addition to clocks that bear his name on the dials, a number of watches have watch papers in their cases showing Troup's name as repairman. One watch paper, dated only a year before his death, bears his name and gives his occupation as clockmaker.

The name Alexander Troup appeared on several occasions in newspapers, in spite of the fact that he did not advertise his business frequently. In 1828 he advertised that he had lost a silver hunting case watch and offered a reward. In 1845 he was listed, along with Judge Haliburton, Dr. Cogswell, the Hon. Michael Tobin, John Stairs, and eighteen other citizens, as one of the founders of the Nova Scotia Horticultural Society, later expanded to the Halifax Public Gardens by John McCulloch. In 1848 he signed a petition, with many other Haligonians, for the "amelioration" of tax on manufactured articles.

Several references state that Alexander Troup Sr. married Mary Black (b. 1788 – d. 1827) of Westmorland, New Brunswick. However, records of

Fig. 75a

Fig. 75a *Clock sold by Alexander Troup.*

St. Matthew's Church, Halifax, confirm that Troup married Mary Lees on 1 April 1805. There was a son, Alexander Jr. (b. 26 Jan. 1806 – d. 8 Oct. 1873). After the death of his first wife, Alexander Troup married Elizabeth (b. 1811 – d. 1886.)

Alexander Troup Jr. entered into apprenticeship with his father in 1820 and continued to work with him in his shop. He then set up his own business, and by 1858 was listed in the city directories at 18 Barrington Street. In the 1860s he moved his shop three times: in 1863 to 162 Argyle Street, in 1864 to 235 Hollis Street, and in 1866 to 66 Barrington Street. He continued to be a watch- and clockmaker until his death on 8 October 1873. It is possible that some of the clocks attributed to his father may have been sold by him.

Thomas Troup (b. 1819 – d. 7 Sept. 1877), second son of Alexander and Mary, worked with his father and brother as watch- and clockmaker.

On the death of Alexander Troup Sr. on 30 December 1856, the *Halifax Morning Chronicle* of 1 January 1857 carried the following notice: "On Tues. evening after a tedious illness, Mr. Alexander Troup, age 80 years – Funeral to take place at Half-past two precisely, tomorrow, Friday. Friends and acquaintances are respectfully invited to attend."

To his wife, Elizabeth, he left his business and "all his worldly goods." His will stated that he hoped his wife would carry on his business after his death. Elizabeth did so, hiring as manager James Carr. In 1863 Carr purchased the business at 164 Argyle Street, at the corner of Argyle and Buckingham streets, and advertised that he was the successor of Alexander Troup.

FIG. 75b

FIG. 75b *Dial detail of the clock in Fig. 75a.*
FIG. 76 *Watch and watch papers including eight papers bearing Alexander Troup's name.*

FIG. 76

TULLES, PALLISTER AND M'DONALD

John Tulles, Thomas Pallister and M'Donald were cabinet-makers and upholsterers on Barrington Street in Halifax, Nova Scotia. They were in business in 1810 and 1811, and during this period they made cases for several tall case clocks. One such clock is on display at the Nova Scotia Museum in Halifax; it was presented to the Province of Nova Scotia by the estate of A. J. Parker of Shubenacadie.

The clock case is made of mahogany with the interior of secondary pine and stands nearly 7 feet, 6 inches. It has reeded quarter columns, a broken-arch cornice, and maple string inlays forming an oval in the mahogany-veneered door and a circle in the base.

The movement is English. A mark impressed on the back of the dial indicates that the dialmaker was James Wilson of Birmingham, England. Wilson made false plates and dials from September 1777 to sometime between 1808 and 1812. The bell of the clock was made by George Ainsworth, a pinion maker who worked in Warrington, England in the early 1800s.

The other known labelled piece of furniture made by Tulles, Pallister and M'Donald is a drum table, location unknown.

D.O.L. WARLOCK

Daniel O'Leary Warlock (b. 1819 – d. July 1901), known as "Doll," was a clock- and watchmaker and silversmith in Saint John, New Brunswick, for over fifty-six years. D.O.L. Warlock was born in Killarney, Ireland, and came to New Brunswick, commencing business in 1840. From 1857 to 1864 his shop was located first at 42 and then at 49 King Street. It was thought to be during this period that he made clocks. One such clock is in the collection of the New Brunswick Museum. It is a rosewood wall clock about 34 inches long. The museum believes it to have been the first clock in the Intercolonial Railway station at Newcastle.

On 20 June 1877 he lost his shop and all its contents in the fire of Saint John, but he re-established his business at the corner of King and Charlotte streets in the Trinity Block. In the 1880s D.O.L. was the agent for E. Howard and Company, Boston, Massachusetts, and for the American Watch Company of Waltham, Massachusetts. He remained active to within a year of his death. In fact, at the age of eighty, he repaired the C.P.R. clock in Saint John station.

D.O.L. Warlock married Mary (b. about 1833 – d. about 1903) after he came to New Brunswick. Their son Henry (b. about 1855 – d. about 1910) was apprenticed to become a watchmaker in 1871. Henry went into business with his father, and after D.O.L.'s death, Henry continued the business under his father's name. Other children of D.O.L. and Mary were Emily (b. about 1857 – d. 1914), Josephine (b. about 1857), Edward C. (b. 1861 – d. before 1892), and Alice (b. 1865 – d. about 1922.)

FIG. 77a *Tall case clock made by Tulles, Pallister and M'Donald, Halifax. – Courtesy the Nova Scotia Museum, Halifax*

WILLIAM W. WELLNER

William Wright Wellner (b. 20 Feb. 1844 – d. 11 Sept. 1907), born at Crapaud, Prince Edward Island, was a watchmaker and silversmith in Charlottetown, Prince Edward Island. William is thought to have learned the watchmaker's trade from John Page and the silversmith trade from Edwin Sterns, both of whom are related to Wellner by marriage. Several silver watch cases made in the 1860s bear his mark, "W.W.," with "Leopard's mask and Lion Passant," and several watches are known to exist with "W.W. Wellner" on the dial.

William W. Wellner established a business of his own in 1868 on Upper Great George Street in Charlottetown. By September 1876 he was listed in the *Mercantile Agency Reference Book* as having in his shop watches and jewellery worth between $5,000 and $10,000, and his credit was "good." By 1881 he had moved his place of business to Grafton Street. An advertisement appearing in the *Maritime Business Directory* in 1900 stated that the business, then at 103 Grafton Street, was the oldest and largest in the province. In the October 1921 issue of the periodical *The Busy East,* a picture is shown of the Wellner store and of W.W. Wellner and L.E. Wellner Sr., proprietors.

FIG. 77b

William W. Wellner was the son of W.B. Wellner (b. 1810 – d. 1892), a merchant who was born in Nova Scotia (as was his father) and came to Prince Edward Island and married there. W.B. Wellner and his family lived in Crapaud and he was active in various civic matters. (In 1839 he was secretary treasurer of the Crapaud Agricultural Society.) In 1848 he left Crapaud for Charlottetown. W.W. Wellner's mother was Mrs Jane Ann (Wright) Dickery (b. 23 May 1821 – d. 28 Apr. 1876). Of six children, only William W., Frances B. (b. 24 Jan. 1850 – d. 28 April 1927), and Warren L. (b. 14 Nov. 1853 – d. 3 Mar. 1889) lived to adulthood. Warren worked in the business with William W. until his death.

William W. Wellner married Rebecca A. MacDonald (b. 20 Aug. 1850 – d. 7 Aug. 1913) on 9 September 1880, and they lived for several years with William's family. Their son, William T. (b. 25 Sept.1882 – d. 12 Feb. 1942), became a watchmaker and engraver and worked with his father. When William W. Wellner died, William T. carried on the business, keeping it in the name of his father. In 1909 Beulah (b. about 1888), sister of William T., was also working for the company. By 1913 Lloyd E. (b. 1 Apr. 1893 – d. 5 Apr. 1971), another son of William W. had also entered the firm. Lloyd saw active service in the War of 1914–1918. At the end of the war, he returned to the business and his brother retired. The business was incorporated in 1919 and between 1920 and 1930 Lloyd E. was president of W.W. Wellner Company. Another daughter, Abigail Jane (b. 17 Sept. 1885 – d. 1 Jan. 1913), did not enter the business.

FIG. 78

FIG. 77b *The label from the tall case clock in Fig. 77a.*
FIG. 78 *W.W. Wellner.*

Lloyd E. Wellner's son Arthur Woodford studied horology at Ryerson Technical Institute in Toronto, and in the 1940s and 1950s was the watchmaker for the firm. Lloyd E. retired in 1959 and Arthur continued the business until 1960.

The family was Wesleyan Methodist and its members were active in the First Methodist Church in Charlottetown. Many members of the family are buried in Charlottetown's People's Cemetery. Descendants of William W. and Rebecca live in Charlottetown today.

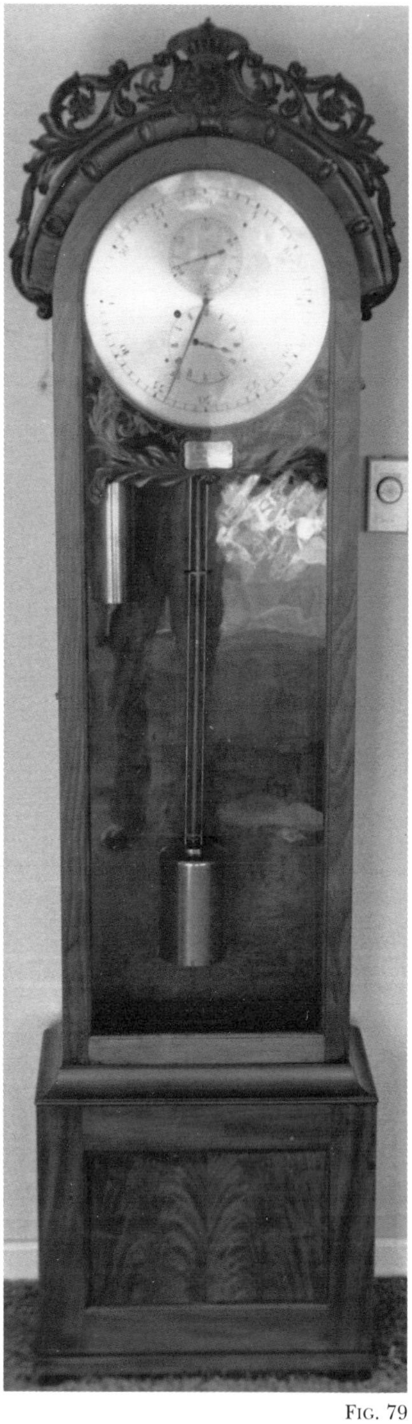

FIG. 79 *The James White clock.*

JAMES WHITE

James White (b. about 1825 in New Brunswick – d. 9 April 1894) was a watch- and clockmaker at Fredericton, New Brunswick. He can be included in the very small group of true Canadian clockmakers who were both born and trained entirely in Canada. Family members recall that he was apprenticed to Justin Spahnn, a Fredericton watchmaker. He later established a shop on Queen Street, Fredericton.

During his lifetime, White was closely involved with the clock in Fredericton's Christ Church Cathedral. This clock is one of the most interesting and historically important clocks in Canada. The Fredericton cathedral itself is a true cathedral – a rarity in Canada, being the paramount church of the Fredericton diocese rather than a simple parish church. It was constructed during the early 1850s and opened in 1853. A clock was purchased in 1853 for £150 and installed in 1854. James White was responsible for its installation and adjustment.

About this time in England, the well-known amateur horologist Edmund Beckett Denison (later to become Lord Grimthorpe) became involved in the lengthy and controversial process of designing a tower clock for the newly completed British Houses of Parliament. Denison's design was eventually accepted and the famous Westminster clock, sometimes erroneously referred to as Big Ben, was the result. The Westminster clock was constructed by the firm of E.J. Dent and installed in 1859. The Fredericton clock was also built by E.J. Dent. To this day, it continues to run well. In the 1970s the church verger commented that the clock varied less than a minute a week.

Prior to 1859 Denison had constructed an early experimental clock, working out his design for an improved gravity escapement. This clock was eventually installed in a church at Cranbrook in Kent. The clock for the Christ Church Cathedral in Fredericton may well be considered the second experimental Denison clock. In 1853 the local Bishop at Fredericton specified that the clock movement must be able to withstand a temperature of minus 40 degrees Fahrenheit, a condition that would seriously hamper, or stop, most clocks. In Denison's book, *Clocks, Watches and Bells,* he comments about the design of a special escapement for the Fredericton church clock. Denison also commended James White, saying he "has shown considerable judgement in making various little alterations which were naturally required in a machine of new construction…and I have heard from him, very satisfactory reports of its performance even while the oil had been frozen as hard as tallow."

The Christ Church clock, incidentally, is not a true tower clock in the sense that it has no outdoor display dials. It is mounted high in the nave of the church. For many years its importance was overlooked. It was not until the 1960s that the clergy and people of Fredericton realized its historic value.

James White was a clockmaker of great talent. His involvement with Edmund Beckett Denison and the Fredericton Cathedral clock led him to develop an interest in precision clocks and the gravity escapement. This interest reached a peak in the early 1850s. Unfortunately, in one sense his talent developed too late. Mass production had largely swept aside the market for hand-crafted precision timepieces and White constructed very few clocks. In 1851, however, James completed an astronomical clock for which he won an award in the Industrial Exhibition in Saint John the same year. The *New Brunswick Courier* of 13 September 1851 comments about the Exhibition that "These rooms contain among other articles a beautiful astronomical clock, the invention and workmanship of Mr. James White of Fredericton." The exhibition was a great success and on one day of that week attracted 50,233 visitors.

A regulator exists that has been a family possession for about 130 years, and may very well be the prize-winning clock described in 1851. The clock is pictured in Figs. 79, 80, and 81. It can be described as follows:

CASE

The clock stands 76 inches high and is 17 inches wide and 9½ inches deep at the waist. The base measures 20 inches in width. The wood used in its construction is mahogany. The case itself has simple lines, with a rounded top surmounted by ornate carved fretwork. There is a large, one-piece glazed door 51½ inches in height. Behind the door the silvered dial is bordered at the bottom by a semi-circular carved fretwork panel, which balances the cresting at the top of the case. In the centre of this lower fret is a silvered plate inscribed "James White Maker Fredericton, NB."

MOVEMENT

The movement plates are 9 inches tall and triangular in shape with rounded tops; they are made of brass ³⁄₁₆ of an inch thick. The plates are held together by five brass posts, the top post being fitted with a ring. Affixed to the front plate is a rectangular brass false plate to which the dial is attached. The escapement is a rare five-legged gravity type, which appears to have been perfected by White himself. The movement is mounted on two heavy cast-iron brackets, which are bolted to the back board.

The dial plate is silvered brass. To minimize friction, as in most fine regulators, there is no motion work for the hands. As a consequence three separate dials are required for the hour, minute, and seconds indicators. These are located on the dial directly over the appropriate movement arbors. There is a single brass weight with a six-spoke pulley. The winding hole is located at approximately the eight o'clock position.

The pendulum rod is temperature-compensated and is constructed from a zinc tube flanked by two smaller steel rods. The bob is of cylindrical shape, 7 inches high and made of a bright metal, possibly steel. Details of the weight and composition of the bob are not known. The pendulum rod hangs from a large brass suspension arm, which is bolted to the back board. There is a silvered beat plate.

White evidently put a great deal of thought into the design of this regulator. It has been reported by a family member that, in addition to his contacts with E.B. Denison, White also obtained the advice of the great British scientist Lord Kelvin, in accurately determining the length of the zinc pendulum rod to provide satisfactory temperature compensation.

A second clock attributed to James White has been reported. This is an octagonal wall clock with elaborately carved case. The movement is described as "heavy, rectangular, brass frame plate with fusee," so evidently it was spring-driven. It is said at one time to have hung on a wall of the governor's building in Fredericton and was evidently retrieved at auction by a White descendant. James White's name is engraved on one of the plates.

The clocks of James White were few in number, but of exceptional quality.

FIG. 80

FIG. 81

FIG. 80 *Movement detail showing five-legged gravity escapement.*
FIG. 81 *Name plate detail from the White regulator.*

Fig. 82 *Benjamin Wolhaupter.*
– Courtesy the York-Sunbury Historical
Society, Inc.
Fig. 83a *A Wolhaupter tall case clock.*

James White married Elizabeth Harned around 1850. The 1851 census records that he and his wife were living with her mother, Mary Harned, and her sister Euphemia. James and Elizabeth had two daughters, Emily Jane B. and Bessie Frances. A son, Henry David, became an electrician and watchmaker and worked with his father, continuing the business into this century. At his death, James left his son, Henry D., all his clock- and watchmaker's tools and workshop "appliances and materials and cabinet containing them as well as all other articles and parts of watches used in the watchmakers' art; also my regulator clock now used and going in my shop on Queen Street in the said city of Fredericton: all counters and show cases and also all my scientific and horological books of every kind."

To his grandson Stewart White he bequeathed his "other clock still unfinished."

JOHN WOLHAUPTER AND SONS, BENJAMIN AND CHARLES WOLHAUPTER

John Wolhaupter (b. Sept. 1771 – d. 13 Jan. 1839) was a silversmith and watchmaker of New York City in the U. S., and Saint John, New Brunswick. He was the son of Gottfried Wolhaupter, New York (originally of Brocken near Schneesburg, Saxony). In 1795 John moved to Saint John and opened a business as a watchmaker and jeweller. However, he travelled often to New York for supplies for his business.

In 1799 he was listed as a "Freeman in the City of Saint John." On 21 May of that year, he announced in the *Royal Gazette* that "John Wolhaupter has taken house on Germain St. opposite Rev. Dr. Byles."

In May 1803 John Wolhaupter decided to leave Saint John, New Brunswick, for New York and requested that all outstanding accounts be settled. He remained in New York until 1805 and on his return to Saint John bought a house from Daniel Micheau on Prince Street, a few doors north of the post office, and opened a shop on Market Square. In 1810, intending to retire, he sold his property and moved to Sheffield on the Saint John River, but remained there for a short time only. In 1813 he purchased a shop on Queen Street, Fredericton, where he was again a watchmaker and jeweller. In 1819 he moved to "Camperdown Street near Mr. Bradley's," but by 1821 and for the next ten years, he continued his business on Queen Street, Fredericton. In 1831 he retired to Richmond, New Brunswick (near Woodstock), to a home purchased for him by his son Phillip David.

John Wolhaupter was an active Mason. In New York, in 1790, he was a member of Hiram Lodge. Shortly after moving to Saint John, New Brunswick, he was a founder of Saint John's Masonic Lodge and became its treasurer.

On 17 May 1795 he married Mary Payne Aycrigg, daughter of Dr. John and Rachel (Lydeker) Aycrigg, and moved to Saint John. John and Mary had at least three sons: Benjamin, Charles, and Phillip David. In 1822 he was married for a second time to Mary Connell at Woodstock. On 13 January 1839 John Wolhaupter died in Richmond, New Brunswick.

Benjamin Wolhaupter (b. 10 July 1800 – d. 27 Jan. 1857) was born in Saint John, the son of John and Mary Wolhaupter. Benjamin apprenticed with his father and in 1813 moved with the family to Fredericton. By 1820 he was in business with his father on Camperdown Street. In that year he married Catherine Phoebe, daughter of Matthew Brennen of Fredericton, and bought a house from Mrs. Bradley, next door to his father's house. There were six children.

In 1825 he moved to a new house on Regency Street "directly above the Gaol," which was shortly demolished; Benjamin purchased part of the property. He advertised as watchmaker and silversmith for the next ten years. Also in or about 1825, Wolhaupter is thought to have made the movement and possibly the case for a tall case clock that bears his name, now at the York-Sunbury Museum. Another tall case clock and also silver pieces were, in 1959, in the possession of the family.

Benjamin took at least two apprentices to train as watchmakers and silversmiths: Benjamin Tibbets in 1827 and his second son, Charles John, in 1833. He was involved in other pursuits beside his trade. In 1839 he and his brother Charles spent time in active military service, during the Maine boundary dispute; he was Captain and Quartermaster, 71st York-Sunbury Militia. In 1837 he was appointed Magistrate in and for the County of York, New Brunswick. From 1840 to 1847 he was president of the Central Fire Insurance Co., a director of the Bank of New Brunswick, a vestryman of the Cathedral, and Commissioner of Public Buildings for New Brunswick. In 1847 he was appointed high sheriff and remained in public office until his death.

FIG. 83b

Benjamin Wolhaupter died on 27 January 1857 in Fredericton at the age of fifty-seven. His portrait may be seen in the York-Sunbury Museum. His obituary in the *Head Quarters* stated that "no man we know of had more milk of human kindness in his breast...always merry...always kind and humane...we do not believe that he left behind a single enemy."

Charles John Wolhaupter (b. 1822 – d. June 1858) was born in Fredericton, the son of Benjamin and Phoebe Wolhaupter. After his apprenticeship to his father from 1833 to 1839, he commenced business in Chatham, New Brunswick, in the store "formerly occupied by Mr. J.S. Samuel nearly opposite to Mr. Joseph Cunard's store." He advertised watches, jewellery, and musical instruments for sale and that he was qualified to repair watches, jewellery, and navigational instruments. In October 1842 he prepared to go to Australia and requested settlement of accounts. Watches were to be picked up from Mr. James Hea.

FIG. 83b Dial detail from the tall case clock by Benjamin Wolhaupter in Fig. 83a.

In 1851 Charles John finally left for Australia, where he spent seven years during the Australian gold rush. He returned to Fredericton in April 1858 with enough gold to make a ring and a cross for his family. In June 1858, he accepted an appointment with the railway company in Saint John. However, during the passage by steamboat from Fredericton to Saint John, he fell overboard and drowned.

Charles left his widow, Harriet, daughter of William Comon and his son, Benjamin Wolhaupter.

FIG. 84

FIG. 85

FIG. 84 *Clock sold by B. & W. Young & Co.*
FIG. 85 *Label of clock in Fig. 84.*

THE YOUNG FAMILY

Members of a family named Young sold clocks in the Maritimes during the 1840 to 1850 period. The clocks have OG cases and are fitted with 30-hour weight-driven brass movements from several makers. The authors have been unable to find any specific personal information about the Youngs. They seem to have been successful businessmen, since their clocks exist in considerable numbers.

As pointed out by Dr. Snowden Taylor, there is a reference to a family of Youngs in the introduction to a reprinted edition of *The Clockmaker, or the Sayings and Doings of Samuel Slick of Slickville* by T.C. Haliburton, issued in 1958 by McClelland and Stewart. This reference comments on the activities of four Young brothers who sold large numbers of clocks in New England and eastern Canada. They eventually became quite wealthy and their fortunes were founded on their clock-peddling success. The only brother named in this account was Samuel Jackson Young, who died at age eighty-seven in 1908. However, his initials do not appear on any known clock labels, but based on his age he could have been a member of the Young family. The reference also mentions the possibility that he could have been the original Sam Slick. It has also been suggested that Moses Barrett was Sam Slick, as noted in the Barrett section of this text. However, it is generally believed that Sam Slick was a composite of several peddlers known to Judge Haliburton. All the Young clocks seen by the authors have brass movements dating from several years after the Sam Slick book was first published.

The Youngs sold clocks in Amherst, Nova Scotia, and St. Stephen, New Brunswick. Several label variations are known, including: B. Young and Brothers, Amherst, N. S.; B. Young & Co., St. Stevens, N. B.; B. & W. Young & Co., St. Stephen, N. B.; and B. & W. Young, St. Stephen, N. B. Three of these labels are shown in Figs. 85, 86, and 87. St. Stevens was an incorrect spelling of St. Stephen.

Several clocks with Amherst labels were fitted with movements by Atkins and Porter (Taylor type 19.1). One clock with a St. Stephen label also contains this movement. The Atkins and Porter company was known to be in business from 1840 to 1846, which helps to confirm the period when the Youngs were active, but this does not identify which address was the earlier. Another Amherst clock has a movement by Chauncey Boardman (Taylor type 2. [10] 12), which was made for some years after 1837.

It is not known if Francis Young, watch- and clockmaker, was related to the Young brothers. Francis Young lived in Saint John, New Brunswick, in the 18th century. He came from London, England, and by 1766 was working in Philadelphia, Pennsylvania. After a brief stay in South Carolina in the late 1770s, he opened a shop in New York City in 1780. He then went to New Brunswick as a United Empire Loyalist, obtaining land in Saint John, where he had a shop at the corner of Saint James and Charlotte streets. The name of Francis Young, watchmaker, was included in a list of Freemen of Saint John in 1785.

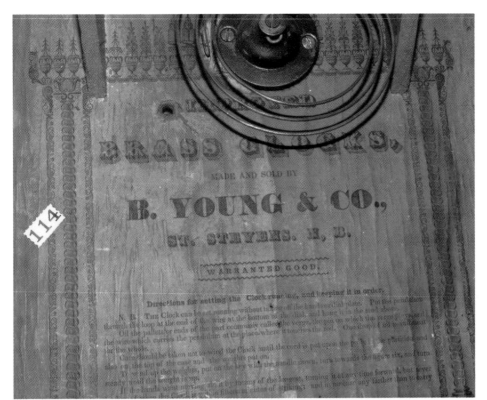

FIG. 86

Fig. 86 *Label of a clock sold by B. Young & Co. with incorrect spelling of St. Stephen.*
FIG. 87 *Clock and label by B. Young & Brothers, Amherst.*

FIG. 87a

FIG. 87b

CHAPTER 3
LOWER CANADA, CANADA EAST AND QUEBEC

FIG. 88

FIG. 89

THE ARDOUIN FAMILY

Members of the Ardouin family served the residents of Quebec City as watch- and clockmakers from about 1820 until about 1865. Five clocks are known to exist that bear the initials C.J.R. or C.J. Ardouin, Quebec, on their dials. A shelf clock, pictured in this section, is 17 inches high and 10 inches wide and is marked "C.J.R. Ardouin, Quebec" on the dial. It is in the Canadiana collection of the Royal Ontario Museum, Toronto, Ontario. The clock is of simple design, made of mahogany veneer over pine and is decorated with a narrow string border rope inlay. The movement is English. A similar clock, also with C.J.R. Ardouin on the dial, spells Quebec incorrectly as "Quebeck."

Charles James Robert Ardouin came to Lower Canada from England with a number of family members and asked for land. In addition to being a watch- and clockmaker, he was also a silversmith and an engraver. He learned his trade in London, England, and worked at various locations on St. John Street, Upper Town, Quebec City from 1820 until at least 1826. In 1822 an advertisement in the *Quebec Gazette* gave his address as 16 St. John Street and stated that he "returns warmest thanks to his friends and generous public for the liberal encouragement they have given him in the line of his profession since his arrival in Quebec and flatters himself by the extensive practise he has had in some of the finest shops in London and his usual assiduity to merit a continuance of their favours." He had for sale 8-day clocks, table clocks, etc., and welcomed watches and clocks of every description for cleaning.

He was also a charter member of the Literary and Historical Society of Quebec. In the 1820s he was a subscriber to the Quebec Fire Society. The exact date of death of C.J.R. Ardouin is not known. However, he died between 1826 and 1844, the year the next Quebec City directory was published. His wife died in the 1860s.

In the city directory of 1844, M. Ardouin, possibly a brother of C.J.R. Ardouin, was listed with his son as watch- and clockmaker and silversmith at 60 St. John Street. In an advertisement the business claimed that, in addition to "keeping on hand a good supply of watches, jewellery and music boxes," they paid particular attention to repairing watches. In the 1850s the business was listed as a goldsmith firm, as well as a watch and clock business.

FIG. 88 *Clock sold by C.J.R. Ardouin, Quebec City.*
– *Courtesy the Royal Ontario Museum, Toronto, Canada*
FIG. 89 *A C.J. Ardouin clock.*

66

FIG. 90
FIG. 91

M. Ardouin's son, Charles James (C.J.) (b. 1820 – d. about 1877), worked with his father, and after M. Ardouin's death, C.J. continued the business at 60 St. John Street until about 1857. He then moved to 25–28 Fabrique Street. By 1863 his shop was at 17–19 St. John Street, and he remained there until 1865. Children of C.J. Ardouin included Mary (b. 1853), Charles (b. 1854), Francis (b. 1860 – d. before 1871), and Edward (b. 1865.)

One tall case clock bears the name C.J. Ardouin, Quebec, on the dial, and another with the Ardouin name, reported to be imported from France, has been seen in the visiting room of the Hôpital-Général, Quebec City. Parts of an Ardouin clock are illustrated in Figs. 90 and 91. All that has survived of this clock is the dial and the movement, but it appears to have been an 8-day striking tall case clock. The dial is white, painted on iron. It is typically British and is inscribed "—douin Quebeck." The same spelling error occurs in a bracket clock by C.J.R. Ardouin, as noted previously, so it is likely that this clock was also sold by C.J.R. The movement is typically British as well, and it has a false plate with a faint inscription "O BORNE" or "OSBORNE". Sometime between the years of 1865 and 1871, Charles J. Ardouin left the clock- and watchmaking business and took the position of chief office clerk in the legislative assembly. He remained in that position until his death about 1877.

FIG. 90 *Ardouin clock dial. Name is partly obliterated. Note spelling of "Quebeck."*
– *Courtesy Kenneth D. Roberts*
FIG. 91 *Typically English 8-day tall case movement from Ardouin clock in Fig. 90.*
– *Courtesy Kenneth D. Roberts*

CHARLES ARNOLDI

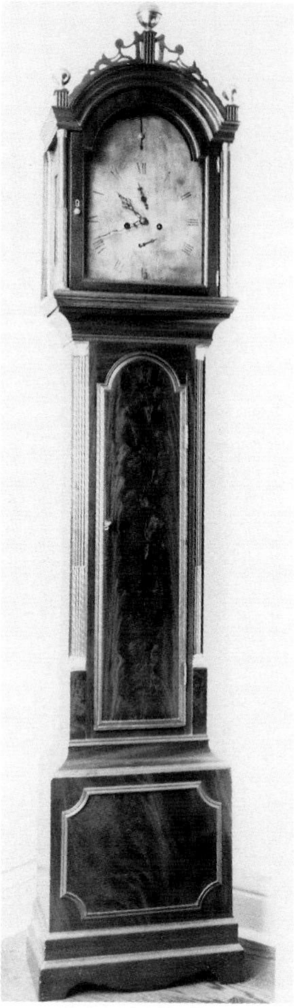

Charles Arnoldi (b. 23 Sept.1779 – d. 17 Dec.1817) was a watch- and clockmaker and jeweller in Montreal, Lower Canada. Two tall case clocks of high quality are known to exist with the name Charles Arnoldi engraved on the brass dials. Charles Arnoldi was also a practising silversmith and his initials, followed by "Montreal," can be found on several pieces of silver.

One impressive tall case clock bearing his name has been reported to the authors and is pictured in Figs. 92 and 93. As can be seen, it has a case with very fine workmanship in figured mahogany. The hood is surmounted by delicate fretwork and three brass finials. It is typical of the top-quality cases being offered by other clockmakers in Montreal at the time.

The dial is solid brass, plain in design, with very little decoration. The mechanism was not examined by the authors, but appears to have typical late 18th-century English configuration, with strike/silent, seconds bit, and calendar.

Charles Arnoldi was associated with Benjamin Comens in 1806 and 1807. The business was at 16 Notre Dame Street, Montreal. Comens came from Windsor, Vermont, and made silver for the Indian trade. However, in November 1811 the Montreal property was sold by the sheriff.

The brief relationship between Arnoldi and Comens seems typical of the way businesses were formed and later disbanded among silversmiths and clockmakers in Montreal during the early 1800s. Of necessity the authors have limited themselves to reporting on the careers of known clockmakers whose products are documented. Many of these men were active silversmiths, which raises the possibility that other silversmiths may also have sold clocks. It is entirely likely that additional clocks and clockmakers in early Montreal will be authenticated in future years. Readers wishing to learn about the activities of these other men are directed to a useful reference, *The Old Silver of Quebec* by Ramsay Traquair.

Traquair, in commenting on John Lumsden, observes that he was in business before 1804. Lumsden is known for both silver and clocks (see p. 103). In 1804 Lumsden's business was taken over by Charles Irish, described as "clock- and watchmaker, jeweller and silversmith." No "Irish" clocks are known. Irish, in turn, was succeeded by Charles Arnoldi on 5 October 1806. As noted above, Arnoldi formed a partnership that year with Benjamin Comens. Comens had just arrived in Montreal from Windsor, Vermont, the home of yet another well-known clockmaker and silversmith, Martin Cheney, who set up shop in Montreal in 1809. Cheney formed a partnership with James Dwight. Dwight had later partnerships with George Savage and Austin Twiss. All these activities are both interesting and confusing, but certainly indicate that the silversmiths and clockmakers of Montreal were closely associated.

Charles Arnoldi left Montreal when he was appointed postmaster at Lavaltrie on 12 June 1812. He also advertised that he would convey travellers to and from Lavaltrie and Montreal. However, he returned to Montreal in 1813.

Charles had married Anne Brown at Montreal on 5 October 1805. Seven children all died at an early age. Arnoldi was a member of a large and well-known family. He was the son of Peter Arnoldi (b. 10 Mar. 1734 – d. 6 Aug. 1801). Peter came to the new world from Forback, Germany, in the first year of the English Regime. He became a merchant and mar-

FIG. 92

FIG. 93

FIG. 92 *Charles Arnoldi clock, Montreal.*
 – *Courtesy Barbara and Henry Dobson*
FIG. 93 *Dial detail of Fig. 92.*
 – *Courtesy Barbara and Henry Dobson*

ried, in Montreal, Johanna Jacobini Philip from Heidelburg, Germany. The family lived in Montreal, St. Jean, Chateauguay, and Three Rivers, where Peter died in 1801. Peter and Johanna had fourteen children, ten of whom reached maturity. Charles's oldest brother, Michael (b. 19 June 1763 – d. 27 Aug. 1807), was trained as a watchmaker and silversmith. Michael was in a partnership until 1784 with Robert Cruikshank. However, ill health prevented him from working at his trade beyond 1792 and he leased his shop to his brother, John Peter, and John Oaks "in return for board, lodging, laundry, and a suit of fine cloth each year for a period of two years." Several pieces of silver made by Michael have been identified.

John Peter, brother of Charles, (b. 29 Apr. 1769 – d. 1801) was a working silversmith in Montreal and made silver for the Indian trade. After leasing Michael's shop with John Oaks, he had several apprentices, among them John Glatter and H. Morand.

The most famous of Charles Arnoldi's brothers was Daniel (b. 4 Mar. 1774 – d. 19 July 1849 of cholera). He was a doctor in Montreal and Riviere-du-Loup. A number of his descendants are well known in Toronto.

ATKINSON AND PETERSON

Two clocks sold by Atkinson and Peterson are in Canadian museums. One clock, illustrated in Fig. 94, is in the Royal Ontario Museum collection, and the other is at Chateau de Ramezay in Montreal, Fig. 95.

The Chateau de Ramezay clock bears the inscription, "Atkinson and Peterson Montreal 1781" on the dial. The dial is white-painted with a seconds bit and calendar. The movement has not been examined, but the dial configuration is typically English. This date is relatively early for a painted dial. The case is very simple in style and may well be of local origin.

The Royal Ontario Museum clock is a much more elegant specimen. Once again, however, the case could be of local origin. The hood is surmounted by delicate fretwork and three brass finials. The hood has free-standing columns on each side of the dial and there are also quarter columns at the waist. Both sets of columns have brass tips, top and bottom. There is a somewhat cryptic provenance for this clock: "To Joseph Mas — (who owned Manor House Terrebonne) and Sophie Ray —, married April 1818 and given to M. Moody. This manor house is now Terrebonne Museum." The date 1818 is relatively late for the elegant brass dial and in view of the 1781 date on the other clock, the authors are tempted to think that this clock was in existence before 1818. Perhaps the provenance should state "given by M. Moody."

The dial is a fine brass specimen with applied spandrels. A disc is attached to the arch portion, bearing the names Atkinson and Peterson. A seconds bit and calendar dial are also present. There is some engraved decoration on the dial. The bell has cast-in ornamentation. The movement itself is of typically English configuration for an 8-day, tall case striking clock.

To date no information has been uncovered on these men or their activities.

FIG. 94

FIG. 94 *Tall case clock sold by Atkinson and Peterson.*
– Courtesy the Royal Ontario Museum, Toronto, Canada
FIG. 95 *This Atkinson and Peterson clock was seen in the Chateau de Ramezay, Montreal.*

FIG. 95

EDWARD P. BAIRD

Edward Payson Baird (b. 21 Jan. 1860 – d. 23 Oct. 1929) was the proprietor of a clockmaking establishment in Montreal, Canada, from 1887 to 1890 when the operation was moved sixty miles south of Montreal to Plattsburgh, New York, and later to Evanston and Chicago, Illinois. The clock cases were manufactured for the purpose of advertising and a wide variety of products were promoted. Over thirty items have been noted (see p. 72).

The parts of the cases that hold the movements were made of wood, but the fronts of the clocks, including the doors that carry advertisements, were made of papier-maché. This material is easily damaged and cannot be treated by chemical paint removers, as more than one collector has learned to his dismay. Baird clocks in mint condition are still to be found, but collectors are cautioned not to be hasty in attempting repairs to any cases in need of restoration.

Although a few gallery clocks exist, the great majority of the cases were "figure eight" in style. The early cases had a rosette on either side of the narrow part of the clock between the top and bottom doors. Rosettes were replaced in later models by plain brackets, allowing for more economical production. In the last period of manufacture, the style of the round bottom door was replaced by a "school clock" type of drop that continued to carry advertising. In this style of clock, instead of advertising on the top door of the clock, the dial itself was painted or embossed with the message outside of the chapter ring.

A detailed study of many of Baird's clocks and his activities can be found in *Edward Payson Baird, Inventor, Industrialist, Entrepreneur* by Leonard L. Schiff and Joseph L. Schiff. Included in the book are pictures of Baird and his family, advertisements, pictures of the Plattsburgh factory, and many illustrations of his clocks, some of them rare.

According to Schiff, Baird purchased his movements from the Seth Thomas Company. With few exceptions, the movements were variants of the Seth Thomas number 97. One variant advertising "Coca Cola" has a 15-day movement using one of the number 50 types. Movements stamped "F. Kroeber, New York," were used in several clocks advertising Venus soap. Seth Thomas is known to have made some movements for Kroeber.

Edward P. Baird was born in Philadelphia, United States of America. His father, James H. Baird, was a Presbyterian minister and his mother, Adeline W., was a member of the Torrey family of New York. In 1875 Edward went to work for William L. Torrey and Company, a firm that made packing boxes. In 1879 he left to join the Seth Thomas Clock Company and became a close friend of Seth E. Thomas, Jr. Edward left this company in 1887 and went to Montreal, Canada. There he was listed in the city directory of 1887–88 as general manager of the Electro-Mechanical Clock Company with offices at 260 St. James Street, Montreal. No additional information about this company has been found, with the exception of a reference in the Schiffs' book to the fact that in 1886 Chester H. Pond of New York, who was associated with the company, was granted a Canadian patent for a synchronizer for electrical mechanical clocks.

In 1888–1889 Baird was listed as a clock manufacturer in the Edward P. Baird and Company at 30 Lemoine Street, Montreal. George E. Baird was also a member of this company. The following year the address of the company was 112 Queen Street. Both men gave, as their home address,

FIG. 96a

FIG. 96b

FIG. 96 *Baird advertising clocks. The clock in Fig. 96a was said to advertise Corticelli (possibly Courtauld) thread before being altered in Picton, Ontario.*

Fig. 97

Fig. 98

"Turkish Baths, 140 St. Monique." By 1891 Edward Baird was living in Plattsburgh, New York, and George was not listed in Montreal.

However, in 1890 Edward did not move his entire operation to Plattsburgh; for some time, advertising clock doors continued to be made in Montreal. It is not unusual to find "Edw. P. Baird & Co., New York and Montreal" on the clocks. At the time that Baird was making clocks in New York, the place of manufacture could be spelled Plattsburg or Plattsburgh and both spellings are found on the clocks.

In Plattsburgh the first and second floors of the Hartwell building were leased. Within a short time, because of the large orders, they moved to 18 Bridge Street, where they employed thirty-five people. A 300-ton hydraulic press, capable of giving a pressure of 1,000 pounds per square inch for moulding clock fronts was purchased. The firm expanded and, in addition to clocks, wood fibre signs, novelties, and time stamps were included in their inventory.

The business survived several fires in the factory. At the time of the second fire, on 24 March 1892, the factory was producing 300 clocks a week.

Baird did not patent his clock innovations in Canada. In June 1894 he was granted an American patent for a clock case made of one piece of moulded glass. In December of that year a patent for the time stamp was granted. Also during this period, he travelled extensively in the United States and Europe, promoting his advertising clocks.

On 30 October 1895 he married Cora Lee Cox of Chicago, and seven months later the couple moved to Evanston, a suburb of Chicago, Illinois, and the Plattsburgh Clock Company ceased to exist. There was insufficient clockmaking work to keep the factory busy, and efforts to diversify did not attract enough supplementary business. Baird began making advertising clocks in Evanston and Chicago. These clocks were different in style from those manufactured previously. The advertising material was embossed or painted on the outer portion of the dial, surrounding the chapter ring on which arabic numbers were used. The

FIG. 97 *Dial detail of a Baird clock, during his years in Montreal.*
FIG. 98 *Seth Thomas movement marked "Baird Clock Co Plattsburg, NY" from the Picton Times clock, Fig. 96a. The movement from the Montreal period clock (Fig. 97) is unmarked, but similar.*

lower portion of the clock, often similar to a "school clock" drop, was also used for advertising.

In 1900 Baird established the Baird Manufacturing Company to make the telephone equipment that he had invented and patented. Between 1900 and 1929 he also patented time-stamp recorders and locks. By 1915, and until his death, locks were of primary interest to him, and he established the Chicago Cabinet Lock Company in order to manufacture them. But his inventions did not occupy all his time. He was a bicyclist, winning many medals. When the Plattsburgh Bicycle Club was organized in May 1891, he was elected its captain. The meetings took place in the Baird clock factory.

Edward P. Baird and Cora had three children: Coreta Cox Baird (b. 31 Aug. 1896), Edward Payson Baird Jr. (b. 16 Jan. 1904 – d. 1923), and Helen I. (Baird) Lang (b. 11 Nov. 1907).

Some of the firms known to have advertised on Baird clocks are Aspinall's Enamel, J.P. Beckers Department Bazaar, Buffos Cigars, Chamberlain Cigars & Cigarettes, Clapperton's Cotton Cord, Clarke's Dog Cakes, Coca Cola, Crown Brand Extracts, Diamond Black Leather Oil, Duffy's Pure Malt Whiskey, El Gaza & Honeymoon Cigars, John Finzer & Brothers, Tobacco, J.M. Fortier Créme de la Créme Cigar Factory, Globe Oil Company, Harrison's Coffee & Spices, Hovis Bread, Jackson Square Cigars, Jesse Joseph & Co. Crown Brand Extracts, Jolly Tar Pastime Old Honesty Plank Road, Jonas' Flavoring Extracts, Henri Jonas & Co., LaAmericana Cigars, Levy's Clothiers & Hatters, Lion "L" Brand Mixed Pickles & Vinegars, Mayo's Tobacco, Mellin's Food, Milkmaid Milk, Monell's Teething Cordial for Children, Nichol's Oriental Balm, Okonite Insulated Wires, Perfection Leather Oil, E. Schmidt & Co., U.S.G. Harness Oil, Snow Drift Baking Powder Co. (Brantford, Ont.), Spanish Blacking, J. Stern & Sons, Clothiers, Vanner & Prest's Molliscorium, Venus Soap, Walkdens Inks Est'd. 1735, Waters' Cola Wine & Quinine Wine, Woolf Hats Clothing, Shoes.

FIG. 99

FIG. 99 *Store in Montreal about 1905 displaying a Baird clock.*
FIG. 100 *Dial detail of clock by J. Balleray & Co., Longueuil.*
FIG. 101a *J. Balleray clock with R. Whiting movement.* – *Courtesy Donohue and Bousquet Antiques, Ottawa*

FIG. 100

FIG. 101a

JOSEPH BALLERAY

Joseph Balleray sold tall case clocks with wooden movements in Long-ueuil, Lower Canada, in the 1830s. The clock illustrated in Fig. 100 combines a wooden dial and movement with a walnut case. The case was evidently made by a local country cabinetmaker who followed the popular style, but did not make an exact copy. The overall effect is pleasing, with a nicely proportioned hood and an exceptionally narrow waist, although the base is heavy and unattractive. The wood-en dial is marked "J. Balleray & Co. Longueuil." Little remains of the original wooden movement, but it appears to be of New England origin.

A second tall case clock made of pine in the style of Twiss clocks is known to exist. Unfortunately, the case has been stripped. The door of the case has a small opening, covered with glass, through which the movement of the pendulum can be checked. The origin of the wooden movement in this clock is not known.

A third Balleray clock is shown in Fig. 101. It is interesting to note that the pine case of this clock seems identical to those sold by the Twiss brothers, who were active in the Montreal area in the same time period (see Fig. 203). The Balleray clock exhibits all the details of decoration, artificial graining, coiled wire hinges, and general outline that are found in Twiss clocks. The dial, too, is similar, with imitation winding holes and floral decoration in the arch. As in Fig. 100, the dial is marked "J. Balleray & Co. Longueuil."

This clock's movement is in good condition and appears to be the work of Riley Whiting of Winchester, Connecticut. Balleray clocks are much less common than those of the Twiss brothers, but on the basis of the two exam-ples seen here, it seems likely that Balleray, like the Twiss brothers, imported movements from Con-necticut and cased them locally at Longueuil.

Another wooden-movement, 30-hour tall case clock has been reported in the collection of the American Clock and Watch Museum, Bristol, Con-necticut. The wooden dial of this clock is marked "W. Balleray Chambly." The authors have been unable to find a connection between W. and J. Balleray, although both communities were close to Montreal.

Little is known about Joseph Balleray. His name was on the list of persons qualified for jury duty in Montreal during the years 1832 to 1835. Also, according to *Le Bulletin des Recherches Historique* and quoted in *Canadian Clocks and Clockmakers* by G. Edmond Burrows, Joseph Balleray, clockmaker, of Blairfundy married Miss Charlton of Montreal on 24 Feb. 1834.

FIG. 101b *Dial detail of clock by J. Balleray, Longueuil.*
FIG. 101c *Views of the R. Whiting movement.*
– *Courtesy Donohue and Bousquet, Antiques, Ottawa*

WILLIAM BAXTER

FIG. 102a

William Baxter (b. 1812 – d. about 1878) was a jeweller, and a watch- and clockmaker in Quebec City from about 1835 to about 1878.

A tall case clock bearing the inscription "W. Baxter Quebec" has been examined by the authors. This clock exhibits a number of interesting features, plus a few inconsistencies.

The case itself, as can be seen in Fig. 102, is constructed in the characteristic tapered-waist style, which, as noted in Chapter 1, seems to be unique to the Quebec area. The principal case woods are pine and mahogany. All visible surfaces are of solid mahogany or fine mahogany veneer. The case stands eight feet tall and is surmounted by three wooden finials. The hood has a characteristic carved crest in the form of leaves. The dial is painted white.

The movement is typically English, being 8-day and weight-driven, with hourly strike by rack and snail. There is a seconds and a calendar dial. The movement is fitted with an iron false plate marked "E. Howell Birmm" (i.e. Birmingham).

The case bears a number of marks, dates, and initials, all from the years 1822, 1823, and 1824, which are well before the time when Baxter opened his shop. The earliest inscription reads "Joseph Marcoux, Septembre 1822, Quebec, Rue Defossés [?] No 27, St Roch." St. Roch was one of the wards or districts of early Quebec City and Defossés Street was the location of many artisans, including cabinetmakers. The inscription is located in an awkward area of the lower case. This suggests that Marcoux may well have made the case and added the inscription during its manufacture.

Another inscription can be found, crudely scratched on the back of the dial and on the rear plate of the movement, reading "G. Maxwill April 4, 1824." Both inscriptions date from a time well before Baxter's arrival in Quebec and indicate that the clock had an earlier existence. Old clocks often suffer many changes and modifications over the years. It is entirely possible that Baxter added his name to the dial and resold the clock at a later date. The full story may never be pieced together, but the clock survives as an imposing specimen bearing the Baxter name.

William Baxter was born in Fife, Scotland. He and his wife (b. 1810 – d. after 1885), a native of Arbroath, Scotland, came to Lower Canada and opened a shop on Buade Street, Quebec City. By 1847 the business had been moved to St. John Street. With the exception of eight years from about 1857 to 1865, when the shop was moved to St. Olivier Street and then to St. George Street, his business remained on St. John Street until the shop was closed in the early 1880s. William and his wife had eight children: James (b. 1836), John (b. 1840), Jean (b. 1841), George (b. 1842), Agnes (b. 1844), Douglas (b. 1845), Rebecca (b. 1846), and Christian (b. 1849.)

His son James joined him in the business and by 1862 had his own shop on Richelieu Street which stayed in business until the mid-1860s, when he was no longer listed in the Quebec City directory. However, when William Baxter died James returned to Quebec City to help his mother manage the St. John shop until it was closed.

FIG. 102 *Clock by Wm. Baxter, Quebec, fashioned in the characteristic Quebec style.*

FIG. 102b

74

G.S.H. BELLEROSE

G.S.H. Bellerose established a factory to make tall case clocks in Three Rivers, Lower Canada. Jean Palardy, in *The Early Furniture of French Canada* credits him with being in business from the end of the 18th century until 1843. There is no doubt that Bellerose was a prolific clockmaker and the *Revue d'ethnologie du Québec /9* suggests that he sold several hundred clocks. Many of his clocks have survived in private collections and museums.

Generally speaking, Bellerose seems to have catered to the market for higher-quality clocks. American wood-movement clocks reached a peak of popularity during the years when Bellerose was active. Although he faced strong competition from such clocks, made by the Twiss brothers and others, he appears to have avoided the use of wood movements, preferring the more expensive English 8-day brass mechanisms. His clock cases too, for the most part, were elegant mahogany specimens, with carving, banding, and fancy inlays. All surviving Bellerose clocks known to the authors are fitted with English 8-day brass movements, are weight-driven, and furnished with painted dials. A tall case clock (Fig. 103) on display in the Provincial Museum in the city of Quebec has a false plate bearing the names "Walker & Hughes." Walker and Hughes were false plate and dial makers in London, England from 1815 to 1835.

As noted above, Bellerose specialized in high-quality cases. One example is shown in Fig. 103. Two other excellent clocks are in the collection of the Royal Ontario Museum. The clock in Fig. 104 exhibits inlays made of mahogany, maple, and satinwood. Both Bellerose clocks are illustrated in *English–Canadian Furniture of the Georgian Period* by D.B. Webster. Interestingly, these three clocks share an identical design characteristic: in each clock, the hood is exceptionally tall in relation to its width, giving the clock a distinctive and not entirely graceful appearance. In other respects the cases are of excellent quality with fine veneers and inlays executed in a very professional manner.

Two clocks are known that are totally unlike the examples discussed above. One is shown in Fig. 536 of *The Early Furniture of French Canada* by Jean Palardy, and the other here, in Fig. 105. These clocks are similar to each other, sharing an unusual tapered shape with a drum-head treatment around the dial, surmounted in one clock by a dome top and in the other by rather unusual clusters of carved fruit. The clocks are simple in construction and the wood used is solid butternut. Both clocks are decorated with "seashell" inlays. These two clocks are distinctive and look more like the work of a country cabinetmaker. However, the existence of these unusual cases, as well as the distinctive tall hoods in the other clocks, are indicative of a developing sense of local styling in Quebec, which has been commented upon elsewhere in this book.

Another area where Bellerose has left a distinct mark is in the production of a few custom-built "famous name" clocks. One of these has been documented in *American Clocks and Clock Makers* by Carl W. Drepperd and bears the name of General Sir Isaac Brock, who was a war hero in Canada and who died in the War of 1812. It is not known whether the clock was made for Brock or as a memorial to him.

Another clock, illustrated in Fig. 107, is known as the "Earl Dalhousie" clock. The Earl of Dalhousie was a British nobleman who played an important role in British North America in the early 19th century. He was first appointed to the post of Governor of Nova Scotia. After a successful term of office in Nova Scotia, he was appointed Governor General of Upper and Lower Canada, succeeding the Duke of Richmond. He

FIG. 103a

FIG. 103b

FIG. 103 *Tall case clock sold by G.S.H. Bellerose. Collections d'ethnographie du Québec*

FIG. 104

arrived in Quebec on 18 June 1820. Unfortunately, in Quebec he became embroiled in local politics, had frequent confrontations with local politicians, and became a decidedly unpopular figure. Unrest in Quebec reached a climax with the rebellion of 1837. At the expiration of his term in 1828 the Earl of Dalhousie left Canada, accepting the position of commander-in-chief in India.

However, he evidently did have time to commission a clock from the Bellerose factory. This is an attractively made clock with clean lines, fine veneers, and good inlays including the typical Bellerose shell motif. The arch of the dial contains the inscription, "Earl Dalhousie," and the illustration in the arch depicts the noble Earl on horseback, leading a group of soldiers. The Dalhousie story is further told in *The Dalhousie Journals*, edited by Marjory Whitelaw.

Documentation of only one Bellerose family has been found in the Three Rivers region and, although it includes information about a well-known priest and a Member of Parliament, a search of the records for the name G.S.H. Bellerose has been unsuccessful. Also, no primary reference has been found for the Bellerose clock-making factory. At least three generations of Bellerose men in the Joseph Hyacinte Bellerose (b. 1720) and Michael Hyacinte Bellerose (b. 1762) families have a name beginning with the initial H. The fact that G.S.H. Bellerose, the clockmaker, also has the initial H may link him to this family.

FIG. 106

FIG. 104 *Bellerose clock in the Royal Ontario Museum.* – *Courtesy the Royal Ontario Museum, Toronto, Canada*
FIG. 105 *This Bellerose clock has a very distinctive case in butternut.*
FIG. 106 *Portrait of the Earl Dalhousie from a book cover.*
 – *Courtesy Barbara and Henry Dobson*

FIG. 105

HENRY BIRKS

Henry Birks (b. 30 Nov. 1840 – d. 16 Apr. 1928) of Montreal was the founder of the Birks jewellery empire that, by its centennial year in 1979, controlled almost 25 percent of the Canadian market. Birks continued to prosper until the recession of 1992 when, like other companies, it experienced financial difficulties. At the time of writing in 1993, some reorganization had taken place and the ownership had changed, but the firm continued.

The story of the family and its business has been published by the company on a number of occasions. "House of Birks" booklets appeared in 1946 and again in 1967. A book, *The First Century*, by Kenneth O. Mac-Leod, was written in 1979 as a tribute to this Canadian company on the occasion of its centennial. This book gives a sensitive and detailed account of the members of the Birks family and their involvement in the business. It is illustrated with many family pictures. The book sketches the history of the business from its beginning to the opening of its ninety-fifth branch store.

Throughout the years, many clocks were sold in Birks stores. Most of the pendulum clocks carrying the Birks name on the dial were purchased from the Seth Thomas Clock Company in the United States. One of the clocks sold in Montreal between 1879 and 1893 is owned by Point Ellis House Museum in Victoria, British Columbia.

In addition to his business interests, Henry Birks was a military man. He was an original member of the "Victoria Rifles," and in 1866, during the Fenian raids, he served at Hemmingford. He and his family also took an active part in church affairs. He was a member and an elder in the American Presbyterian Church in Montreal, where, beginning at the age of eighteen, he taught the infant class in Sunday school for a period of twenty-one years. He was a generous supporter of the YMCA, and according to the articles and books published about him, he led a carefully regulated life, and allowed himself only six hours of sleep a night. He disallowed tobacco and liquor. He was considered to be a cautious man, carefully surveying a situation before he made a move, but his moves were determined by his vision of things to come. Despite his contention that "We are making all the money that is good for us," Henry enjoyed being a successful businessman. Each day, including his eighty-seventh birthday, he walked up and down the aisles of his Montreal store, greeting customers and chatting with the staff. Of himself, he said, "I'm just an old eight-day clock and I need to be wound up once a week, to true my balance and be set in beat."

The three sons of Henry Birks took an active part in the operations of the business. William Massey Birks (b. Oct. 1868 – d. 4 July 1950) entered the business in 1885 and was a buyer, purchasing goods for the store in Europe. It was largely due to his foresight that branch stores were established, the first being opened in Ottawa in 1901. W.M. Birks was president of the company from 1928 until 1938 and chairman from 1938 to 1950.

John Henry Birks, "Harry," (b. 1870 – d. 24 Mar. 1949) entered the firm in 1891 after graduating as a mechanical engineer from Massachusetts Institute of Technology. His interest was in the manufacturing end of the business and the factories were under his control. He succeeded his elder brother as president of Henry Birks and Sons in 1938, remaining president until 1944.

Gerald Walker Birks (b. 1872 – d. 13 Oct. 1950) entered the business

FIG. 107a

FIG. 107b

FIG. 107 *The "Earl Dalhousie" clock by*
G.S.H. Bellerose.
– Courtesy Barbara and Henry Dobson

in 1888 and was in charge of advertising until the First World War. He was responsible for the development of the Birks catalogue. During the war he was in command of the Canadian Military YMCA in London, England, and he afterwards devoted his life full-time to the work of the YMCA.

Further contributions of Henry Birks's sons and grandsons, their many interests, and the honours conferred upon them are documented in *The First Century* by Kenneth O. MacLeod.

In addition to the sale of clocks, Birks stores sold watches bearing the Birks name. Henry Birks established a "watch factory" where movements purchased primarily from Switzerland were finished by watchmakers. In 1906 the company employed a dozen watchmakers. In 1925 a watch case factory was purchased, and it remained in existence for about a decade.

FIG. 108

FIG. 108 *Family portrait taken in 1880. Henry Birks and wife with sons Gerald, William and Harry. – Courtesy Henry Birks and Sons Limited, Montreal*

The ancestors of Henry Birks had worked for centuries, as silversmiths. Before the 10th century, early members of the family arrived in England from Scandinavia. (The name is derived from a Scandinavian word for the birch tree.) Beginning in 1550 the Birks freemen were listed on the court rolls of the Cutlers' Company of Sheffield, and many members of the family became master cutlers. The Birks family settled in Yorkshire, and there, in the 16th century, Richard Birks handed on his knowledge to his son.

Henry's father, John Birks, was born at Wombwell Hall near Darfield and Barnsley in the Sheffield and Doncaster district. On 1 September 1825 in St. Mary's Church, Barnsley, he married Anne Massey, daughter of Richard Massey of Monk Bretton. In 1832 the family arrived in Montreal at the height of the cholera epidemic, to which they lost two of their three children. A daughter survived. The family moved from St. Paul Street to St. James Street, where Henry was born, one of seven children born in Canada. In 1852 the family lived at "Burnside," a house built by James McGill located near the present-day McGill University gates.

Henry Birks came to the attention of Theodore Lyman, partner in the firm Savage and Lyman, watchmakers and jewellers. Lyman was impressed by Henry's excellent attendance record and punctuality at the Sunday school where Lyman was Superintendent. Henry joined the firm on 22 April 1857. One of his duties was to sleep above the store with a pistol under his pillow to ward off break-ins.

A year later Savage and Lyman moved their business to the Cathedral Block on Notre Dame Street. In 1868 Henry Birks was made a junior partner at a salary of $1,000 per year and the business moved to St. James Street at the west corner of Dollard Lane. The name was changed to Savage, Lyman, and Company. Henry Birks was forced in 1877 to sell his interest in the business in order to "make good" a note he had endorsed for one of his brothers. This sad experience persuaded him to later institute the policy whereby no member of the Birks family could sign a negotiable paper without agreement of the company's directors.

The prosperous 1860s gave way to the world-wide depression of the 1870s. A severe drop in annual turnover at Savage, Lyman and Company compelled them to go into liquidation in 1878. Henry Birks was placed in charge of the liquidation of the stock.

In March of 1879 Henry Birks purchased the Savage and Lyman stock with $4,000 of his own savings and $1,000 borrowed from his wife, Harriet. He then opened his store at 222 St. James Street under the name of Henry Birks and Company. Having observed customary business practices during his twenty-two

FIG. 109

FIG. 109 *Clock sold by Henry Birks & Co. – Courtesy Point Ellis House Museum, Victoria, B.C.*

years with Savage and Lyman, Henry Birks decided on different policies. His business would be based on cash for all purchases and one selling price for all customers. He met with immediate success, turning over his stock seven times during the first eleven months. His business continued to expand over the next four years at a rate of more than 25 percent per year. Cramped for space, he moved to 232 St. James Street.

In 1885 he moved across the street to bigger quarters at 235–237 St. James Street. Henry realized that, with the growth of the city of Montreal, the business section would of necessity move northward. In 1893 he began building a larger store on Phillips Square at the corner of St. Catherine and Union streets on land which originally belonged to Joseph Frobisher, explorer and fur trader. He moved into these premises in 1894. The Henry Birks building was enlarged in 1902 and again in 1907. A further ten-storey building, built in 1913 and enlarged in 1926 on the south side of Cathcart Street, was known as the New Birks Building (later Birks Building).

Shortly after his move to Phillips Square, Henry Birks arranged for factories to supply much of the goods that he sold in his store. He had opened his first jewellery factory above his St. James Street store in 1887 and then moved it to Phillips Square. In 1898 he acquired the manufacturing firm of Robert Hendery and John Leslie, which had produced silverware for him, and moved its silver factory to Phillips Square in 1900. In 1913 it was moved to the New Birks Building, but the jewellery factory remained at Phillips Square. A watch case factory, purchased in 1925, lasted about a decade.

In 1893 Henry Birks changed the name of the firm to Henry Birks and Sons, reflecting the excellent relationship Henry enjoyed with his three sons. He gave them early responsibility and treated them as equals, dividing profits, salary, capital, and interest in the business into four equal parts. In 1905 the firm was incorporated into a limited company when Henry Birks and Sons consolidated with Ryrie Brothers. However, the name of the business was not changed until 1924.

FIG. 110a

FIG. 110b

FIG. 110 *Case and dial of a fine tall case clock by Thomas Cathro. The case is in the "Quebec" style except for the fretwork at the top.*

THOMAS G. CATHRO

One fine tall case clock that has come to the attention of the authors bears the name Thomas G. Cathro, Quebec. Directories list Thomas George Cathro at 15 Notre Dame Street, Quebec City, during the period 1822 to 1844.

The clock case is constructed in the distinctive "Quebec" style. The primary wood is mahogany. Most of the characteristics of this unique style are present: a rounded top to the hood, small columns beside the dial, tapered-waist and door, gothic ornamentation on the door, quarter columns in the waist, and a scrolled base with integral feet. One noticeable departure from this style is the elaborate fretwork cresting on the hood in place of the usual carved leaves.

The dial and movement of this clock are typically English of the first half of the 19th century and correspond in age to the period Cathro was in business.

Thomas G. Cathro came to Lower Canada with his brother J. Cathro as "military" settlers. They applied in 1822 for land in the Township of Buckland. Although they received the land, Thomas settled in Quebec City. In the census of that city in 1825, Thomas and his wife were between twenty-five and forty years of age, and they had one daughter younger than fourteen.

R. CATTON

Richard Catton came to St. John's, Newfoundland, from England in 1815. He advertised on 12 November 1816 in the *Royal Gazette* that he was conducting a "Lottery of jewellery and trinkets now on hand, Water Street." On the twenty-sixth of that month he advertised that he was a watchmaker and goldsmith and that his shop was on "Lower Water Street nearly opposite premises of Messrs. Robertson and Mortimer." By 4 October 1817, as indicated by an advertisement in the *N.B. Courier,* Catton had relocated to Saint John, New Brunswick, on the corner of Prince William Street and Market Square.

In 1819 Catton moved to Quebec City with his brother Thomas, and they applied for land in Lower Canada. However, Richard opened a shop at 13 Mountain Street and in November 1820 placed an advertisement in the *Quebec Gazette* which read: "Richard Catton, chronometer, watch- and clockmaker, 13 Mountain Street has received by the 'St. Lawrence' a neat and select assortment of superior London-made gold and silver watches, 8-day spring dials, gold seals, keys, ear-rings, brooches, finger-rings, silver spoons, plated goods, ladies' dress lace, clasp. Also on hand a few handsome alabaster vases."

As well as being a silversmith and a watch- and clockmaker, Richard Catton also was a dealer in furs.

The bracket clock pictured in Fig. 111 has a late 18th-century movement, with engraved back plate, verge escapement, and bob pendulum. The case is of the late 18th- or early 19th-century break-arch style. The finish is ebonized veneer. The dial is painted steel, very simple in style.

On 8 February 1821 a notice appeared in the *Quebec Gazette* asking all persons who had claims against Richard Catton to submit them. It is not known if he had retired from business or died; however, no death notice was found. His name did not appear in the Quebec City directory of 1826.

It is known that John Bean (b. 1785 – d. 4 Feb. 1830) succeeded Richard Catton at 13 Mountain Street and advertised his watchmaker's trade on 8 March 1821.

LOUIS FOUREUR *DIT* CHAMPAGNE

Louis Foureur or LeFoureur, (b. 2 June 1720 – d. 16 April 1789) of Montreal was known as "Champagne." Early in his life he learned the trade of cabinetmaker and wood carver under the guidance of J. Bte. Filiau *dit* Dubois and his brother François Filiau. Examples of his work include six wooden church candles and a number of church chapels in which his carvings are exhibited. According to the *Dictionary of Canadian Biography, Volume IV*, Louis Foureur *dit* Champagne became interested in clockmaking and studied all aspects of the work of Filiau *dit* Dubois and decided to surpass him as a clockmaker. To accomplish this, he secretly studied clocks made in France and proceeded to make a clock so excellent, elaborate and complicated that it established his reputation in Montreal. He continued to make clocks until his death, but apparently none of his clocks have survived. In spite of this rivalry, Champagne and François Filiau *dit* Dubois became close friends. Dubois attended Champagne's wedding and was godfather to his eldest son.

In addition to making clocks, Champagne took in at least six apprentices. One began his apprenticeship in 1744 for four years. The last apprentice started his six-year apprenticeship in 1775.

Champagne bought a house in 1747 on Notre Dame Street, Montreal, where he lived until his death. He was considered wealthy and owned other houses and land that were sold after his death.

Louis Foureur *dit* Champagne was the son of Pierre Foureur, blacksmith and locksmith, and Ann Céleste Desforges. He was married on 9 November 1744 to Catherine Guertin of Montreal, and there were seven children. Two sons, Jean Louis and Pierre, were silversmiths in the 1780s and 1790s.

FIG. 111

ERNEST CHANTELOUP

The name "E. Chanteloup, Montreal," appears on the small interior dial of a tower clock originally installed in the post office of Cornwall, Ontario. The post office was demolished in 1955, but the clock was saved and awaits restoration. The time-and-strike movement of the clock has a type of double three-legged gravity escapement.

Relatively little information has been uncovered on Chanteloup's business activities. However, as the illustrations in this section confirm, he was capable of supplying tower clock movements of excellent quality. His advertisements claimed that, as an iron and brass founder, he was able to provide a wide variety of cast materials. Therefore, it is reasonable to believe that he could have constructed part or all of the tower clock illustrated here. This tower clock follows the general practices of the time quite closely. In the opinion of one authority, the escapement is similar to that used in certain Seth Thomas tower clocks. However, sufficient differences were noted to suggest that Chanteloup prepared his own castings and installed his own version of the running gear. It is not known whether he did his own gear cutting.

FIG. 112

FIG. 111 *Bracket clock by R. Catton, Quebec. – Courtesy Serge L'Archer*
FIG. 112 *A general view of a Chanteloup tower clock.*

FIG. 113

Unfortunately, this is the only Chanteloup clock available for study. It is to be hoped that in future more examples will become available, making it possible for more definite conclusions to be drawn.

Ernest Chanteloup began his career in Montreal around 1860 as a lamp-maker on Craig Street. By 1869 he was manufacturing bronze goods, tin ware, chandeliers, and gasoliers; he also served as a gas fitter and plumber. In the early 1870s he was also manufacturing telegraphic and electrical instruments, copper work, locomotive and car fittings, cooking ranges, and church ornaments. By 1878, in addition to his other activities, he was the agent for the Union Water Meter Company. He eventually manufactured gas and oil burners and easy chairs for parks.

In 1879 Chanteloup began to advertise that he was also a manufacturer of tower clocks. Advertisements similar to that shown in Fig. 115 appeared in the directories until 1894. In 1895 Chanteloup's factory was purchased by James C. King and David Yuile.

Ernest Chanteloup boarded in Montreal for about ten years at the Cartier Hotel on Cartier Square. In the summer he lived at the summer residence he built in 1884 on the Upper Lachine Road, Lachine, Quebec. By 1894 no mention of E. Chanteloup was made in the city directory; it is assumed that he died about that time.

E. CHANTELOUP,
Iron and Brass Founder,

LIGHT AND ORNAMENTAL CASTINGS.

VAULTS, DOORS, IRON COLUMNS, WROUGHT, CAST AND RIVETTED BEAMS AND GIRDERS.

Locomotive and Car Fittings, Telegraphic and Electrical Instruments and Supplies, Fire Alarm Apparatus, Artistic Bronzes, Church Ornaments, Lamps and Gasaliers, Copper Works, Handsome Fire Grates and Chimney Fronts.

MANUFACTURER OF

CHURCH ☆ TOWER ☆ CLOCKS,

— AND —

☆ LOCOMOTIVE BELLS ☆

INCLUDING CHIMES AND PEALS.

Gas and Coal Oil Burners of all Descriptions.

Steam Fittings, Coal Screens, Hot Water and Steam Apparatus, French Window Fasteners and Espagnolettes, Railway Supplies of all descriptions. Fire Iron and Dogs and Fenders, Builders' Hardware.

OFFICE AND WORKS.

587 TO 593 CRAIG STREET,
MONTREAL.

FIG. 115

FIG. 113 *Close-up of the gravity escapement used by Chanteloup.*
FIG. 114 *Details of the tower clock movement.*
FIG. 115 *Chanteloup advertisement found in the Montreal city directories of the 1880s.*

FIG. 114

MARTIN CHENEY

The Cheney family name is one of the earliest and best known in American clockmaking. The activities of Benjamin Cheney (b. 1725 – d. 1815), son of Benjamin and Elizabeth Long Cheney have been extensively documented. He was known in Connecticut as a maker of brass and wood movement clocks from about 1745. He was the first maker to produce wood movements for tall case clocks in any quantity, although all his clocks were hand-made, one at a time.[1]

The name of Benjamin Cheney has also been associated with that of Benjamin Willard, senior member of the famous Willard family of clockmakers. Willard had a brief business association with Cheney. This was not an apprenticeship, but the close proximity of the clock shop may have influenced Willard to enter into clockmaking several years later.

Benjamin's brother Timothy Cheney (b. 1731 – d. 1795) was also a maker of wood and brass clocks. Benjamin trained several apprentices, including his sons Asahel, Elisha, Martin, and Russell. Each of these sons became a well-known maker. Olcott Cheney, son of Elisha, carried the family trade through a third generation. Asahel, Russell, Elisha, and Olcott all spent their lives entirely in the United States. Martin worked as a clockmaker at Windsor, Vermont, for about six years, but moved to Montreal in 1809, living the rest of his life in Canada.

Martin Cheney (b. 1778 – d. about 1855), after completing his apprenticeship, set up shop in Windsor, Vermont. His activities there have been documented in several places. He made and sold a variety of tall case and shelf clocks, including a number with complicated mechanisms. Several years ago, a musical tall case clock by Martin Cheney, Windsor, was sold at auction in New York. The authors have also seen two fine "Massachusetts shelf" clocks by Cheney bearing his Windsor address. One of these clocks is illustrated in Figs. 117 and 120.

Insight into Martin Cheney's Windsor activities can be gained from the following references. In the first, Sewell S. Cutting discussed him in a speech on 4 July 1876 at Windsor Vermont:

> Here in Windsor in early times lived and worked Martin Cheney, famous here and famous in Canada for nearly half a century. His clocks were the pride of Vermont households and today, are precious heirlooms. Some of them told not only the hours, but the days of the month and changes in the moon, and some of them had musical accompaniments with a tune for each day of the week, piously inclined to a meditative air on Saturday, playing St. Mary's on Sunday, but at Sunday midnight as if impatient at so long restraint, starting off in joyous glee with "Over the Water to Charly."

The second reference is an advertisement Cheney himself placed in the *Windsor Gazette:*

> Martin Cheney informs his customers & the public that he has just received from Boston, a number of warranted silver cased WATCHES of an excellent quality which will be disposed of very low for Cash. – Likewise – A general assortment of the most fashionable Watch Chains – Silver Thimbles – plated Sugar Tongs – Scissors Chains – Crystal Beads, Gold Earnubs and Finger Rings, Silver Pearl and common Sleeve buttons, Silver Plated and Polished Steel Knee-buckles, Toy Watches – Violin Strings – Fancy Beads – etc.

FIG. 116

FIG. 116 *This presentation banjo is marked "Martin Cheney Montreal" on the dial. A hand-written inscription in the case states that it was purchased in 1825 for $40.*

Musical, Alarm, Moon and plain Eight Day Clocks and Time-pieces made and warranted. Watch-repairing performed with assiduity and dispatch. He returns his thanks to those who have favored him in business and assures them that it shall be his pleasure to execute all orders in a manner which may merit a continuance of their patronage.

N.B. Cash given for old silver copper and brass.
Windsor, May 31 1802.

Some years ago an article entitled "Musical Clocks of Early America and Their Music" by Kate Van Winkler Keller[2] provided some interesting background on these musical tall case clocks. The tune, "Over the Water to Charly [sic]"[3] was used by several makers, including Thomas Harland and Daniel Burnap. The article confirms that Asahel Cheney, Martin's brother, had advertised at Putney, Vermont, in 1798 that he could manufacture musical clocks. The above quotations confirm that Martin offered a similar product line and leave little doubt that he was a skilled and versatile maker.

His offer to purchase old brass refers to a practice common among clockmakers of the period. They often found it necessary to make their own clock parts from scrap brass, melted down and recast in moulds. Cheney also advertised in 1804 and 1805 "for a lad of 14 or 15 years" to become his apprentice.

In politics at this time, Martin Cheney supported the American Federalist party. This party lost popularity in the years following 1801, which led to much dissatisfaction among its followers. Beginning in 1806 there occurred a small wave of migration to Canada from Windsor and other areas. Included in their number were a few jewellers, silversmiths, and clockmakers (see reference to Benjamin Comens in the section on Charles Arnoldi). In 1809 Martin Cheney joined the exodus. His business in Windsor was taken over by Isaac Townsend, a silversmith and clockmaker from Boston.

After his arrival in Montreal, Cheney formed a partnership with James A. Dwight on St. Paul Street. They placed an advertisement in the *Canadian Courant* and the *Montreal Gazette* on 20 March 1809, announcing the opening of their business. The advertisement stated that "they manufacture and expect shortly to have for sale, elegant house clocks and timepieces which will be warranted good." They also offered watches, watch furniture, and jewellery.

The partnership of Cheney and Dwight lasted only a short time, but Cheney continued his business for at least two decades on St. Paul Street. Inscriptions in two of Cheney's Montreal clocks refer to sales dates of 1824 and 1825. In June 1827 Cheney advertised in a Vermont newspaper, for a journeyman clockmaker "to whom good encouragement will be given if application is made soon. One should be preferred that wishes further knowledge of the business." Applications were screened either by Cheney in Montreal or by the staff of the *Northern Sentinel* in Burlington, Vermont. It is not known how much longer Cheney remained in Montreal, but he was present at baptisms of his grandchildren well into the 1830s, as indicated by church records.

Martin Cheney moved from Montreal to the Home District (York County), Upper Canada, where he owned property in Whitby Township, now part of Ontario County. He was listed in 1846–47 and 1850–51 direc-

FIG. 117

FIG. 117 *Massachusetts Shelf Clock by Martin Cheney at Windsor, Vermont. – Courtesy the Henry Ford Museum, Dearborn, Michigan*

tories for the County of York as owner of Lot 10, Concession 1, Whitby Township. His son Frederick (b. 1804 – d. 1848) moved to Whitby Township and purchased Lot 11 beside his father. Martin's wife, Fanny Patrick (b. 1787 – d. 30 May 1855), and son Frederick are buried in Union Cemetery, South Presbyterian Section A, Oshawa, Ontario. There is no cemetery record of Martin, although it is probable that he lies beside his wife. Another son, Edward, and his family lived in Montreal.

The major portion of Martin Cheney's working life was spent in Montreal. The authors have not attempted to examine his output during the early years in Windsor, Vermont, other than to illustrate the Massachusetts shelf clock (Figs. 117 and 120) which is owned by the Henry Ford Museum, Dearborn, Michigan. This clock is a fine and typical example of the style. The movement differs in some respects from the movements found in Cheney's clocks from the Montreal period. There is, however, some resemblance to the type A movement discussed below.

Over a period of years the authors have had an opportunity to study a considerable number of Martin Cheney's Montreal clocks. At the time of writing, some twenty-three specimens are known, consisting of twelve weight-driven wall clocks and eleven weight-driven banjos. All the clocks run for approximately eight days, time only. Several other clocks attributed to Cheney have been reported, but since neither the dials nor the movements have been marked, they have not been included.

There has been a question in the past whether the clocks sold by Martin Cheney in Montreal were partially or entirely made in Montreal, or whether he merely added his name to dials and movements of another maker. A few persons, with little evidence, have been quick to say that he "must have brought them in from the States." One or two have noted a resemblance to the movements of Joseph Nye Dunning. However, a careful comparison of published examples of Dunning movements with Cheney movements, shown here, does not support this. Dunning had a practice of mounting the pendulum suspension bridge well above the top edge of the front plate. This can be seen in certain earlier Cheney movements (i.e. type A). Most type-B and all type-C movements differ completely from Dunning's work. These movement types are described more fully in the pages that follow.

In an article on Dunning,[4] Lester Dworetsky has commented that "it is extremely hard to distinguish brass banjo clock movements from one another without some good starting point – e.g. a distinctive type of case used only by one maker, clocks with dials signed by the maker, unique design of the movement, etc." Dworetsky also comments: "...the old clockmakers had very little interest in duplicating movement after movement in exactly the same proportions. This disinterest would explain in part, why it was never done. Another source of variability is found in the differing abilities among apprentices who carried out the casting, filing, wheel cutting, plate planishing, etc. These two points could account for this writer's observation that no two movements (earlier than rolled brass, stamped movements) that he has examined are exactly the same." These comments are appropriate to the movements found in Martin Cheney clocks.

The authors are of the opinion that Martin Cheney made the movements of the clocks that bear his name either on the dial or movement plates. Recognizing that conclusive evidence one way or the other does not seem to exist, the authors point to a number of factors supporting their belief:

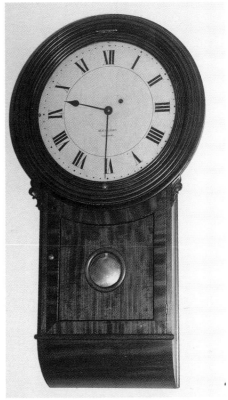

FIG. 118

FIG. 119

FIG. 118 *A typical wall clock by Martin Cheney Montreal. This was his most popular style.* – Courtesy Robert Phillip, Canadian Museum of Time

FIG. 119 *Original reverse glass painting on a banjo clock by Martin Cheney.*

FIG. 123

FIG. 120

FIG. 122

FIG. 124

FIG. 120 *Movement from shelf clock in Fig. 117 made by Martin Cheney at Windsor, Vermont. – Courtesy the Henry Ford Museum, Dearborn, Michigan* FIG. 121 *Cheney type-A movement, marked "Cheney" at upper left and numbered "133" upper right.* FIG. 122 *Martin Cheney type-B movement. The Cheney stamp is in the upper right corner.* FIG. 123 *Cheney type-C movement not stamped, but found only in some Cheney clocks. Compare with variant in Fig. 125. Movement illustrated here is from clock in Fig.130 and has separate pendulum post.* FIG. 124 *Martin Cheney's characteristic silversmith stamp, upper left corner of type-A movement.*

1. Martin Cheney was a recognized clockmaker trained in his father's shop.

2. He is credited with making clocks for sale in Windsor, Vermont.

3. Upon his and Dwight's arrival in Montreal, he advertised that "they manufacture and expect shortly to have for sale elegant house clocks...."

4. Cheney advertised for and probably employed apprentices and, in 1827, was looking for a journeyman clockmaker.

5. Three movement styles occur in the Montreal clocks. Some evolutionary trends can be seen in these movements. At the same time, they do not exactly resemble the work of contemporary U.S. makers. Cheney, however, as an expatriate American, no doubt kept in touch with his brothers and other clockmakers who, after all, were not far away. He would have been reasonably familiar with American developments.

6. There would have been economic incentive to make clocks locally. The compounded costs of makers' mark-up, transportation, and import duties would be minimized. American merchandise would have been unavailable during the War of 1812 period and no doubt anti-American sentiment lingered for some time afterward.

7. Cheney was a recognized Canadian silversmith. Silver tableware bearing his distinctive stamp is known. He used the same silversmith's stamp to mark type-A and type-B movements. It is unlikely that he would have dismantled and marked movements unless they came from his own shop.

MOVEMENTS

As noted above, three distinctive movements are found repeatedly in Cheney clocks of the Montreal period. These movements have been designated type A, type B, and type C. No two examples are exactly the same, but there are obvious similarities in each group. The designation of the three groups presumes that they were produced in that chronological order. References to Cheney in Montreal become scarce after the early 1830s. It is unlikely that many movements would have been produced after the early 1830s in any event. By this time, the far-reaching effects of the mass production of clocks in Connecticut would have reached Montreal, making it difficult or impossible for hand-crafted clocks to compete. The characteristics of the three movements can be summed up as follows:

TYPE A

The assumption that these movements were the first to be produced is based on some similarities to the Windsor, Vermont, movement shown in Fig. 120. The shape of the pendulum suspension bridge is similar. The hour wheel in each clock is solid without spokes, a condition seldom seen in banjo movements. The position of the idler wheel, to the left of the centre arbor, at the nine o'clock position is also unusual. The Windsor movement has its idler wheel in an even less common position, about two o'clock relative to the main arbor. This may not have much relevance except to suggest that Cheney had little preference about its placement. The later Cheney movements all show the more common positioning, to the right of the main arbor.

Type-A movements seem to have been sequentially numbered. In Fig. 121, the number 133 can be seen in the upper right corner of the front plate. In the clock in Fig. 129, which also has a type-A movement, the number is 137. These numbers also appear on the pendulum rod and, in the latter clock, the number is also written inside the case. A third clock bears the number 111 on the case, but not on the movement. Another

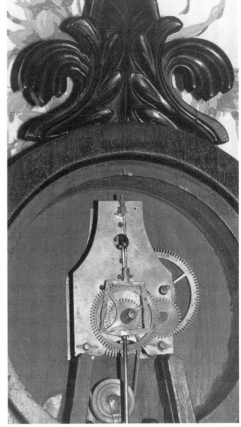

FIG. 125

FIG. 125 *Type-C movement variant from clock in Fig. 133. The upper movement post also serves to support the pendulum rod.*

seemingly unique feature of the type-A movement is the shape of the "keystone" on the pendulum rod. The three examples of this movement, all have rectangular keystones with no taper. In two of these movements, Cheney's silversmith stamp appears on the upper left corner of the front plate. A single screw with machine screw thread fastened the movement to the backboard of the case.

TYPE B

This movement is slightly narrower and somewhat taller than type A. One wall clock similar to the example shown in Fig. 118 has a specific sale date of 1824 and is fitted with a type-B movement, which places it in the middle-to-later years of Cheney's time in Montreal. These wall clocks exist in larger numbers than any other style of clock sold by Cheney, and any that the authors have examined are fitted with type-B movements.

The idler wheel has been moved to the right of the centre arbor and the hour wheel now has filed out spokes. The pendulum suspension bridge is mounted in a lower position and has a somewhat different shape. The "keystones" found with these movements are tapered, very similar to those in U.S. clocks. The type-B movement is usually marked with Cheney's stamp, which has now moved to the upper right corner. The movement is usually held to the back board with a single screw with machine thread, although one movement has been reported with two screws and a shorter vertical plate dimension.

The authors have no explanation for the differences noted above. There are several possibilities, including changing apprentices or other workers in the shop or a desire to conform more closely to American trends. Cheney would have been out of touch with U.S. makers for a number of years during and after the War of 1812. After the war he may have been able to purchase ready-made brass castings for clock parts from U.S. sources, rather than depending on his own resources.

TYPE C

This movement is noticeably different from its predecessors. It is found in wall and banjo clocks that show more "Empire" styling influence, suggesting that they are later. The most obvious difference is that the plates now have a unique bell-like shape. This plate shape has only been seen in clocks sold by Martin Cheney. There are a few "A-frame" movements produced in the U.S.A., but a comparison shows many differences. The type-C movement has only three posts rather than four. There is no longer a suspension bridge. This has been replaced by a simple post in the front plate. The "keystone" is again tapered, but now has a circular cutout, filed in the lower left corner. One clock has been reported without a keystone. The pendulum rod was merely bent to avoid contact with the centre arbor. The wheels themselves appear to be similar to those in the type-B movement, but their position on the arbors has changed. All wheels except the escape wheel are now at the back of the movement, whereas in type A and type B, the third wheel and escape wheel pinion were near the front.

A variant is illustrated in Fig. 125. In this movement simplification has been carried further. The plates again are bell-shaped, but without the decorative arched top. The separate pendulum mounting post is gone. In its place, the upper movement post has been lengthened to serve a dual purpose. It is slit at the end to accept the suspension spring. The plate is held between a pin and a shoulder on the post. The movements are usually fixed to the back board by two screws, again with machine thread.

FIG. 126

FIG. 126 *Unusual Martin Cheney banjo. The base portion is more typical of his office-type wall clocks.*

The reasons for the changes in the type-C movement may have been partly economic. These movements contain less metal and would have been a little easier to make. By the late 1820s Cheney was no doubt beginning to feel the pressure of competition from cheap mass-produced clocks from Connecticut.

In summary, as noted previously, the authors have knowledge of at least twenty-three authentic Montreal Cheney clocks, plus one loose type-B movement. Fourteen of these movements have been available for study: three type A, six type B and five type C. It is quite possible that other variants exist. However, there is enough distinctiveness to the movements to make a strong case that they were produced by Cheney or under his direct supervision.

TABLE 1
MOVEMENT DIMENSIONS

	WIDTH	HEIGHT	DEPTH
TYPE A	2¾"	3¹¹⁄₁₆"	1½"
TYPE B	2⁹⁄₁₆"	4¹⁄₁₆"	1⁹⁄₁₆"
TYPE C	2½"	4⅛"	1½"

Variations of ¹⁄₁₆" have been noted in these dimensions.

TABLE 2
TOOTH COUNTS

Great Wheel	96
2nd Wheel / Pinion	64/8
3rd Wheel / Pinion	60/7
Escape Wheel / Pinion	30/7
Cannon Pinion	36
Minute Wheel / Pinion	36/6
Hour Wheel	72

These counts are identical for types A, B, and C movements.

THE CASES

When the cases of Cheney clocks are examined, there can be little doubt of their uniqueness. Cheney would have found it advantageous to employ local craftsmen to make his cases. By the early 1800s Montreal had become a fairly large and affluent population centre. It was a seaport and therefore had access to tropical woods. There was already in Lower Canada a sizeable group of skilled cabinetmakers.

The presentation banjos in Figs. 116, 127, and 128 are similar over-all to American products, but some differences can be seen in the shape of the "chimney" surmounting the dial and in the carved "boss" at the very bottom. The two later banjos shown in Figs. 132 and 133 are unlike any other known clocks. They are similar to each other and bear some similarity, in their "keyhole" shape, to the wall clocks in Figs. 130 and 131.

The rather unusual banjo in Fig. 126 has a base section that looks as though it belonged on a wall clock. The overall result is ungainly, but unique.

Another Cheney banjo is illustrated in *English–Canadian Furniture of the Georgian Period*, by Donald Blake Webster, Fig. 141. This clock, too, illustrates the distinctive design characteristics which are often found in Cheney's Montreal clocks. The finial is of unusual style with a central "pinecone" flanked by carved scrolls. The entire crest is flat, about a half-inch thick, and in the authors' opinion could be original, since one other Cheney clock is known with a somewhat similar ornament. The extensive

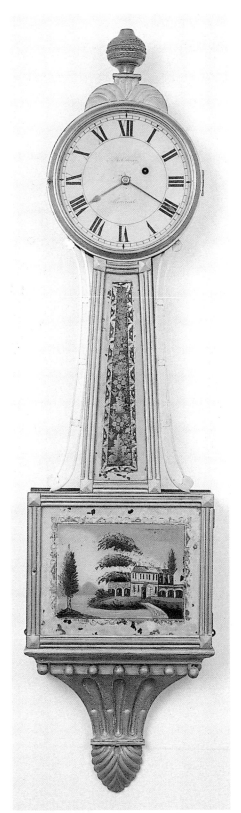

FIG. 127

FIG. 127 *Another Martin Cheney Montreal presentation banjo, with original glass.*
– *Courtesy Stan Kirschner*

FIG. 129

FIG. 130

FIG. 128

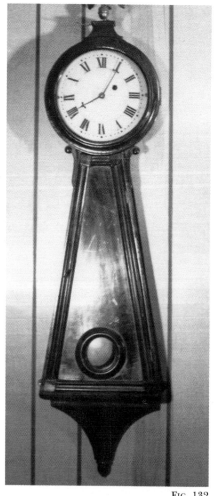

FIG. 131

FIG. 132

FIG. 133

FIG. 128 *Martin Cheney Montreal banjo clock (replacement glass). Note similarity of case details to Fig. 127.*
FIG. 129 *Attractive M. Cheney wall clock with type-A movement and brass dial.* FIG. 130 *M. Cheney wall clock exhibiting "keyhole" shape.* FIG. 131 *Another "keyhole" Cheney wall clock without the ornamentation seen in Fig. 130.* FIG. 132 *Martin Cheney banjo with type-C movement and showing distinctive "keyhole" design.* FIG. 133 *Another M. Cheney banjo with "keyhole" shape.*

use of rosewood in the case is unusual, with mahogany and pine as secondary woods. The bezel is of heavy brass and the white painted metal dial is slightly convex.

The movement is type B with Cheney's characteristic stamp on the left side. Case, dial, and movement are marked with the number 121. The head has very thick, circular walls, as has been noted in other Cheney banjos and in the Dwight and Griffin banjo described in this book. This appears to be typical of Montreal cases. The weight is lead.

The wall clocks exist in two styles. The clock shown in Fig. 118 is the most common style of Cheney wall clock to be found, with ten being known at time of writing. All have a generally similar configuration, but many minor variations from clock to clock can be found in the dimensions of the rectangular "drop" portion of the case, in the size of the little observation window over the pendulum bob and in the presence or absence of carved ornaments below the bezel. This little window seems to have been a trademark of Cheney's, since it appears in many clocks, including several banjos. The bezels on these wall clocks are always fitted with a single hinge at the top and a latch or knob at the bottom. Most dials have a chapter ring diameter of 12 inches, but one clock with a 15-inch ring is known. The two wall clocks in Figs. 130 and 131 are probably later. The "keyhole" shape is unique.

Martin Cheney may well have produced clocks in other styles. The authors have been told of a Cheney bracket clock, but were unable to verify its existence. Unfortunately, a century and a half later, the researcher must rely primarily on surviving clocks to piece together the story of Martin Cheney. As other clocks become known, a more complete picture may emerge.

The reader's attention must also be drawn to the tall case clock in Fig. 135. This elegant specimen was on public display several years ago. Its label, "Cheney Le Mecanisme, J. Bte Sancer la boite, 15 Juin 1816," combines Cheney's name as supplier of the movement with that of cabinet-maker Jean Baptiste Sancer. The movement and dial currently in the clock are English. The dial bears the name "J.N. Gibson, Saltcoats." It is not clear whether this is a later substitution for an earlier movement by Cheney, or whether Cheney imported the movement. Cheney probably did import English watches and, like his competitors in Montreal, may have imported tall case movements as well. The case itself is an excellent example of the high-quality craftsmanship available in Montreal at the time.

Some readers are no doubt familiar with *The Banjo Timepiece* by Chipman P. Ela. In a second printing of this book a few years ago, Mr. Ela included an addendum on Martin Cheney containing data and illustrations provided to him by one of the authors of this book. This information includes a lengthy listing of statistical information for the banjo clock in Fig. 128, which may be of interest to some researchers. In his comments, Ela gives Martin Cheney credit for originating a "Grand Piano" movement which he had previously credited to Samuel Abbott. The authors feel this statement is unwarranted since there is no evidence to link Cheney to the "Grand Piano" movement. Ela was influenced by the unusual bell shape of the Cheney type-C movement, but the overall resemblance is minimal.

Fig. 134

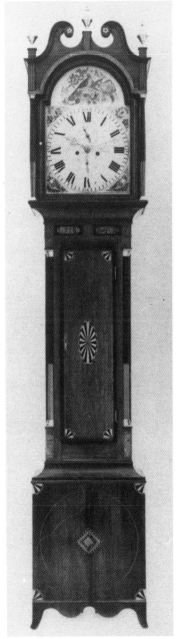

Fig. 135a

Fig. 134 *Cheney label from clock in Fig. 132.*
Fig. 135a *Martin Cheney tall case clock.* — *Courtesy Barbara and Henry Dobson*

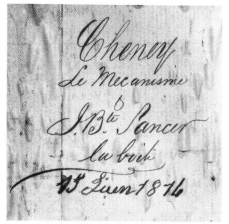

FIG. 135b

FIG. 136

FIG. 135b *The label of the Cheney clock in Fig. 135a. – Courtesy Barbara and Henry Dobson*

FIG. 136 *Dial of a Morbier clock inscribed "J.H.D. Codère à Sherbrooke Qué 1893."*

J.H.D. CODÈRE

Joseph H.D. Codère (b. 1864) was a watchmaker and jeweller in Sherbrooke, Quebec.[5] His name appears on a Morbier clock dial.

Several large Codère families lived in Sherbrooke before Confederation. J.H.D. Codère was the son of François and Emily Codère. François was one of the partners in a hardware and leather business. Joseph H.D. became a watchmaker and jeweller and opened a business on Wellington Street, Sherbrooke, in the 1880s and was still in business in 1925. His name was not in the Sherbrooke telephone book of 1928.

J.H.D. Codère married Marie (b. 1865) and Marguerite was born in 1890.

THOMAS DRYSDALE

Thomas Drysdale was a watch- and clockmaker in Quebec City in the 1840s. He left Edinburgh, Scotland, in 1832 with his wife and several small children and settled in New York. In the late 1830s he moved to Quebec City, and the family was recorded in the census of 1842. He was listed as a jeweller in the city directory of 1844–1845; at that time his shop was at 5 St. Joseph Street. By 1847 he had moved his business to 13½ Buade Street. It is probable that he died about 1849 as the 1850 Quebec City Directory listed only his wife.

During the 1840s Thomas Drysdale sold tall case clocks. "Thomas Drysdale, Quebec City," appears on the dial of the clock that is in the entrance hall of Dundurn Castle in Hamilton, Ontario. The name on the movement of the clock is Howell and Cooke, who made clock movements in Birmingham, England, in that time period. The case is a fine example of the typical "Quebec" case styling in mahogany. The movement, painted dial, and hands are typically English of the period (see Fig.137).

By 1842 the family had grown to sixteen members, including two servants and two other individuals who lived with the family. Sons Thomas and John and a daughter, Rebecca (b. 1829), were born in Edinburgh. Christopher (b. 1832) was born in New York, and the Drysdales' last two children, Mary (b. 1840) and Walter (b. 1842), were born in Quebec City. In the late 1840s, Mrs. Drysdale (b.1806) opened the Drysdale Ladies School at 13 Hope Street. After the death of her husband, the school, at that time called St. Andrews School, was moved to St. Ann Street. There was no directory listing for the school in 1852.

Two sons, John and Thomas, worked in Quebec City until 1873. After that date, no members of the family with the name Drysdale appeared in the city directories.

JEAN-BAPTISTE FILIAU *DIT* DUBOIS

Dubois was a clockmaker in Montreal during the French Regime (about 1730–1760). No example of his clocks is known. However, Dubois was an "horologeur" in New France by 1729, as evidenced by the church records of Notre Dame Cathedral, which gave his occupation as such when his daughter Marie-Louise was born. He also gave horologer as his occupation when the general survey of all houses of Montreal took place in 1740.

Available records show that Dubois was born in Quebec City and contracted smallpox as a child. During the epidemic, all schools were closed and he received little formal education.

He apprenticed to the master carpenter and sculptor Jean-Dauphin in Quebec City and then left that city to work in Montreal. After being unable to repair several clocks for his wealthy patrons, Dubois became interested in the mechanisms and about 1730 he returned to Quebec City to study with a clockmaker thought to be Pierre-Henri Solo.

On his return to Montreal, Dubois commenced repairing all the clocks and watches that were presented to him to fix. At the same time, he studied each carefully. He then proceeded to build clocks of his own invention, making tools as they were needed. Word soon spread about his skill. Writing in a book about Filiau *dit* Dubois, Emile Falardeau recorded that all Montreal was following "Dubois Time." He was responsible for timing court cases, religious services, and the time of the angelus. He also continued to make clocks. In 1825 Michel Bibeau wrote an article saying that he had seen clocks made by Dubois (translation) "which were very good and the workmanship was of fairly good taste, considering resources he had at his disposal at the time."[6] An inventory of 1783 mentions "a clock with cords and weights which does not chime, made by Dubois, the clockmaker of this town…with its pine case."

In 1749 a Swedish scientist named Kalm, after a trip to Montreal, reported that he had observed "excellent clocks that were being made in Lower Canada by a man without formal education." He had apparently seen Dubois at work. It is also stated that an advertisement appeared in 1813 in *Le Spectateur*, a Montreal newspaper, asking for information about M. Dubois, a carpenter who resided in Montreal before 1760 and who had mastered the clockmaker's trade.

In spite of his popularity as a clockmaker, Dubois was forced to continue his trade of carpentry and cabinetmaking. It was impossible to make a living in Lower Canada as a horologer. His craftsmanship in working with wood has been documented in at least one church in a suburb of Montreal. He passed on his knowledge to L. Foureur, known as Champagne. L. Foureur then made and sold clocks as an independent craftsman.

However, at the end of the French Regime, the British government sent General Murray in December 1760 to establish a government in "Canada" favorable to Britain and his instructions were to "not for any reason whatsoever give your assent to any laws establishing factories or businesses that could prove harmful or prejudicial to this Kingdom and that you make every effort to discourage, hinder and restrict any attempts that may be made to establish such factories or businesses…." It was therefore impossible for French-Canadian clockmakers and other craftsmen to develop their crafts for at least two generations. Although several men continued to work as clockmakers, they had no successors.

Dubois died around 1770. His wife, Geneviève Végès, born in 1703, was buried on 10 May 1775. Her death certificate stated that she was the widow of "Dubois, the clockmaker."

FIG. 137

FIG. 137 *Clock in Dundurn Castle, sold by Thomas Drysdale.*
 – Courtesy Dundurn Castle, Dept. of Culture and Recreation, City of Hamilton

FRANÇOIS DUMOULIN

François Dumoulin was a French Canadian who worked in Montreal from about 1775 to 1810. A rare bracket clock in the Canadiana collection of the Royal Ontario Museum has the name "Fra. Dumoulin à Montreal" engraved on its brass dial. The movement is definitely English. The case is British in style and resembles clocks sold by Charles Geddes in New York and François Valin in Quebec City. The case is mahogany, veneered over secondary pine. There are no separate inlays. The height of the clock is 18¼ inches and the width is 10¾ inches.

François Dumoulin was a merchant in Montreal in the late 1770s. During the 1780s he took at least three people to court, seizing their properties. However, in September 1795 he was found guilty of perjury and sentenced to imprisonment and a fine. By 1798 he had closed his business and moved to Three Rivers. His second wife brought a suit against François in November 1807, and his property was sold by the sheriff.

François Dumoulin was the son of Jean-Pierre Dumoulin. His mother, Louise Chevalier de Verney, came from the Canton of Berne, Switzerland. François's first wife was Margaret Baby-Chenneville, whom he married in August 1776. His second wife, whom he married in February 1784, was Louise-Charlotte Cressé-Poulin (b. 24 March 1751). Her father, Louis-Pierre Poulin, was part of the horologer Poulin family.

François Dumoulin had three sons: Louis François, John Emmanuel, and Pierre Benjamin.

JAMES A. DWIGHT

James Adams Dwight was a watch- and clockmaker and silversmith from the United States who was in business in Montreal by 1809 and until about 1846. Several wall clocks exist with the name "Dwight" on the dial. However, his importance in Canadian horology is augmented by the fact that, during his career, he was a partner with each of three important clockmakers: Martin Cheney, George Savage, and Austin Twiss. The partnership in each case lasted only a short time.

In an advertisement in the *Montreal Gazette* of 27 March 1809, the Cheney-Dwight partnership was announced. The advertisement informed the public that Cheney and Dwight, horologers, had commenced business on St. Paul Street, a few doors south of the Old Market, where "orders in their line will be thankfully received and punctually attended to…. They manufacture and expect shortly to have for sale elegant house clocks and timepieces which will be warranted good." In addition the advertisement stated that they were also repairing all kinds of watches and offered for sale watches, jewellery, etc.

In an advertisement in the 25 December 1809 issue of *The Courant and Montreal Advertiser* Cheney and Dwight offered for sale watches, clocks, and jewellery "at the sign of the Gold Watch, a few doors south of the Old Market." Although its present location is unknown, they apparently sold a musical clock with seven tunes that changed for each day of the week.

Another of J.A. Dwight's partners was George Savage. The partnership began in 1818 and dissolved in 1819. Silver bearing the mark used during their partnership has been found. The location of their shop was 56 St. Paul Street.

FIG. 138

FIG. 138 *Bracket clock with the name François Dumoulin on its dial. – Courtesy the Royal Ontario Museum, Toronto, Canada*

When Austin Twiss came to Lower Canada in 1821, Dwight was his partner for a short time.

During the 1820s Dwight continued to conduct business on St. Paul Street. However, several land deals he attempted to conclude were unsuccessful, and a lot on Notre Dame Street was seized and sold in 1821. In 1823 he and his partner, James C. Pierce, lost land in the Township of Stanstead.

It is possible that a further partnership existed between Dwight and a Mr. Griffin. A banjo clock inscribed "Dwight & Griffin Montreal," illustrated in Fig. 141, was acquired by the Canadian Decorative Arts Department of the Royal Ontario Museum in 1981.

The clock has been tentatively dated by the museum at 1840–1850. James Dwight died about 1846, so if he was involved, this partnership must have existed before that date. There is no mention of such a partnership in city directories, so it may have been of short duration. Examination of the clock, however, suggests another possibility: that the partnership existed around 1817. There are several inscriptions on the case, in two forms. The first is either "212D" or "2/2D." The second is the date 26/7/17. Each inscription occurs in two places – on the throat panel and on the base – suggesting that they may have been placed there during the assembly of the case. The movement has been stamped with the figure 2 on the hour wheel and on the idler wheel. It is possible, of course, that some of these are jewellers' marks for repair dates (1917?). However, their location and frequency are not typical of repair marks.

The clock is an attractive banjo with a carved, gilded wooden finial. The dial has a wooden bezel, is flat in form and made from iron and painted. It is held by three machine screws that fit into threaded brass inserts in the head. The weight is of lead, and there is a continuous sheet of metal between the weight channel and the pendulum, covering the full base area and the throat. The throat panel and lower door are covered in flame mahogany veneer over pine.

The movement has not been identified as the product of any known maker. The hour wheel and idler wheel are both solid, without spokes. This is uncommon but has been seen in clocks by Martin Cheney, also in Montreal. Otherwise, the movement is not similar to other Cheney movements.

The base of the case has curved sides. A similar case by Martin Cheney is known, probably from the 1820-to-1830 period. The Dwight and Griffin clock has a small circular window in the lower door, a detail that is also seen frequently in Cheney clocks. The occurrence of these details in the clocks of two Montreal makers suggests that the style was popular in Montreal and the cases may have come from local cabinet-makers.

There were several families named Griffin in Montreal during the first half of the 19th century. In 1816 Robert Griffin (baptized 1786), a banker, was at 33 St. Paul Street, and James Dwight had his establishment at 57 St. Paul Street. Martin Cheney, clockmaker, was just a short distance away at 107 St. Paul Street. Because none of the Griffin men was in an occupation associated with horology, a partnership may have been formed for financial reasons.

In *The Early Furniture of French Canada* by Jean Palardy, there is mention of a William Griffin, 1799, in Montreal in a list of Scottish woodworkers and cabinetmakers who came to Quebec in the late 18th and early 19th centuries.

FIG. 139

FIG. 140

FIG. 139 *Clock sold by James Dwight in Montreal.*

FIG. 140 *This dial with outside diameter of 16 inches is all that remains of a clock by James A. Dwight.*

FIG. 141

FIG. 141 *This banjo clock labelled "Dwight & Griffin Montreal" may have come from another of Dwight's partnerships. – Courtesy the Royal Ontario Museum, Toronto, Canada*

Additional information about Dwight in the census records of 1825 suggests that, at that time, Dwight was less than forty years of age. By the time the Montreal City Directory was published in 1842, he had taken his son, James A. Dwight Jr., into his business, which was listed as "James A. Dwight & Son." The business was located at 151 Notre Dame Street at the corner of St. François Xavier Street. Their home was on St. Antoine Street. The city directory indicated that they were importers of clocks, watches, jewellery, plate and plated ware, cutlery, and fancy articles. Watches, clocks, and jewellery were "neatly repaired" in their store.

Apparently in 1845 or 1846 James Dwight Sr. died and his son took over the business and lived on Brunswick Street. The exact date of death of the elder Dwight is not known. This may be attributed to the fact that the Protestant cemetery on Dorchester Street was closed and, although some of the bodies were transferred to the new cemetery, the records of many persons buried in the old cemetery were lost.

James Dwight Jr. continued in business for a very short time. By 1852 no Dwight was in business in Montreal and William Creyk had taken over the clock, watch, and jewellery business at 151 Notre Dame Street.

After an unsuccessful business venture in New York, James Dwight Jr. returned to Montreal in the 1850s. A letter written by John Wood (see p. 139) to his son in August 1855 relates the following incidents concerning James Dwight:

> You have probably heard of the very notorious James Dwight. After his failure in New York, he got into business again and a very few months later absconded with a young lady of a respectable family. He was advertised and the two were taken. The woman was placed in the lunatic asylum; from him they took what money he had and he came to Montreal. His wife is suing him for a divorce, and as soon as that is obtained he intends to marry the woman. I cannot say that Learmont[7] is wrong in giving employ, though I should find some trouble at present in adopting the same course.

The records of the American Presbyterian Church indicate that James Dwight Jr. and his first wife, Phoebe, had two children, both of whom died as infants, in 1832 and 1833.

JEAN FROMENT

Jean Froment was a goldsmith, silversmith,[8] and clockmaker[9] in Quebec City and Acadia. According to the *Revue d'ethnologie du Québec/9*, he worked as a clockmaker about 1750. Very little information has been found about the man and no clocks made by him have been found. However, he is considered to be an important clockmaker of New France by the Quebec Department of Cultural Affairs.

In 1751 Jean Froment signed power-of-attorney to his brother François before leaving for Acadia as an interpreter of English. His skill as a silversmith is attested to by the existence of a silver missionary pyx, now in Moncton Cathedral.

On 23 July 1751 the instrument of power-of-attorney was probated. This could indicate the year of Froment's death; it might also indicate that he was deported from Acadia at that time.

JACQUES GOSSELIN

Jacques Gosselin is listed in the records of the Provincial Museum of Quebec as a clockmaker in Quebec City in 1758. It is possible that N. Gosselin, mentioned in *The Early Furniture of French Canada* by Jean Palardy, worked with him. It is interesting to note that the Gosselin name is also associated with repairs to the St. Sulpice Seminary clock, described later in this book (p. 115). Barnabé Gosselin and his brother Gabriel (a gunsmith) made and installed parts on this ancient clock. The installation was done in 1823, so these men may have been related to Jacques Gosselin.

Jacques was the son of Jean-Baptiste Gosselin and Françoise Boutonne of St. Eustache, Lower Canada. He married Genevieve L'Enclus (b. 1734). One child, Charlotte, was born 8 May 1757 and died the next year.

JAMES GRANT

James Grant was a clockmaker and engraver in Lower Canada during the last part of the 18th century. A bracket clock, now in the Canadiana collection of the Royal Ontario Museum, has "Jam's Grant Montreal" engraved on the back plate of the brass movement.

The clock has an arched top and the left side, as well as the front door, is hinged. It is made of mahogany veneer over butternut and the base on which the movement sits is pine. The case is ebonized with polished black stain. This finish and the feet are suspected by the museum of being added some time after the clock was made. The height of the clock is 14¾ inches and the width is 10⅜ inches.

Although nothing has been found by the authors about the personal life of James Grant, other information indicated that he was working in Quebec City by 1783. A publication called *Horloges et Horlogerie* by Pierre-Georges Roy,[10] reveals that James Grant engaged a fourteen-year-old apprentice for seven years. The young man "committed himself to stay with his master, to serve him faithfully, not to be absent without permission, not to participate in any games of chance, not to frequent taverns and inns and not to marry."

In June 1785 Grant's name was among other signatures on a letter published in the *Quebec Gazette*. The letter was presented to the Hon. Thomas Dunn, lamenting his departure to England and extolling his many virtues. A similar letter was published, addressed to the Hon. Henry Hamilton, in November 1785. Also in November of that year, an advertisement appeared in the *Quebec Gazette*. This advertisement (Fig. 143) also revealed his other talents.

From Quebec City, Grant moved to Montreal and re-established his business. The date of his departure from Quebec City could have been August 1790 as, according to the *Quebec Gazette,* a letter was waiting for him in the Quebec City post office. By October 1790 he was among those persons who signed a letter addressed to Lord Dorchester, concerning making Montreal a customs port. Further letters to various public officials included the signature of James Grant, Montreal. These letters appeared in the *Quebec Gazette* up to 1799.

In 1795 Grant asked for and received land near Montreal. Between 1792 and 1799, he advertised in the newspapers as a clockmaker and engraver.

A record of his death has not been found.

FIG. 142a

FIG. 142b

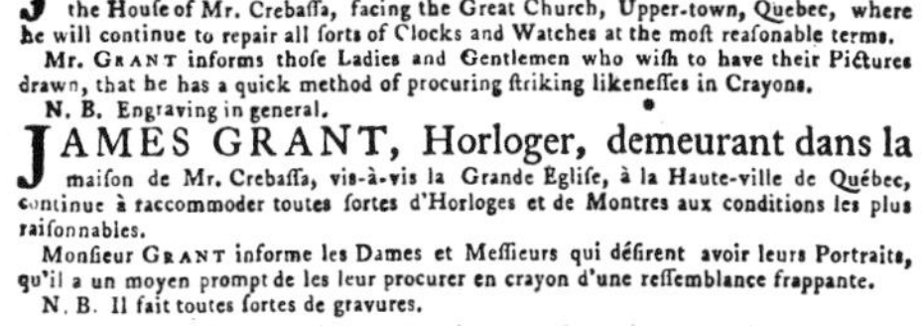

JAMES GRANT, Clock and Watch-maker, at the House of Mr. Crebaffa, facing the Great Church, Upper-town, Quebec, where he will continue to repair all forts of Clocks and Watches at the moft reasonable terms.
Mr. GRANT informs thofe Ladies and Gentlemen who wifh to have their Pictures drawn, that he has a quick method of procuring ftriking likeneffes in Crayons.
N. B. Engraving in general.

JAMES GRANT, Horloger, demeurant dans la maifon de Mr. Crebaffa, vis-à-vis la Grande Eglife, à la Haute-ville de Québec, continue à raccommoder toutes fortes d'Horloges et de Montres aux conditions les plus raifonnables.
Monfieur GRANT informe les Dames et Meffieurs qui défirent avoir leurs Portraits, qu'il a un moyen prompt de les leur procurer en crayon d'une reffemblance frappante.
N. B. Il fait toutes fortes de gravures.

FIG. 143

FIG. 142 *A James Grant bracket clock and its back plate. – Courtesy the Royal Ontario Museum, Toronto, Canada*
FIG. 143 *An advertisement in the Quebec Gazette.*

FIG. 144a

JAMES HANNA AND JAMES G. HANNA

James Hanna (b. about 1737 – d. 26 Jan. 1807) and his son, James Godfrey Hanna (b. 1788 – d. 1851), were watch- and clockmakers in Quebec City in the 18th and 19th centuries. The father, James Hanna, came from Dublin, Ireland, to Quebec City around 1760 and opened a shop in the house of the merchant John McCord, at 15 Fabrique Street. In 1764 he could be found at the shop with the sign of the "Eagle and Watch." He later purchased a property from the Jesuits on Fabrique Street for £360 where he lived and had his shop. In 1805 he made this address his permanent residence. In January 1794 he advertised that the shop was located in the "Corner House facing the Post Office." For forty years James Hanna supplied the people of Quebec City with silver and gold items, plated wares, clocks, and watches. A typical advertisement appeared in the *Quebec Gazette* of 10 July 1794.

The name "James Hanna" is found on the dials of high-quality tall case clocks. One such clock is described in the book, *Revue d'ethnologie du Québec / 9*, published by the Université du Québec at Trois-Rivières. This clock, which is in a private collection, had the date 1760 on the dial in addition to the name "James Hanna, Quebec," thus establishing the date that he was in Canada. A second clock, also in a private collection, has the name of the place spelled "Quebecc."

Another clock with the name of James Hanna on the dial is pictured in Fig. 148.

The movement of a Hanna clock, recently examined, has the name "Quebec" struck twice with a die stamp into the front plate near two of the corners. The impression is fairly deep, and the letters are raised. This impression could indicate that the movement was made in Quebec. However, in his many advertisements in the *Quebec Gazette*, Hanna never indicated that the movements of any clocks that he sold were made in Lower Canada. In fact, all his advertisements announced with pride that he "imported from London and Scotland a variety of watches, clocks and other goods." Frequent trips to Britain by both James Hanna and his son were referred to in the *Quebec Gazette*. Therefore, the importing of goods from Britain by the Hannas was a frequent occurrence. Also, a commercial venture into clockmaking in Lower Canada after 1760 would have caused great displeasure to the government of that day, which was promoting British goods and actively discouraging manufacturing in Canada.

In addition to advertisements offering goods for sale, the *Quebec Gazette* mentioned other aspects of Hanna's life. In 1771 he offered for sale a farm in the parish of St. Ours and petitioned for and received more land in 1773 "situated on the River Schesvonstek." He continued to buy houses which he then rented. Also, his signature added weight to those of other Quebec City inhabitants in letters to the newspaper. One such letter was addressed to the Honourable Thomas Dunn in 1785. His signature was one of a number accompanying an address to Lord Dorchester on his departure in 1791, and to the Prince of Wales on his departure in 1794. James Hanna contributed to and was a member of both the Quebec Fire Society and the Quebec Benevolent Society. With other citizens, he petitioned King George III for land to build a Presbyterian Church and a university. In 1795, along with six others including James Orkney, James Hanna petitioned for exemption from a law regulating forges.

In 1803 James Hanna placed an announcement in the *Quebec Gazette*

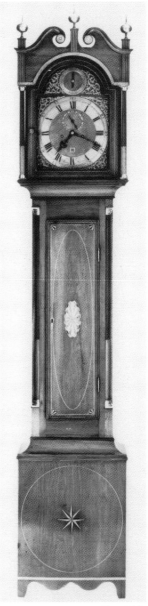

FIG. 144b

Quebec, July, 1794. L. FROMENTEAU.

JAMES HANNA,

Corner House, facing the Post-Office:

HAS imported from LONDON and SCOTLAND, a variety of watches, clocks, and time-pieces, watch-chains, seals, trinkets, and watch-glasses.

A VARIETY OF LATEST FASHIONED JEWELLERY.

Silver Work.—Tea pots, candles, cream pots, salts and spare glasses, table spoons, desert, tureen, gravy and salt spoons, tea spoons and tongs, dish crosses with lamps, shoe buckles, knee and stock ditto, childrens clasps, thimbles, ear-picks, bodkins, buttons, rings, shirt buckles, Freemasons medals, punch spoons, coral and bells, pap spoons, sword and hanger tips, cups, Mr. Hubert's portrait in lockets and pins.

Plated Work.—Tea urns, tea pots, candles, sugar dishes, slop bowls, cream pots, pint cups, tumblers for canteens pint and half pint, candlesticks, crewet stands, shoe and knee buckles, tureen, table and tea spoons, bottle stands, childrens clasps, patent spring and Newmarket spurs, cocking and ... pieces, pistols, swords and hangers, Highland dirks, ditto snuff mills horn and hoof, sword belts, hooks and chains, regimental sword knots, powder horns and shot bags, sportsmens bottles, knives and cork screws, pen knives, gardeners pruning knives, and pocket ditto, travelling knives and forks, counting house knives, sets of gold grains, and odd grains to replace lost ditto, cocks spurs, nail nippers, nut crackers, pliers, pencil cases, ... crayon, ... spectacles for all ages, concave and reading glasses, goggles, dog chains, collars, joint leather ditto, brass ditto with locks, boot garters, steel gallows buckles, shoe and knee buckles, black ditto, stock buckles and buttons, cucumber slicers, tweezers, lancets, fishing rods, and a variety of fishing tackle, baskets, landing nets, cabbage ditto, hair combs, soldiers crooked ditto, pocket compasses, mathematical instruments, hydrometers barometers, and thermometers, glaziers diamond pencils, table bells, leather ink pots, fountain pens, spy glasses, Reeves's patent colours, ivory for portraits, hair pencils, Crayon's Indian ink, lemon squeezers and strainers, watch stands, back-gammon tables and dice, chess and draught boards, sportsmens pocket horns, memorandum day of the week books, Ladies pocket books, ditto purses, snuff boxes newly invented by a Canadian, strong and best for preserving the snuff, Irish bag pipes, ...

JAMES HANNA,

Dans la Maison du coin faisant face au Bureau de la Poste,

A Importé de Londres et d'Ecosse, une variété de Montres, Horloges, et Pendules, Chaines de Montres, Cachets, Colifichets, et Verres à Montres.

UNE VARIÉTÉ DE BIJOUTERIE DES PLUS NOUVELLES MODES.

Ouvrages d'Argent.—Theyeres, Candies, Crémiers, Salieres, et Verres de relais, Cuilleres de table, et à dessert, Soupieres, Cuilleres à ragout et à sel, idem à thé et Pincettes, Croix à plats avec lampes, Boucles à Souliers, à Jartieres et à Col, Agraffes pour les Souliers d'Enfans, Cure-oreilles, Dés, Poinçons, Boutons, Anneaux, Boucles à chemises, Médailles de Francs-maçons, Cuilleres à punch, Corail et Clochettes, Cuilleres à bouillie, Bouts d'Epée et de Coutelas, Coupes, Portraits de Mr. Hubert en Loquets et Epinglettes.

Ouvrages plaqués—Urnes à thé, Theyeres, Candies, Sucriers, Boles, Crémiers, Pocines de chopine, Goblets pour Cantiner de chopine et demiard, Chandeliers, Porte-fioles, Boucles à Souliers et à Jartieres, Soupieres, Cuilleres à soupe et à thé, Porte-bouteilles, Agraffes à Souliers pour les Enfans, Eperons à ressorts et de New market, Fusils de chasse, Pistolets, Epées et Couteaux de chasse, Dagues de Montagnards, Tabatieres, de corne, Ceinturons d'Epées, Crochets et chaines, nœuds d'Epées régimentaux, Cornes à poudre et Sacs à plomb, Bouteilles de chasseurs, Couteaux, Tire-bouchons, Canifs, Couteaux à Jardiniers et à poche, Couteaux et Fourchettes de voyage, Couteaux de comptoirs, des assortimens de grains à peser l'Or et pour remplacer ceux qui manquent, Eperons de coqs, Pincettes à cloux, Tenailles, Etuis de crayons, Ciseaux et Etuis, Lunettes pour tous âges, Verres concaves et à lire, Lorgnettes, Chaines de chiens, Colliers, idem de cuir à joints, idem d'airain avec serrures, Jartieres de bottes, Boucles d'acier pour bandes, Boucles de Souliers et à Jartieres, idem noires, Boucles et Boutons de col, machines pour tailler les concombres, Pincettes, Lancettes, Perches à ligne, et une variété d'apparats de pêche, Paniers, Filets, Peignes à cheveux, idem croches pour les Soldats, Compas de poche, instrumens de Mathématique, Hydromètres, Barometres et Thermomètres, Diamans de vitriers, Clochettes de table, Ecritoires de cuir, Plumes de cuivre, Longue-vues, Couleurs fines de Reeves, Platins d'ivoire pour Portraits, Pinceaux, Crayons, Encre de la chine, Presseurs de Citrons et couloirs, montures de Montres, Tables de back-gammon, et Dés, Damiers et échequiers, Cornes de poche pour les chasseurs, Livres de ...

FIG. 145

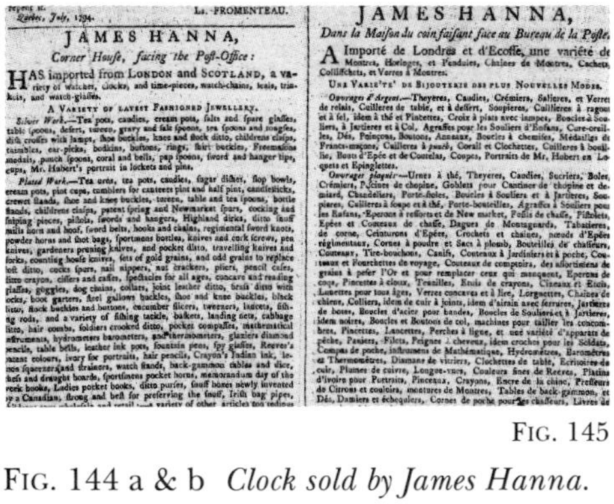

FIG. 144 a & b *Clock sold by James Hanna. – Courtesy Barbara and Henry Dobson*

FIG. 145 *Advertisement in July 1794* Quebec Gazette. *– Courtesy National Archives of Canada*

FIG. 146

FIG. 147

FIG. 146 *Clock by James Hanna, in Quebec tapered-waist case.*
FIG. 147 *Dial detail in clock shown in Fig. 146. Note spelling of "Quebecc."*

thanking his clients for their support and soliciting continued support for his son who was joining the business. The son, James G. Hanna, would have been about fifteen years old at the time.

The similarity of names might have led to confusion between father and son. However, son James always used his middle initial, G., in his name, and his father signed his name using just initials. In any case, their activities are easy to separate, because their signatures were distinctive, as can be seen from church records where both signed a document.

The two Hanna men, father and son, continued the business until 1807, in which year, on 26 January, James Hanna Sr. died. He was buried in Mount Hermoun Cemetery. His son took over the business and continued to advertise in the *Quebec Gazette*. His association with his father's British suppliers also continued. In 1809, for example, three British ships are known to have brought goods from London and Liverpool.

Following the example of his father, James G. Hanna sold clocks carrying his name on the dial. A clock owned by the Provincial Museum of Quebec is pictured in the book by J. Palardy, *The Early Furniture of French Canada*. This clock is signed James G. Hanna. The movement of the clock was made by Robert Wood, London, England. This clock has a painted dial.

In 1816 James G. Hanna moved to 16 Mountain Street, but the business was soon in difficulty. He took a partner, François Delagrave, a silversmith. In November 1817 he announced in the *Quebec Gazette* that the "subscriber gives notice that he will petition the Provincial Legislation at its next session in order to obtain the privilege of building a Toll Bridge over the River Famine in the Parish of St. Francis, Nouvelle Beauce." However, by 19 February 1818 he had declared bankruptcy. Delagrave continued in business until 1832. The bankruptcy was announced in the *Quebec Gazette:*

> Notice – the subscribers, elected trustees and managers of the bankrupt estate of Mr. James Godfrey Hanna, of this city, merchant, request who have claims against him to produce them, duly authenticated and those who are indebted to the said James Godfrey Hanna either by account, notes, obligations or otherwise are requested to pay immediately to Louis Gauvereau, one of the undersigned at his present address and to no other person, otherwise they will be prosecuted according to the law.
> *Signed,* Louis Gauvereau, John Reinhart and Anthony Anderson, Trustees.

By April 1818 the sheriff gave notice that the stock of James G. Hanna would be sold in May of that year. In April 1821 the sheriff offered the property for sale. James G. Hanna left Quebec City before the property sale took place. He moved to Saint-Charles-de-la-Belle-Alliance and brought about twenty-five Irish families to the region. They manufactured cloth and sewing thread.

It is not known if James Hanna Sr. was already married when, at the age of twenty-six, he came to Lower Canada from Dublin. It is known that two girls born in Lower Canada had the name Hanna and lived in his house. Jane (b. about 1769 – d. 1817) married James Orkney, watch- and clockmaker of Quebec City. Mary Hanna (b. about 1778 – d. 1800) married John Macnider, merchant of Quebec City, in 1794. Only one daughter of Mary and John Macnider, named Margaret, survived.

There were close family ties between James Hanna and the Orkney

FIG. 148

FIG. 149

FIG. 150

FIG. 152

FIG. 151

FIG. 153

FIG. 154

FIG. 148 *A James Hanna tall case clock.*   FIG. 149 *View of movement of clock in Fig. 148.*   FIG. 150 *This James Hanna tall case clock shows different styling.* *– Courtesy Barbara and Henry Dobson*   FIG. 151 *James Hanna bracket clock. Hanna's name is not presently on the dial, but is scratched into the reverse side.*   FIG. 152 *Movement marked "Rob't Wood, London" in Hanna bracket clock in Fig. 151.*   FIG. 153 *Signature of James Hanna and signature of his son.*   FIG. 154 *Dial of a tall case clock by James G. Hanna, Quebec.*

FIG. 155

FIG. 156a

FIG. 156b

FIG. 155 *A N. Jalbert, Montreal, clock.*
FIG. 156 *The clock and label of
a Porter Kimball clock.*

and Macnider families. James Hanna witnessed births, marriages, and deaths, and the children were named in honour of members of each other's families.

James Hanna was married for a second time on 9 December 1787 to Elizabeth Saul. On 9 November 1788 James Godfrey was baptized. Over the next ten years, James and Elizabeth's family continued to grow. Children of James and Elizabeth were James Godfrey (b. 1788 – d. 1851), Margaret (b. 1792), Anne (b. 1795), William John (b. 1797), Elizabeth (b. 1799), Samuel George (b. 1801), Jackson Moore O. (b. 1803), and Amelia, born 25 July 1807 to Elizabeth Hanna, widow of the late James Hanna. It is thought that Elizabeth died about 1817, as a case was brought to trial involving James G. Hanna in a suit against John Macnider for the custody of the sisters of James G. Hanna who were still minors.

NAPOLEON JALBERT

Napoleon Jalbert was a watchmaker and jeweller in Montreal, Quebec, from 1884 to about 1918. Two tall case clocks have been reported to the authors, both with the inscription "N. Jalbert Montreal" engraved on the brass dials. The examples appear to be similar in style and stand 7 feet tall. The clock in Fig. 155 has a single brass finial and originally was fitted with a rocking ship which moved in a position above the chapter ring. The case is cherry. The second clock, while similar, is said to be of mahogany construction and has three wooden finials. It also has a moving device on the dial, a rocking ship, which may or may not be original. The movements are said to be of English origin and are fitted with both calendar and seconds dials.

Napoleon Jalbert's business was located on Notre Dame Street until about 1894, when he moved to St. Hubert Street. In 1894, one Fortunat Jalbert, whose relationship to Napoleon has not been established, but who shared the same address, opened a watch repair and jewellery store on St. James Street. Early in the 20th century, Fortunat changed his business to a nickel-plating operation. Napoleon Jalbert's name was listed in the Montreal City Directory up to the 1919–1920 issue.

PORTER KIMBALL

Porter Kimball sold clocks in the Eastern Townships of Lower Canada, from the town of Stanstead. Stanstead is located about 20 miles south of the present city of Sherbrooke, Quebec, a few miles north of the border with the American state of Vermont.

The clocks, of which perhaps a dozen are known, are all transition or pillar-and-splat style. They are fitted with typical wood movements, 30-hour Terry-type. Two clocks are illustrated here. Others are shown in Burrows, *Canadian Clocks and Clockmakers,* and another is owned by the Stanstead Historical Society. The labels examined are all similar and bear the legend "Walton and Gaylord Printers, Stanstead, Quebec." This firm went out of business in 1837, suggesting that Kimball was active in that decade. One movement style has been identified, namely Taylor-type 8-211 (later amended to 8-221) by an unidentified maker. The labels state that the clocks were "manufactured and sold by Porter Kimball," but the authors have not been able to confirm that he actually made any part of the clocks.

Pillar-and-splat cases are commonly found in two slightly different styles. In one style, the half columns are attached to the front of the case on each side of a separate door. In the second style, the columns are nailed directly onto a wider door, which then forms the entire centre front of the case. Kimball's clocks follow the latter style, with some of the doors being hinged on the left side.

Little definite information has been found on Kimball himself. Other Kimballs, however, are known to have been active as clockmakers at Boston, Massachusetts, and Montpelier, Vermont. It is possible that Porter was part of one of these families and sold clocks for a few years north of the border.

JOHN LUMSDEN

John Lumsden was a silversmith, jeweller, and horologer in Montreal, Lower Canada, from the late 1700s to 1802. Three Lumsden clocks are known: two tall case clocks and a bracket clock. All clocks are fitted with brass dials and have the information "John Lumsden, Montreal" engraved in the arched portion of the dial. The dials are relatively plain with no applied ornamentation. Each chapter ring is simply engraved and a calendar dial is located below the main central arbors. One tall case clock, grain-painted and of basic design, is pictured in Fig. 158.

This clock has a typical English tall case striking movement. It is weight-driven and includes a seconds dial, a calendar dial, and a strike/ silent indicator. The movement has had a rather crude repair to the escapement in which the anchor has been moved to one side along the arbor. Extended pallets are bolted to it, reaching back to the escape wheel.

A second Lumsden tall case clock, reported recently, has a brass dial similar to the previous clock. It has a more elaborate case with a large head and base and a narrow waist. The lower door has a somewhat unusual glass panel. The bracket clock has typical late 18th-century styling. The primary wood is mahogany. The case has a domed top with a handle, simple wooden feet, and a door fitted with arch-top glass. There are also arched side windows. The 8-day repeating movement has verge escapement and bob pendulum. The back plate is attractively engraved (see Fig. 162).

Information about John Lumsden's personal activities and whereabouts is relatively limited. Available records revealed the following information:

1. John Lumsden married Mary Logoterie in Christ Church, an Anglican Church in Montreal, in November 1793. He lived on Notre Dame Street.
2. John and Mary had two children: son William, who was born on 13 August 1794, baptized in September of the same year, and died on 18 May 1795; and daughter Margaret, who died as an infant on 30 July 1796.
3. In 1794 Lumsden and other Montreal residents supported the idea of building a court house and signed a document asking for vacant land on Notre Dame Street to be designated for this purpose.
4. In July 1799 his name appeared in the *Quebec Gazette* as one of the many who signed the address to Lord Robert Prescott on his departure.
5. His name is listed in the *Quebec Gazette* in 1802 upon his return from a trip to London, England.
6. Lumsden died on 9 December 1802. His burial service took place in the same church in which he was married.
7. His business was sold to Charles Irish in 1804.
8. Charles Arnoldi purchased the business from Irish in 1806.

FIG. 157

FIG. 158

FIG. 157 *This clock by Porter Kimball has lost its splat and corner "chimneys."*
FIG. 158 *John Lumsden tall case clock.*
 — *Courtesy Royal Ontario Museum, Toronto, Canada*

FIG. 160

FIG. 161a

FIG. 159

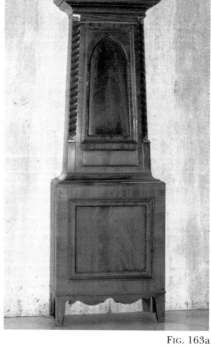

FIG. 161b

FIG. 162

FIG. 163a

FIG. 163b

FIG. 159 *Lumsden tall case clock.*   FIG. 160 *Movement of clock in Fig. 159.*   FIG. 161 *Bracket clock by John Lumsden.*
FIG. 162 *Back plate detail from Fig. 161.*   FIG. 163 *McMaster clock in the lobby of the Quebec Hilton Hotel.*

WILLIAM McMASTER

William McMaster (b. about 1793 – d. 21 Sept. 1854) left Scotland for Quebec City in 1827. There, he set up a clock- and watchmaking shop at 46 St. John Street, Upper Town. William was in business there until his death, at which time his wife and his two sons, William and Thomas, continued to operate the store. According to the Quebec City directory, the shop was still in existence in 1863. However, by 1868, the directory no longer listed the McMaster shop.

Several clocks sold by William McMaster are known. One tall case clock has been displayed in the lobby of the Quebec Hilton Hotel, Quebec City, for a number of years. This clock is an excellent example of the unique slope-waisted Quebec style, which the authors have commented upon several times. It is a handsome clock in fine figured mahogany veneer with carved "twist" columns. It is decorated with brass fittings and finials. The white-painted dial appears to be typically English of the period, with seconds bit and calendar display. The movement has not been examined by the authors, but the dial arrangement, again, suggests an English origin.

A second McMaster clock in a private collection is illustrated in Fig. 164. The dial and movement appear to be similar to that of the preceding clock. The movement has a false plate with the name "Birmingham" engraved on it. The case, however, is of interest. While it is more conventional than the Hilton Hotel clock, it has, nevertheless, enough unique characteristics to suggest that it, too, was made by a local craftsman. The curved cresting at the top of the hood bears some resemblance to the typical Quebec style and there are attractive fan-shaped carvings above and below the dial.

In January 1852, when the 1851 census was taken, an apprentice, Gordon Sprunt, a nineteen-year-old Scot, was living with William McMaster.

William McMaster married Margaret Bell (b. 1817) by special licence on 19 September 1843 at the Wesleyan Methodist Church. There were three children: William Bell (b. 6 June 1844) Thomas Chalmers (b. 22 Mar. 1847), and Margaret Walker (b. 9 Dec. 1848 – d. 26 May 1855).

THE MONTREAL WATCH CASE COMPANY

The Montreal Watch Case Company was situated at 125 Vitre Street in Montreal. The factory commenced operation in 1887 and continued until 1909. They manufactured gold watch cases and in addition carried on an extensive watch repair trade.

According to a special Montreal edition of *The Dominion Illustrated,* the president of the company was Mr. Moses (Moise) Schwob, who was a manufacturer and an importer of watches and diamonds on Notre Dame Street in Montreal. Mr. Schwob was one of the Schwob brothers[11] who were creditors of the Canada Clock Company, Hamilton, Ontario, when it went bankrupt in 1886. In 1888 they offered to buy the remaining stock of the company, worth $30,000, for fifty cents on the dollar.

Another person involved in the Montreal Watch Case Company was C.H.A. Grant, one of the largest stockholders. Grant was the general manager of the company. In 1897 the city directory listed W.J. Stewart as secretary-treasurer.

Schwob, Grant, and Stewart were highly respected and well known in the business community. By the turn of the century, according to the city directory, C.H.A. Grant owned the company.

In September 1907 the business was moved to 63 Alexander Street. No listing was given for the company in the 1910–1911 city directory.

FIG. 164

FIG. 165

FIG. 164 *A McMaster Clock.*
FIG. 165 *The Montreal Watch Case Company, from* The Dominion Illustrated, *1891.*

JAMES ORKNEY

FIG. 166

FIG. 167

FIG. 166 *Orkney clock owned by the Canadian Museum of Civilization, Hull, Quebec*
FIG. 167 *Tall case clock sold by James Orkney, Quebec City.* – *Courtesy the Royal Ontario Museum, Toronto, Canada*

James Orkney (b. about 1760 – d. 24 Jan. 1832) was a clockmaker, silversmith, and jeweller of Quebec City, Lower Canada, from about 1786 until 1826. Tall case clocks bearing the name "James Orkney, Quebec" on the dial are found in private collections and museums. According to a survey conducted by J. Varkaris in 1978, more "Orkney" clocks exist than clocks of any other Lower Canada maker, with the exception of the Twiss brothers.

The Orkney clocks are quite distinctive, as the majority seen by the authors have hoods decorated with three finials and open fretwork, a feature that originated in New England. The Orkney clock owned by the Canadian Museum of Civilization, Hull, Quebec, has a unique pendulum door, in that the door frame is fitted with bevelled glass that appears to be original. Using glass in the door is unusual in other clocks of this era. The case is mahogany with some mahogany veneer over secondary pine. The shell-shaped inlay, and inlays of other designs were imported. The clock is 90 inches in height and 20¾ inches in width.

A similar clock, owned by the Royal Ontario Museum in Toronto, is 92½ inches in height and 20½ inches in width. The clock case is made of mahogany and it is decorated with imported fan-shaped inlays. Crotch mahogany veneer was used on the lower door. The case has fluted brass-topped quarter columns and bracketed base and feet.

Pictured in *English-Canadian Furniture of the Georgian Period* by Donald Blake Webster is a clock from a private collection that has a broken-arch cornice instead of fretwork on the top of the hood. The case is mahogany with secondary pine. It stands 91⅛ inches high and is 17⅞ inches wide. It is decorated with maple and mahogany fan and oval inlays, thought to have been made by Orkney rather than imported. However, imported inlay was used to decorate the lower part of the case. Another clock in a private collection has a brass dial and a rocking ship mechanism. The clock case is without fretwork. A clock also without fretwork on the case is in the Collection d'ethnographie du Québec. The painted dial has a picture in the arch. Several clocks, one of which is known to have fretwork on the hood, exhibit the phases of the moon on their dials.

The one known bracket clock is in the collection of the Royal Ontario Museum. It is Chippendale in style, with mahogany and curly maple veneer. Secondary parts are of oak. The clock case lattice and fret are made of brass.

The movements of the Orkney clocks are British. Two clocks use movements that bear the name "Russell, London." On several false plates of the movements is the name "Nicholas Birmingham." The names on the false plate of the clock owned by the Canadian Museum of Civilization are W.C. and J. Nicholas. William worked from 1793 to 1822. Joshua was associated with the business from 1797 to 1801.

James Orkney was born in Scotland and learned his trade there. He arrived in Quebec City around 1786 and established a business. His first advertisement appeared in the *Quebec Gazette* in July 1787, offering for sale clocks, watches, and jewellery. In the census records, James Orkney was at 13 Mountain Street in 1792 and at 26 Mountain Street in 1795. Although the 1822 Quebec City directory gave 34 Mountain Street as his address, the business was again listed in 1826 as 26 Mountain Street.

For a short time around 1800 James Orkney had a silversmith named Joseph Sasseville as a partner. After leaving Orkney, Sasseville and his sons owned prosperous businesses in Quebec City. Joseph, among other items, made the silver armbands that were worn by the Indian chiefs who visited England in 1824.

In addition to advertisements of his business, Orkney's name appears in the *Quebec Gazette* along with other Quebec City citizens who signed petitions to Lord Dorchester in 1791 and the address to H.R.H. Prince Edward in 1794. Five years later he signed an address to Lord Robert Prescott. In 1794 he also signed a declaration of loyalty toward the constitution and the government. He subscribed to various relief funds and was a member and contributor to the Quebec City Fire Society. In 1797 he was a juryman at the trial of David McLane for high treason. In 1802 he signed the memorial to King George III, asking for land to build a Presbyterian church. A number of documents recording Orkney's other business activities can be found in the Provincial Archives of Quebec. For example, one such document of July 1799 records the purchase of property from a blacksmith, François Bedouin of Montreal. Another document written in 1808, concerns a contract with stonemasons who had been hired to construct the jail of Quebec City. James Orkney assured the commissioners that the work would be properly done.

FIG. 168

Orkney frequently visited Great Britain. In 1803, he was listed as a passenger on the brig *Ann*, returning from Greenock, Scotland. At other times, he arrived at Quebec City, having sailed from London or Liverpool.

After his wife's death James Orkney gave most of his estate to his son, Alexander, who was living with him. He gave notice on 1 January 1818 that he had decided to liquidate his business. At that time he had on hand, among other articles, twenty-five tall case clocks. In 1823 he bought the modest seigneury of Île-aux-Ruaux near Île d'Orléans. On 20 September of that year, he engaged James Boyt to "do any tasks assigned." Boyt was legally bound to inform Orkney of anything that could cause him harm.

The following births, deaths, and marriages are recorded in the church records in the Provincial Archives of Quebec. James Orkney married Jane Hanna (b. about 1769 – d. 2 Aug. 1817) on 3 July 1790 in St. Andrew's Presbyterian Church. Jane Hanna was a close relative, probably a daughter, of James Hanna, a clockmaker of Quebec City. There were eight children: James Jr. (b. May 1791 – d. 27 Sept. 1818), Robert (b. Nov. 1792 – d. 23 Oct. 1861), Helen (b. 5 May 1794), Jane (b. 9 July 1798 – d. 25 July 1798), Alexander (b. 17 May 1800), John Hanna (b. 23 Dec. 1804), William (b. 1 June 1806 – d. 20 Dec. 1806), and Maria (b. 31 Mar. 1810).

James Orkney died in Quebec City on 24 January 1832 and is buried in Mount Hermoun Cemetery.

FIG. 169

FIG. 168 *Orkney bracket clock.*
  – *Courtesy the Royal Ontario Museum, Toronto, Canada*
FIG. 169 *Dial detail of a fine tall case clock by J. Orkney, Quebec.*

JOSEPH PETIT CLAIRE

Joseph Petit Claire (d. 11 Feb. 1809) was a watch- and clockmaker in Lower Canada from the 1790s until his death. At least two tall case clocks bear his name on their dials. During the period in which these clocks were made, Petit Claire worked in Montreal. One of these clocks is illustrated in *English–Canadian Furniture of the Georgian Period* by D.B. Webster. The clock is made of mahogany with secondary pine. The pendulum door is shaped from a single board. The clock is simple in design and is without inlays. The lower section of the case has been lowered by 3 to 4 inches, leaving the clock 89½ inches in height. On the brass face is engraved in script, "Joseph Petit Clair, Montreal."

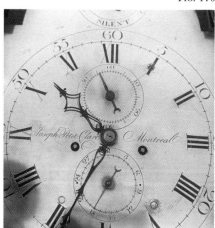

The second tall case clock, in a private collection, has a case of ebonized mahogany with tile inserts. On the dial of the clock is the name "Joseph Petit Clare, Montreal." It was not unusual two hundred years ago to find a number of spellings of surnames. However, a legal document exists in which he signed his name, "Joseph Petit Claire." The clock, illustrated in Figs. 170 and 172, shows clearly the ebonized case and tile inserts. The brass dial appears to be typical English styling of the late 18th century, when Petit Claire was active. The dial itself is simple and devoid of decoration. It has three subsidiary dials for strike/silent, seconds and calendar. There are two circular marks above V and VII. These appear to be caused by a loosening of the riveted mounting posts that attach the dial to the false plate, or movement. The authors were unable to examine the movement.

Petit Claire and five members of his family lived on Buade Street, Lower Town, Quebec City, from at least 1790. By 1795 he had moved to the "Rue sous le Fort."

In May 1797 Joseph Petit Claire moved to Montreal from Quebec City and announced his arrival in the 29 May issue of the *Montreal Gazette*:

> JOSEPH PETITCLAIRE [*sic*] Clock and Watch Maker lately from Quebec, informs his Friends and Public in General that he has taken a House in St. François Street, facing Mr. Blondeau, where he will repair and clean Clocks and Watches in the best manner, and exercise all other branches of his profession on the most reasonable terms – Old Gold and Silver bought or exchanged – Also Good Watches for Sale.

In October 1801 Joseph Petit Claire took an apprentice, Louis Baron, aged seventeen, for a period of four years. The contents of the contract for the apprenticeship are partially quoted below:

> Louis Baron would accomplish within legality and honesty all that is asked from him, safeguarding his interests and would see that no harm is done to his master.

In return, Mr. Petit Claire agreed to "treat Louis in a human and honest manner and to train him as a clockmaker." Louis's father assumed the cost of lodging and his son's various needs.

It is believed that Joseph Petit Claire was the son of Claude Petit Claire (b. 1741 – d. 28 June 1785) and Marie-Anne Baron-Lambert (b.1738) of Quebec City. Joseph Petit Claire married Marie Louise Delciat in Quebec City on 12 June 1794. He was her second husband. Records of the Roman Catholic Notre Dame Cathedral indicate that they had a son, also called Joseph, on 17 November 1805. The father, Joseph Petit Claire, died on 11 February 1809. After his death, Marie Louise married for a third time, in October 1809, Pierre Huguet *dit* Latour, a wigmaker, silversmith, and merchant (b. 1749 – d. 1817).

FIG. 170

FIG. 171

FIG. 172

FIG. 170 *Tall case clock by Petit Claire, Montreal, with brass dial. Case is decorated with inlaid ceramic tiles.*
FIG. 171 *Dial detail of clock in Fig. 170.*
FIG. 172 *Detail of tile set into the base of the clock in Fig. 170.*

PETER (PIERRE) POULIN

Peter Poulin (b. 1809 – d. mid-1870s) was a watch- and clockmaker in Quebec City. A tall case clock bearing his name is illustrated in *Canadian Clocks and Clockmakers* by G. Edmond Burrows. Over the years the authors have seen several clocks bearing Poulin's name, but unfortunately did not have the foresight to photograph them. Poulin appears to have imported clocks, primarily from England, and to have added his name. One such clock was a wall clock, probably fusee, typically English in appearance and bearing his name. The clock illustrated by Burrows (Fig. 227) has a painted dial, typical of English manufacture in the early decades of the 19th century. The case, however, has some unusual decoration, suggesting that it may have been made locally in Quebec.

Peter Poulin went as a young man from Quebec City to New York, and it may have been there that he apprenticed as a watch- and clockmaker. He married Delana (b. 1809) in the United States and did not return to Quebec City with his family, Peter E. (b. 1837), Sarah (b. 1840), James D. (b. 1842), and Helena (b. 1845), until the late 1840s. In 1851, he and his family were living with his father, Pierre Poulin (b. 1784), a native of Quebec City.

Peter established a watchmaking and jewellery business on St. John Street around 1850 and the next year, also on St. John Street, he opened a homeopathic pharmacy where he sold "Humphry's specific homeopathic medicines."

His son, Peter E., apprenticed as a watchmaker and joined his father in the business at 49 St. John Street, then known as Peter Poulin and Son. Peter E. Poulin expanded the business, and after the death of his father, became a manufacturing jeweller. In this century the business was carried on by Peter E. Poulin's son.

MICHEL ROUSSEAU

Michel Rousseau (b. 1805 – d. 1853) was a cabinetmaker who lived in Lotbinière, Quebec, a town located on the St. Lawrence River about 35 miles west of Quebec City. Family history credits Rousseau with having made three tall case clocks. The authors are fortunate in having received information on one of these clocks, which is in the possession of descendants living in the United States.

The clock, illustrated in Figs. 173, 174, 175, and 176, is of particular interest because it is a fine example of the local Quebec, tapered-waist style. This style has been commented upon several times in this book. The clock's history is documented in a book entitled *Rooms* by Corinne R. Curtis, Michel Rousseau's granddaughter. *Rooms* was published in 1979 by Vantage Press Inc., New York City. Some relevant excerpts are as follows:

> David Rousseau, my father, was born in 1844 in Lotbinière, Canada, far up the St. Lawrence River. His father Michel Rousseau and his mother Cecile Lemay had migrated to Canada in the early eighteenth century.... David Rousseau was an inventor and a successful one. Between 1873 and 1901 he took out 31 patents on electrical inventions... He died in 1918....
>
> Grandfather's clock stood at the foot of the stairs. Papa's sister sent it to him from Lotbinière, Quebec. She said that he should have it because he was the oldest child. Papa was very proud of it and told us about it many times. "You will not see many clocks like

FIG. 173

FIG. 174

FIG. 173 *Clock by Michel Rousseau. The stripped pine case illustrates the unique Quebec case style.*

FIG. 174 *Dial of the clock in Fig. 173, said to have been painted by Madame Rousseau.*

FIG. 175

FIG. 176

FIG. 175 *Original wood movement from*
Rousseau clock, front view.
Hour wheel, etc., not present.
FIG. 176 *Rousseau clock movement,*
rear view. – *Photographs courtesy*
Ward Francillon

this one," he said. "The inside works are all wooden. It was made by your grandfather, and the face was painted by your grandmother and there are only three of them." The face was yellowed with age. The numerals were done in yellow paint [*sic*] and there was a pretty pink rose in the centre.

As can be seen in the illustrations, the clock did indeed have an 8-day wooden movement and the rose in the arch still looks fresh after more than a century and a half. The movement itself appears to be of American origin. Mr. Ward Francillon, an authority on early wood movements, comments that it closely resembles movements by Asa Hopkins, although there are unexplained differences.

These tapered-waist clocks were made in fairly large numbers and over twenty are known to the authors. There are two basic types. Those in pine cases are fitted with inexpensive movements, usually American 30-hour or 8-day wooden movements, although one is known with a Black Forest wood-and-brass movement. The pine cases were originally painted with imitation wood-grain finish (by now often stripped) and are not usually marked with any maker's name. They appear to have been made by several cabinetmakers, probably in rural areas. A second group was of higher quality with mahogany cases and English brass movements. They were usually sold by urban vendors in Quebec City and Montreal to a more affluent clientele.

The Rousseau clock is of particular interest because it can be reliably traced to a specific craftsman. The authors have long held the opinion that the tapered-waist style is unique to Quebec and that the pine examples were probably made by country cabinetmakers. This example offers concrete proof.

The case has most of the characteristic features of the style. Construction is entirely of pine. It has been stripped of its original finish. The rectangular base is decorated with moulding and is scalloped at the bottom. The tapered midsection has two quarter columns set into the front corners, and a tapered door. The hood has free-standing columns and is surmounted by a curved crest, carved to resemble leaves or fern fronds. Three brass finials dominate the crest. This clock has a hinged door in front of the dial and two small side windows.

The movement, as noted previously, most closely resembles the output of Asa Hopkins of Northfield, Connecticut. Mr. Francillon has recorded three other tapered-waist pine clocks with probable Hopkins movements. Each of the others was fitted with a rocking ship ornament at the top of the dial, a feature that does not appear in the Rousseau clock. This difference could explain the fact that the Rousseau movement has only two plate pillars at the top where the others have three. The other details of the four movements suggest a common maker. It must be cautioned, however, that many of the early wood movements can only be attributed to a specific maker since they are never marked.

THE ST. SULPICE SEMINARY CLOCK

The distinction of being the oldest public clock in Canada belongs to a tower clock in the Saint Sulpice Seminary, which is associated with Notre Dame Church in Montreal and is now a home for retired priests. The present seminary building was begun in 1685 during the time when the area was known as New France. It is located on Notre Dame Street and is said to be the oldest surviving building in the city. At the outset, it must be noted that the clock one sees today is a modern replacement with an electric movement. Fortunately, the historic original movement has been preserved and is in safekeeping in Montreal.

FIG. 177

Fortunately, too, for many years a diary of the clock's history was kept. Over forty pages of hand-written information are to be found in the seminary archives. Much of the information provided here came from this source.

From the time of its installation, circa 1701, the clock served a dual purpose. Its regular beat and quarter-hour strike controlled the activities of the seminary. In addition, it was for many decades the only public clock in the city. As it grew old, however, the clock suffered from many mechanical problems. Repairs were repeatedly carried out and the diary records the details of thirteen of these between 1751 and 1835.

As time went on, other public clocks were erected in the city and eventually smaller domestic clocks became plentiful and cheap. By the early years of the 20th century, the clock had stopped completely and there seems to have been little incentive to repair it yet again. Finally, in 1966, an electric movement was installed. Several years later the electric movement failed, and for some years there was no clock in the tower. The original movement was moved to the École Polytechnique at the Université de Montréal for storage.

In 1982 interest in the old clock was revived. An engineering firm from Hamilton, Ontario, was engaged to restore the clock tower under the direction of Mr. Ron Cox. Mr. Cox prepared a modern copy of the old dial and hands for installation in the tower. An electronic movement and bell system replaced the original movement. The old movement was taken to Hamilton at the time, but was later returned to Montreal for safekeeping and restoration. Unfortunately, at some time during this period, the original pendulum was lost.

During the time the clock was in Hamilton, the interest of local clock collectors was triggered and they made inquiries about the clock's origins. The authors are indebted to Mr. Charles Murray of Hamilton who has made available the photographs used in this text. Mr. Murray also kindly loaned his correspondence, a copy of the clock's diary, and other data.

Through a contact in the Antiquarian Horological Society in Great Britain, Mr. Murray was able to correspond with M. Yves de Silans of Angers, France. M. de Silans, a recognized authority on early French clocks, had restored a similar clock, dated 1647 that had been converted

FIG. 178

FIG. 177 *General view, Seminary of St. Sulpice, Montreal.*
FIG. 178 *Clock in tower at St. Sulpice Seminary.*

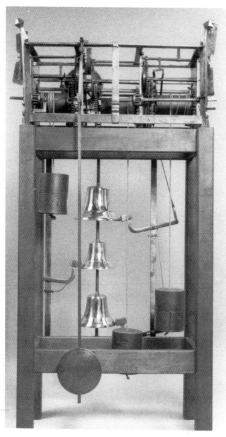

FIG. 179

FIG. 179 *Similar French clock with pendulum, bells etc.*
– Courtesy M. Yves de Silans

in the 18th century to a more modern escapement. M. de Silans restored the movement to a crown wheel and verge escapement, but has commented that his clock bore the marks of an original foliot escapement. The foliot, of course, pre-dated the invention of the pendulum and would be contemporary with 1647. In other respects, the de Silans clock was very similar to the St. Sulpice clock. M. de Silans traced the origins of his clock to Charente-Maritime in the Royan district of France, but cautioned that his clock may have been built in some other area. He was, therefore, unable to make any precise identification of the St. Sulpice clock's origin.

M. de Silans observed that the existing escapement on the St. Sulpice clock was of a somewhat later period and wondered if it was original.[12]

For comparison, a photograph of the restored de Silans clock is shown in Fig. 179. Both the de Silans and the St. Sulpice clock are equipped with a time train and two striking trains, ringing three bells. The construction details of the movement frames are quite similar. M. de Silans was not able to date the St. Sulpice clock exactly, but suggested that it could be as early as 1680.

The mechanical details of the St. Sulpice clock in its present form can be summarized as follows:

1. The mechanism is mounted in a frame of forged iron bars. The corner posts of the frame are formed at the top into faceted "heads." Note the similarity of the "heads" on the de Silans clock.
2. There are three trains of wheels, two of which control striking of hours and the quarters, plus a warning bell five minutes before each strike. The central train is the time train.
3. The three bells were mounted externally in the tower and were rung by linkage to the movement. A fourth bell was added later to ring the Angelus three times daily. At a later date, a small slave clock was installed inside the building, linked to the tower.
4. The escapement consists of a fifteen-tooth wheel and a forged iron anchor. This was originally installed during the second repair, some time after 1751, replacing an earlier system that appears to have been crown wheel and verge. Further refinements were made to the escapement in the sixth repair. Major pendulum modifications were also carried out in the second repair. It is not known whether the clock ever possessed a foliot.
5. Each train was weight-driven. The ropes were wound on wooden drums. A crude maintaining power was achieved by manually hanging a weight on the time train great wheel during the winding operation.
6. There are two marks on the frame. One is the mark of a forge at Malingsbo, Sweden. There is no ready explanation for this mark and it may have been incorporated during one of the many repairs. The other mark is appropriately a small cross.

The diary concerning the clock is a lengthy and significant document. The authors have included a short summary which provides many interesting sidelights on the clock's history. Reading the diary, one is impressed by the care that was lavished on the old clock, and its importance to the daily rituals. One can also sense the frustration the custodians must have experienced in trying to maintain a delicate and somewhat complicated mechanical device in a seminary environment, handicapped by harsh weather, remoteness, and perhaps insufficient knowledge and facilities.

The authors are particularly indebted to Ms. Elizabeth McArton of the Royal Ontario Museum for her patient and conscientious efforts in translating the diary. It is acknowledged that some portions of the photostat were indistinct and some misinterpretations may exist. The authors hope the summary provided will inspire some other researchers, skilled in both 18th-century French language and early clock technology, to attempt a more detailed study – they are satisfied, however, that the following comments reflect the essence of the story.

ORIGINS

The seminary clock was brought to Montreal by Monsignor de Belmont, who became Head (Superieur) of the Seminary in 1701. It was said to have cost 800 francs. It is not entirely clear when the clock was actually installed. One reference[13] has suggested that the clock was not installed until 1740, when a special portal or tower was built to house it. The diary comments on the remains of an earlier clock, which had been seen in the seminary attic, but nothing else is known about it.

By 1751 the clock had ceased working. Monsignor Normant, successor to Monsignor de Belmont, asked M. Guillon, the seminary priest who was responsible for the clock, to petition the government for help in obtaining a clock in better condition. M. Guillon replied that he could make repairs to the present clock. He roughed out a new "balance wheel," i.e. escape wheel, with fifteen teeth. However, he returned to France before it was installed.

FIRST REPAIR

The diary comments briefly on new bushings being installed in the strike mechanisms. All wheels were checked, but could not be straightened. Lead weights were made to replace stone weights.

SECOND REPAIR

The diary states that the "roue de rencontre" which the authors have interpreted as the crown wheel, possessing twenty-nine teeth, was removed and the fifteen-tooth wheel (previously roughed out by M. Guillon) was finished and installed in its place. The old pendulum, measuring 3 feet in length, was removed and a new one, measuring 9 feet, was put in its place. This pendulum had an 80-pound bob. Some changes in the masonry of the tower were made to accommodate the longer pendulum.

THIRD REPAIR

This repair is described as more important, more difficult, and longer than those made previously. It appears that a minute hand was installed and the dial was restored. The minute hand was mounted on a very thin rod, which turned within the iron pipe holding the hour hand. This rod was too light and flexed under the weight of the minute hand. An error of "un demi quart d'heure" (half a quarter hour) was experienced – slow on the up sweep and fast on the down. This caused many problems. The diary does not specify how the problem was resolved, although for some time the clock continued to have only the hour hand.

FOURTH REPAIR

Some modifications were made to the strike mechanisms to improve their operation. The pendulum bob occasionally came into contact with the masonry walls during its swing and the walls were hollowed out to suit.

FIFTH REPAIR

The diary comments that this repair increased the load on the clock. The Superieur, Monsignor Montgolfier, decided that it would be desirable to have a dial and bell system within the building. A small dial and striking system, connected to the big clock, were installed under the direction of M. Poncin. The dial and hands were made and engraved at the seminary, and gilded by

FIG. 180

FIG. 180 *St. Sulpice clock movement as seen in 1982. Some parts are not assembled.*

the Sisters of the Congregation. This slave clock was connected by two rods and a series of "star wheels," i.e. gears, to the tower. The lower bells were actuated by wires attached to the upper striking levers. Around 1800 M. Poncin obtained a lathe with which he was able to replace or repair bushings in the big clock and true up the wheels to make them mesh better. He repaired the bell hammers and attempted to protect them with tin covers, although these were only used for one year because of the difficulty of installing and removing them.

SIXTH REPAIR

The "balance wheel" installed in the second repair was removed and replaced with a new shaft and copper wheel. This escape wheel had fifteen teeth as before. However, the teeth were shaped differently to eliminate recoil, creating a dead beat type of escapement. The escapement shown in Fig. 182 may well be the wheel described here. The term "cuivre" (copper) was used to describe the wheel and several other items in the diary. It is quite likely that the metal employed was actually brass.

During this repair, the pendulum suspension rod was provided with a knife-edge suspension and a screw by which the bob could be raised and lowered. Around 1808 the iron suspension rod was replaced with hardwood.

SEVENTH REPAIR

Some repairs to the strike train were carried out. The indoor clock mechanism was refurbished, as well as the shafts connecting it to the tower.

EIGHTH REPAIR OR, RATHER, IMPROVEMENT [*SIC*]

To protect the clock from dust, both the movement and the pendulum were enclosed in a case. Since the chamber containing the clock was in darkness, it had been necessary to employ a candle or lantern when visiting the clock. This was considered to be hazardous. Monsignor Molin had a window installed, with four panes. Some improvements were made to the floor and to the scaffolding that was set up, when necessary, to gain access to the bell hammers.

NINTH REPAIR OR IMPROVEMENT

Wooden stop devices were installed to control winding. This allowed the weights to be raised to suitable height without over-winding. A container for lead weights was attached to the pendulum. It was believed that adding or subtracting weights would help to regulate the clock. Another device was installed in 1820 to remind the clock winder to use a counterweight for maintaining power. (This is more fully described in the winding instructions below.)

TENTH REPAIR OR IMPROVEMENT

After twenty years the knife suspension (sixth repair) had rusted enough to stop the clock. It was removed. A Mr. Henry, clockmaker, made a spring one inch wide and one foot long, which was used instead of the knife edge to suspend the pendulum rod. The spring was held by a fork at its midpoint. This proved to be very successful.

ELEVENTH REPAIR

A pivoted detent, which sounded the warning strokes before each strike, was abandoned in 1823. Instead, they copied the mechanism of another clock in the "Little Seminary" (Petit Seminaire), which sounded the warning strokes using the same mechanism that sounded the hours. M. Gabriel Gosselin, gunsmith, made a 9¼-inch copper wheel with 126 teeth, which was sufficient to sound all the strikes in a twelve-hour period (forty-eight warning strokes and seventy-eight hour strokes). Gabriel Gosselin was a brother to the late Barnabé Gosselin, who made the balance wheel in the sixth repair. Gabriel Gosselin also altered the pinion of the copper wheel above, from six leaves to eight leaves. The cost of this repair was 16½ louis (French coin).

The wooden winding drums on the three great wheels were reduced in diameter. The weight on the hour strike barrel was cut by a third to twenty-one pounds. The quarter-hour detent was straightened.

TWELFTH REPAIR

A few weeks later, the different parts of the clock (presumably the three trains and the outer case) were marked with red, blue, and black paint to make reassembly easier.

THIRTEENTH REPAIR

The diary comments that the twelfth repair was carried out in 1823 and the thirteenth repair in 1835. The latter repair was made by a French clockmaker, M. Dautel. Rust had been a problem. Unspecified repairs were made to the escapement and several pinions. The cost was ten louis (French coin).

COMMENTS ON REGULATING THE CLOCK

[This section is quoted directly from the diary, in a literal translation.]
How they regulated the clock, now on true time, now on mean time:
In the old days they set the big Seminary clock on true time as it is indicated on a sundial. Later M. Poncin, at the request of the late M. Jean Delisle, Notary, who was well versed in astronomy and physics, set it on mean time which lasted until about 1797 when true time was used again.

In September 1822, to comply with the wishes of several priests of the Seminary, of a few astronomers, of other intelligent laymen, they stopped setting the clock on true time as had been done for about twenty-five years and they set it once again on mean time.

FIG. 181

FIG. 181 *St. Sulpice clock movement.*

They found that this method of setting clocks on mean time didn't give rise to any notable drawbacks, neither from the beginning of matins the next day, nor for the eucharistic fasting which began at midnight. So that, on 2 November when the clock, which had been set to mean time, indicates 11.43 and 45 seconds, then it is exactly noon or midnight by the sun.

Thus it was decided by Monsignour Roux, Head and by several other priests on 31 August 1822, and since then, with regard to Matins, by Monsignor Plessis, Bishop of Quebec, on 1 March 1823. *[This passage graphically shows the importance of the clock in the daily ritual of the Seminary.]*

INSTRUCTIONS FOR PERSON WHO WINDS THE SEMINARY'S CLOCK
[These instructions are quoted completely, again in a literal translation.]
1. He must take care to wind it every day.
2. He winds it at 11 a.m. from the beginning of November up to the time when the priests start to go to dine at La Montagne [The Mountain], that is, about Ascension.
3. During all the time that the priests go once a week to dine at the country house, the clock is wound at 6 p.m.
4. To go and wind the clock, you must start as soon as the time indicated above is reached, that is, as soon as possible after 11 o'clock in winter and 6 o'clock in summer. The reason is that if you are a few minutes off, you risk putting the striking mechanism out of order.
5. If you didn't wind the clock at the time indicated above, you must wind it as soon as you [are able] taking care, however, not to go there until immediately after the quarters, the halves or the three quarters have sounded.
6. As soon as the person who winds the clock notices that it has stopped, or that there is something wrong with it, he notifies the priest responsible for the supervision of this clock.
7. When the person who is going to wind the clock arrives in the study where it is placed, he begins by opening the appropriate doors and panels, then he winds up the middle weight, taking care to suspend from the large wheel the hook which supports the weight, intended to make it continue to run while it is being wound, after which he places the crank and turns it to the left. When the weight is ready to arrive up above, he keeps on his guard so that, when the crank is stopped suddenly by purposefully meeting an arm, he doesn't let it return back, but holds it firmly and gently lets the weight go down to the place where it must stop. Then he takes off this crank and removes from the big wheel the hook indicated above.
8. Then he winds the works which sound the warning strokes and the hours, also turning the crank to the left; finally he winds the quarters striking mechanism, turning the crank to the right. When these two points are almost at the top, he keeps on his guard and holds up the crank as indicated in No. 7.
9. You shouldn't bring light up to the clock except when it is necessary, taking care to have the light firmly inside a lantern.
10. To oil the clockworks of the big clock, you must stand near this clock, your back turned towards the little window which lights it and you must have a vessel containing olive oil or goose fat. You should put it on lightly:

1. on the 3 largest wheels on the front of the pegs bearing the 3 wooden cylinders, by passing a feather (from bottom to top) on the 9 teeth.

2. On the 3 wheels, each one of which is attached to each cylinder, passing the feather from the top to bottom over a dozen teeth.

3. On the 3 wheels, the teeth of which mesh with the 3 wheels indicated in No.1. The feather must also be passed from top to bottom over 8 teeth.

4. Then you must oil a few teeth in the balance wheel or copper wheel; then at both ends of all the axles of the wheels and of the flywheels, the copper dowels of the 3 big wheels, the ends of the 3 detents, the pivots of all the wheels, the sorts of nuts that are in the middle of each fly wheel.

WAYS OF REGULATING THE HANDS
[Also a literal translation]

To regulate the hands of the dial which is outside, after having released the screw which holds the 2 parts of the vertical rod at the height of about 5 feet, you should grab with the left hand the lower part of the rod and hold it solidly stopped by turning this hand from right to left, then with the right hand, you must move the top part of the rod from right to left if you have to make the hands go forward, and from left to right if you have to make them go back, paying attention so that this part of the rod doesn't get lowered when it goes into the pipe where its lower end is. Then you must tighten the screw.

Included in the diary is a lengthy specification prepared by S.T. Lepaute, a well-known clockmaking firm in Paris, France, dated 7 January 1822. It appears to be a quotation for a new clock for the seminary, offered for a price of 2,800 francs, plus shipping. Obviously this clock was never purchased. The article includes details of a more modern clock, which would have provided the bell-ringing sequence with greater accuracy and resistance to the elements.

The diary also includes a lengthy discussion of the mechanical characteristics, dimensions, etc., for both the great clock of the seminary and the other clock referred to in the diary – the later clock of the Little Seminary.

In summary, the authors are pleased to have been able to include extensive detail of the St. Sulpice clock. It is truly one of the great clocks of the Western Hemisphere. Only rarely does such a wealth of year-to-year historical detail survive. It is fortunate, also, that the rarity and historic value of the clock have been recognized. At the same time, one can feel a little saddened to realize that, after almost three centuries of joys and sorrows, its work has been taken over by an electric motor.

FIG. 182

FIG. 182 *Escapement detail,*
St. Sulpice clock as presently set up.

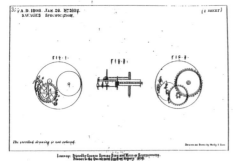

FIG. 183

A.D. 1808 . . . . N° 3102.

Watches and Clocks.

SAVAGE'S SPECIFICATION.

TO ALL TO WHOM THESE PRESENTS SHALL COME, I, GEORGE
SAVAGE, of Huddersfield, in the County of York, Watchmaker, send greeting.
WHEREAS His present most Excellent Majesty King George the Third,
by His Letters Patent under the Great Seal of the United Kingdom of Great
5 Britain and Ireland, bearing date at Westminster, the Twenty-sixth day of
January, in the forty-eighth year of His reign, for Himself, His heirs and
successors, did give and grant unto me, the said George Savage, my executors,
administrators, and assigns, His special licence, full power, sole privilege and
authority, that I, the said George Savage, my executors, administrators, and
10 assigns, and every of them, by myself and themselves, or by my and their
deputy or deputies, servants or agents, or such others as I, the said George
Savage, my executors, administrators, or assigns, at any time agreed with, and
no others, from time to time and at all times thereafter during the term of
years therein expressed, should and lawfully might make, use, exercise, and
15 vend, within that part of His said Majesty's United Kingdom called England,
His Dominion of Wales, and the Town of Berwick-upon-Tweed, my Invention of "A NEW METHOD OF REGULATING OR EQUALISING THE FORCE OR POWER OF
THE MAIN SPRING IN WATCHES OR OTHER SIMILAR MACHINES FOR MEASURING TIME;"
in which said Letters Patent there is contained a proviso obliging me, the
20 said George Savage, by an instrument in writing under my hand and seal,
particularly to describe and ascertain the nature of my said Invention, and
in what manner the same is to be performed, and to cause the same to be
inrolled in His Majesty's High Court of Chancery within one calendar month
next and immediately after the date of the said recited Letters Patent, as

FIG. 184

FIG. 183 *George Savage. – Courtesy
Henry Birks & Sons Ltd., Montreal*
FIG. 184 *Savage watch patent.*

GEORGE SAVAGE

George Savage (b. 1767 – d. 21 July 1845) was a well-known watch- and clockmaker from England. Shortly after his arrival in Lower Canada in 1818 he established a business in Montreal. This business was successful for over sixty years. Clocks displaying the names G. Savage, Savage & Son, and Savage and Lyman are prized possessions of collectors today. In addition to clocks and watches his firm sold jewellery and silverware.

The Savage family had been in England for centuries. LeSieur Thomas Le Sauvage went to England with William the Conqueror in 1066 and settled in Derbyshire. George's father, for some time a soldier in H.M. 22nd Regiment Royal Fusiliers in England, was a watchmaker. Before coming to Montreal, George lived in Huddersfield, Yorkshire, and it was there that he worked to perfect the lever escapement and invented the "two-pin" variety. He applied for and was granted patent No. 3102 in 1808 for "A New Method of Regulating or Equalizing the Force or Power of the Main Spring in Watches or similar Machines for Measuring Time."

According to the booklet, *The House of Birks,* published for Henry Birks and Sons Ltd., George Savage moved to 5 St. James Street, Clerkenwell, London, England, where he was a successful jeweller and watchmaker. His invention "for a detached escapement for watches which was a combination of the lever and the chronometer" won the medal of the Society of Arts.

In 1818 George Savage and his family (except John, who remained in England) came to Montreal, Lower Canada.[14] For a year George was associated with James A. Dwight, who at the time was in business on St. Paul Street, close to Bonsecours Market. Their silver mark during the partnership was "D. & GS".[15]

After the dissolution of the partnership with James Dwight, George Savage established his own business on the corner of St. Peter and Notre Dame streets. His son, Joseph (b. about 1799 – d. 6 Feb. 1859), was his partner and the firm was known as G. Savage and Son.

In 1826 the business was moved to St. Paul Street opposite the church of Hôtel-Dieu. At this time, Joseph announced in the *Montreal Herald* of 6 March that "Business under the name of George Savage and Son will in the future be conducted by Joseph Savage only." The name of the firm, however, was not changed. An advertisement appeared in the *Quebec Gazette* of 6 October 1828 offering for sale by Savage and Son "a splendid selection of gold, silver and plated ware...that will be exhibited for a few days at 11 Market Place, Upper Town, Quebec City." Also, Joseph continued to conduct business on St. Paul Street until mid-1840.

In the late 1830s and early 1840s, clocks were sold by the Savage firms with the name "Savage & Son" on the dial. Some of these clocks carry the place name "London" on the dial, while others that appear to have been made within the same time frame have the name "Montreal" as the place of business. As Savage clocks were, for the most part, imported from Britain, it would appear that clocks carrying either place name were sold in Canada at the time. George Savage Sr. had been in partnership with George Jr. in London, England. The partnership was re-established in Toronto. The "London" clocks may have been sold at the Toronto Savage establishment in the 1840s.

The clock shown in Fig. 187 differs from the other Savage clocks that are illustrated. It has a typical, early 19th-century English 8-day, double-fusee movement, without repeater feature. The plates are not engraved but have shaped upper corners (see Fig. 188). The case is unusual, being of solid mahogany with elaborate scrolled top and side ornamentation. There is applied carving and a scrolled brass inlay on the case front. The painted dial is circular, with a matching circular brass bezel door. The hands are of brass and have a distinctive shape.

Unfortunately, the dial has been repainted, but on the reverse side, there are an amazing number of jewellers' repair inscriptions from firms in Montreal and Ottawa, beginning in 1851 and extending well into the 20th century. The style of this clock suggests that it is of a later period, i.e., the 1840s. The original styling also suggests that the case might have been made by a local Montreal-area craftsman.

A clock with inlaid brass designs and a fusee movement in a fruit-wood case, similar in style to the clock in Fig. 186a, has "Savage & Co. Montreal" on the dial.

In 1835 John Wood entered the employ of George Savage in the St. Paul Street store on the corner of St. Dizier Lane. In the following year George Savage retired from active business. By 1837 the success of the firm and an over-stock of key-wind watches encouraged G. Savage to open a branch uptown on Notre Dame Street, two doors west of St. Gabriel Street. This store, of which John Wood was fully in charge, became the principal business and continued in this location for twenty years. Eventually, the courthouse stood on this location. John Wood left the Savage store in 1839 and opened a business of his own at 130 St. Paul Street (see p. 139). The St. Paul Street store, owned by the Savage family, was in business until about 1841, under the management of William Learmont.

George Savage Sr. advertised as a silversmith as well as a watchmaker. In the first ten to fifteen years, however, he contracted out the silver business to skilled craftsman in Montreal. When business developed to the point that he needed a full-time silversmith on the premises (in the 1830s), he brought Robert Hendery to Canada from Scotland. Although Hendery established a business for himself in 1840, he continued to manufacture silver for G. Savage and Son, as well as for others. Another craftsman making silver articles for Savage & Son and for Savage and Lyman, was Peter Beaulé. In order to defray costs, in April 1853 Joseph Savage bought a machine to make silver spoons.

George Savage died on 21 July 1845 and, as a member of the First Congregational Church of Montreal (Zion), was buried in the English Protestant burial ground south of Dorchester Street near St. Urbain Street. This cemetery served the community from 1799 to 1847. After this date, burials were made in the newly acquired Mount Royal Cemetery, and some of the bodies, including that of George Savage, were moved there from the old cemetery. His removal took place on 18 October 1867. The old cemetery land was expropriated as a park in 1871; the area was then known as Dufferin Square.

After the death of George Savage, the business continued to expand. In 1851 Joseph Savage accepted a partner, Colonel Theodore Lyman (b. 1818 – d. 1901). At this time the name of the firm was changed to Savage and Lyman. Colonel Lyman was the brother-in-law of Joseph Savage, who had married Abigail J. Lyman on 18 May 1829. This firm advertised their business as "wholesale and retail importers of clocks, watches, jewellery and plated ware, etc." According to the *Canadian Illustrated News*, the firm moved in 1856 to the Cathedral Block and remained there until 1872. It was a time of great prosperity in Canada and the store was "one of the finest of its kind in Canada."

FIG. 185

FIG. 186a

FIG. 186b

FIG. 185 *Joseph Savage.* – *Courtesy Henry Birks & Sons Ltd., Montreal*
FIG. 186 *Savage & Son "London" clocks.*

119

FIG. 187

FIG. 188

FIG. 189

FIG. 190a

FIG. 190b

FIG. 191a

FIG. 187 *Savage & Son, Montreal bracket clock with flamboyant styling.* FIG. 188 *Unsigned movement in bracket clock, Fig. 187.*
FIG. 189 *Tall case clock by Savage & Son, Montreal. The unusual hood shape may be of local origin.*
FIG. 190 *Two clocks by Savage & Son, Montreal. The clock on right has a simple walnut case.*
FIG. 191a *Dial detail from the tall case clock in Fig. 191b.* – *Courtesy Collections d'ethnographie du Québec*

FIG. 192

FIG. 193

FIG. 191b

FIG. 194

FIG. 195a

FIG. 195b

FIG. 191b *Savage & Son tall case clock.* – *Courtesy Collections d'ethnographie du Québec* FIG. 192 *Bill of sale.*
FIG. 193 *Advertisement from the Montreal Directory of 1848.* – *Courtesy the National Library of Canada* FIG. 194 *The Savage and Lyman store.* – *Courtesy the National Archives of Canada* FIG. 195 *Clocks sold by the firm of Savage & Lyman.*

FIG. 196    FIG. 197

FIG. 198

FIG. 196 *Interior of Savage, Lyman and Co. store from an engraving in the* Canadian Illustrated News.
– *Courtesy the National Library of Canada*
FIG. 197 *Advertisement in the Montreal Directory, 1869–70.*
– *Courtesy the National Library of Canada*
FIG. 198 *The Savage, Lyman & Co. store on James Street.*
From Canadian Illustrated News.
– *Courtesy the National Library of Canada*

Theodore Lyman remained as a partner in the firms Savage and Lyman and Savage, Lyman and Co. until 1885.

The store on the Cathedral Block, considered extensive when first occupied, proved too small for the increasing business of this rapidly growing city and they moved to 226–230 James Street. The new store was lighted by three "dazzling sperm oil lamps.... The architectural style and beauty of the building was considered unsurpassed by any store in the continent." In front of the building was a clock that stood on an iron pillar, "a great boon to the public of a city where the correct time is most difficult to ascertain."

On 22 April 1857 Henry Birks (b. 1840 – d. 1928) began his apprenticeship with Savage and Lyman. Henry was accepted primarily because Theodore Lyman was impressed with Henry's excellent attendance record at Sunday school, where Lyman was the superintendent. Henry Birks remained with the firm for twenty-two years.

In 1859 Joseph Savage died, leaving no person in the firm with the name Savage. A further change to the name of the business occurred in 1869 when Henry Birks became one of two new partners. The firm was then known as Savage, Lyman and Co.

Unfortunately, due to the depression of the 1870s, the annual turnover of the business dropped to a third of its previous figure of $300,000, and the firm was forced to go into liquidation. In 1878 Henry Birks was placed in charge of liquidating the stock.

On 1 March 1879 Henry Birks bought the stock from the trustees of the bankrupt company and formed a company of his own at 222 St. James Street, thus establishing a business destined to thrive for over a century, as described elsewhere in this book.

A firm called "Savage & Lyman" opened its doors again across from Henry Birks in 1879, but closed permanently in 1885.

GUSTAVE SEIFERT

Gustave Seifert (b. 1813 – d. 1909), born in Prussia, was a watch- and clockmaker in Quebec City. His name appears on an elaborate bracket clock that is on view in Laurier House Museum in Ottawa, Ontario. The English movement is of high quality and the chimes are sounded on bells.

The Seifert business was founded in 1857 at 22 Couillard Street, Quebec City. In addition to supplying watches and clocks to customers, Gustave Seifert advertised as a manufacturing jeweller and silversmith. By 1872 the business, then known as "The European Bazaar," had moved to 26 Fabrique Street. Sons Albert E. (b. 1861 – d. about 1923) and G. Otto (b. 1869) trained with their father, and in 1899 the business was incorporated under the name of Seifert, G. & Sons Inc. In September of that year, the business was moved for the last time to 14–16 Fabrique Street. Early in 1900 Gustave retired, leaving the business under the management of his sons. A third son, Harold L. (b. 1884), joined the business in 1907. By 1920 members of a third generation of Seiferts, Gustave E. and Charles O., were managing the business.

In 1930 Seifert and Sons Inc. was taken over by Henry Birks and Sons and was Birks's first Quebec City store. Harold L. Seifert managed the business from 1929 to 1935. Gus E. Seifert was in charge from 1935 to 1961, when he retired due to ill health. The tradition was continued by Raulin E. Seifert, who managed the store from 1970 to 1973.

FIG. 199 *G. Seifert clock.*
    – Courtesy Canadian Parks Service

FIG. 199

123

FIG. 200

FIG. 201

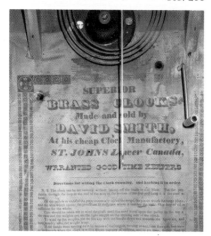

FIG. 202

FIG. 200 *Label of the David Smith,*
Godmanchester OG clock.
FIG. 201 *OG clock by David Smith, St. Johns.*
Fig. 202 *Label from clock in Fig. 201 by*
David Smith "at his cheap Clock
Manufactory, ST. JOHNS, Lower Canada."

DAVID SMITH, GODMANCHESTER

David Smith (b. about 1800) sold clocks in the late 1830s in Godmanchester, Huntingdon County, Lower Canada. Two clocks sold by David Smith have been examined. Both clocks are 8-day OGs and were sold by the Forestville Clock Company,[16] which made clocks in the United States beginning in 1835. A picture of the label of one clock is shown in Fig. 200. There is no printer's name on the label.

According to the records located at the National Archives of Canada, David Smith and his wife, Jane (b. about 1798), answered an advertisement in their homeland, Ireland, offering free land to persons who immigrated to Canada. Their sponsor was James Brown who, on behalf of seventeen people, petitioned on 2 April 1822 for land for these people "to his Excellency, George, Earl of Dalhousie, Governor-in-Chief in and over Provinces of Lower Canada, Upper Canada, Nova Scotia and New Brunswick."

The land allotted to David Smith was located in Godmanchester Township, Lower Canada. Census records indicate that for many years he owned Lot 15, Concession IV of that township, taking possession of the land in 1824.

Unfortunately, the pages that included that area are missing from the census of 1831. However, the 1842 census records show that David ran a successful farming operation, which included production of wheat, barley, oats, potatoes, and maple syrup. Besides these products, large quantities of cloth were manufactured on the farm, which was also stocked with a variety of farm animals.

David Smith was head of a household of five people. In addition to himself and his wife, both of whom belonged to the Church of England, there was one man over the age of sixty and one man between eighteen and twenty-one years of age living on the farm. Both were Wesleyan Methodist. Also included in the household was one "coloured" person over forty-five years of age who was Presbyterian. From these records it is apparent that David and his wife had no children.

The Smiths lived on their farm in Godmanchester until sometime in the 1870s. Neither of them was listed in the 1881 census of Godmanchester Township.

DAVID SMITH, ST. JOHNS, LOWER CANADA

David Smith sold clocks in St. Johns (now St. Jean), County of St. Johns, during the late 1830s. Printed on the label of one of these clocks are the words, "Made and sold at his cheap Clock Manufactory." There is little doubt that the clocks were imported from the United States, as the label was printed by J.M. Sterns, Middleburg, Vermont. This label is different from the label on the clocks sold by David Smith of Godmanchester.

Unfortunately, efforts have failed to prove that the David Smith from St. Johns and the David Smith from Godmanchester were different persons. However, the authors suspect that two persons were involved. David Smith was a well-established farmer in Godmanchester before and after the period during which the clocks were sold, and there would appear to be little reason for him at the same time to establish another location from which to sell clocks about 50 miles away at St. Johns. However, the 1851 census of St. Johns, the first census available for that place, shows no David Smith in its list of residents, so no proof exists that there were two David Smiths, living in two locations.

PIERRE-HENRI SOLO

Solo is considered to be one of the first watch- and clockmakers of "Canada." He had his business in Quebec City shortly after 1730.

Although very little information was found about Solo and his activities, his existence in Quebec has been verified by the following facts: his name as a merchant clockmaker appears in 1729 in an inventory of goods supplied by the court following the death of a patron; a 1923–24 report of the provincial archivist, Pierre Georges Roy, states that in July 1730 Solo certified the value of a watch given in a lottery by Jean-Baptiste Lozeau.

Another small reference to Solo is made in the book entitled *Revue d'ethnologie du Québec /9* in which it is suggested that a clockmaker, Jean-Baptiste Filiau *dit* Dubois, apprenticed with Solo.

THE TWISS FAMILY

Twiss clocks are prized by private collectors and museums. These Canadiana clocks were distributed by the Twiss brothers, Austin, Benjamin, Joseph, Ira, and Russell, and their father Hiram, from the Montreal area to Upper and Lower Canada during the period from 1821 until perhaps as late as 1850.

The clocks are tall case in style and it has usually been assumed that the movements were made in Connecticut and assembled and cased in the Montreal area. The cases were made from Quebec pine and were skillfully painted to imitate expensive woods. When viewed from a distance, the effect of graining and inlay are apparent. The cases were highlighted with gold trim, and the long door of the case was fitted with unusual coiled-wire hinges. Although the size of the cases appears to be identical at first glance, close scrutiny shows minor variations in height, width, door size, hood height, etc. Such variations can be seen among the Twiss clocks owned by the Canadian Museum of Civilization, Hull, Quebec.

As noted above, there has long been a belief, based on a good deal of practical evidence, that Twiss movements were all of Connecticut origin. However, from time to time conflicting opinions have been advanced, suggesting that at least some movement manufacturing was carried out in Montreal. It is difficult to make any final judgement 150 years later, except to agree that most, if not all movements found in surviving Twiss clocks were imported. These movements appear to be the product of Silas Hoadley, with the exception of a few clocks fitted with movements by Riley Whiting. In fact, one clock reported by Burrows (*Canadian Clocks and Clockmakers,* Fig. 217) has a dial on which the Twiss name is painted over that of Whiting.

In support of the belief that the movements came from the Hoadley factory, there are a number of pieces of evidence:
1. There was a long association between the Twiss family members and Silas Hoadley. In fact, several of them learned the trade of clockmaking at Hoadley's factory.
2. Hoadley was credited with being the largest manufacturer of these wooden 30-hour tall case movements. He began manufacture in his own factory around 1815 after a brief partnership with Seth Thomas and continued to make clocks of all styles until 1845 or later. This coincides with the active period of the Twiss brothers in Montreal. No one knows for sure when manufacture of these wood movements ceased, although by

FIG. 203a

FIG. 203b

FIG. 203 *A typical Twiss clock sold by I. Twiss, and showing original imitation grain painting and wooden finial painted to simulate brass.*

FIG. 204

FIG. 205a

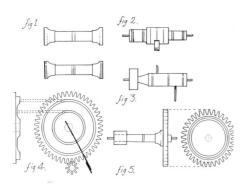

FIG. 206

FIG. 204 *Close-up of graining and hinge.*
FIG. 205a *Label from an Ira Twiss clock.*
For Fig. 205b, see page 263.
FIG. 206 *Line decoration found on component parts of Twiss movements.*
– Courtesy of John A. Crosby

the late 1830s they were definitely being phased out. It is possible that a few clocks continued to be cased in Montreal during the 1840s; these could have been made with stockpiled movements. Hoadley's long period in business is one good explanation for the fact that there is remarkable uniformity in Twiss movements over their long business career.

3. Quoted later in this chapter is a letter by Carrie Twiss Burgess, a descendant of Russell Twiss, in which reference is made to the importation of movements.

4. The authors are indebted to Ward Francillon, noted authority on American wooden clock movements, for comments he has provided on the Twiss clocks in Ottawa museums and elsewhere. The authors are likewise indebted to John A. Crosby of Ottawa for providing photographs of an Ira Twiss movement and for sharing with the authors the results of his study of ten Twiss clocks. He also contributed drawings of constituent parts, which are shown here.

In summary, Mr. Francillon stated, "I do not know of any feature of construction that would make their [Twiss] movements identifiable as unique to them. Everything I have seen, including the photos you have sent, appears to be the product of Hoadley or another identifiable maker such as Riley Whiting." Mr. Francillon also commented on the presence or absence of stiffeners in the Twiss movements. These were half-round wooden blocks glued to the seat board between the plates. Their purpose was to prevent the movement from being pinched if the seat board shrank across its grain. He states that these were "absolute marks of identification" to Hoadley's shop and were used in the later years of Hoadley's production.

Those movements without stiffeners were described by Francillon as "Hoadley-like," being identical in all respects to the preceding movements except for the absence of stiffeners. He felt, however, that there was "a 99% odds-on chance" of these being the products of the Hoadley shop.

There are also several factors that suggest that the Twiss brothers may have been involved in at least some movement production:

1. One of the most significant factors is the statement on an Ira Twiss label (Fig. 205a) that states "...clocks made by Ira Twiss at his carding mill at Côte des Neiges, three miles from Montreal, where are manufactured every part comprising these clocks which are warranted to keep time..." The "Austin Twiss" label[17] in Fig. 205b (see p. 263) made an identical claim in French. Written in French on the reverse of the label is a bill of sale showing that the clock was sold in 1830.

2. The existence of the Twiss carding mill is significant, since Ira manufactured and sold carding mill equipment. In later years, Joseph Twiss engaged in the design and manufacture of fanning mills. With all this mechanical activity, the manufacture of clock movements would have been a distinct possibility. In fact, in the census records taken while any of the brothers were in Canada, the occupation was always given as "horologer."

3. A wooden tower clock attributed to the Twiss brothers is described later in this section. As far as the authors can determine, no other movement with similar characteristics is known. The design is simple, being based entirely on a 30-hour, time-and-strike, tall case movement, except that all dimensions have been doubled. This clock could indeed have been produced at one of the Twiss shops in Montreal.

4. Hiram Twiss received two patents, which indicate that he had an ex-

tensive knowledge of clock movement fabrication. (It must be acknowledged that the only clock ever made according to these patents was actually manufactured by Silas Hoadley.) These patents are discussed again later in this section.

A series of drawings by John A. Crosby is reproduced in Fig. 206. These illustrate some of the decorative aspects of Twiss movement parts, as studied by Crosby. These details appear to correspond to Hoadley production details provided by Ward Francillon, although such decoration was seldom very consistent.

The shells of the Twiss weights were made out of tin and filled with pebbles or sand. Occasionally bits of metal were used, and when necessary a layer of sawdust was inserted to prevent shifting. The weights were topped with wood, to which hooks and cords were attached. The bob of the pendulum was of brass-covered lead and was formed by pouring lead into a pocket of brass. A loop of wire was placed at the top of the bob.

The dials were made of wood and had decorations on the corners outside the chapter ring. The painting in the arch was frequently of flowers, often in a basket. Occasionally, a scene of houses or a landscape was portrayed. Gessoed gilt decoration accompanied the coloured painting. Clock collectors believe that many of the dials were painted in convents by nuns. The temptation to foster the rumour that Cornelius Krieghoff also painted the scenes on the dials must be resisted. Although this famous Canadian painter did work for a period in Lower Canada, he did not arrive there until 1840, after Russell had moved his clock establishment away from Côte des Neiges, which at the time was 3 miles away from Montreal.

The hands of Twiss clocks featured a diamond-shaped opening. The

FIG. 207 *Two views of a "Hoadley" movement in a J.B.& R. Twiss clock showing stiffeners on the mounting board.*

FIG. 207

FIG. 208

FIG. 208 *Two views of a "Hoadley-like" movement in a J.&R. Twiss clock. Stiffeners are not present.*

shaft between the opening and the arbor was either plain or a split elongated oval.

The hoods of the clocks were made with two "free" columns on either side, a broken-arch cornice, and a central post at the top. Twiss clocks do exist with projections of equal height on the top of the hood, but this type of configuration may have been the result of changes made by the owners after the clocks were bought. There are Twiss clocks, known to the authors, which have no broken-arch cornice. Although such clocks have been in the families for generations, it is not known if they were originally made so. An examination of one such clock revealed that the nails used were definitely from a later period.

An I. Twiss clock in excellent condition was recently sold and is now in a private collection (see Fig. 203.). The five-pillar movement is ivory bushed, a feature of all known Twiss clocks. The second wheel on the time side has been doubled by a repair man. It is not known if the metal inserts in the fly are original. The clock has its original weights, although one has been repaired. The lead pendulum bob cannot be adjusted. Unlike other Twiss clocks examined by the authors, this Ira Twiss clock has retained its original finial which is made of wood and painted to simulate brass.

The clock belonged to Esmond Butler and his family. Mr. Butler was aide-de-camp to Governor General Vincent Massey and three successive Governors General. During the time that he and his family lived in the Annex to Rideau Hall, the clock was part of his personal effects.

Original labels are found only rarely in Twiss clocks. Labels by Ira and Austin are illustrated in Figs. 205a and 205b and a J.B. & R. Twiss label is shown in Fig. 212. The "Ira" label is entirely in English while the "Austin" and "J.B. & R." labels are entirely in French. It seems likely that labels were printed in either language to suit the preference of individual customers.

As noted previously, the Ira Twiss label refers to his carding mill in Côte des Neiges. The J.B. & R. Twiss label shows Montreal as the address. The latter label was printed by the Montreal shop of Louis Perrault. This label, among other things, recommends oiling the escape wheel every five or six months with oil of olives or almonds, especially before the onset of cold weather. Perrault's name appears in the jury lists of 1832–1835 and the city directories until 1894. His son Louis (b. 1840 – d. about 1894) continued his father's business, which became a publishing company as well as a printing shop.

In addition to the tall case clocks that were sold in Canada, the Twiss family manufactured shelf clocks in their Meriden, Connecticut, factory. Some of these clocks bear the name Benjamin Twiss and others have the names of Benjamin and Hiram Twiss (B. & H.) on their labels. These clocks are pillar-and-splat in style with the lower glass often portraying a young woman. According to G. Edmond Burrows in his book, *Canadian Clocks and Clockmakers*, "All of the decorating was done by two or three young women in a small shop back of the dwelling house at the south-west corner of North Broad and Brittannia Streets, in Meriden." The movements appear to have been made by several makers. Some clocks are fitted with "groaner" movements made by Chauncey Boardman. Other clocks contain movements of the type made by Ephraim Downs, E.G. Atkins, and possibly Olcott Cheney (subtype 9.225 as designated by Dr. S. Taylor).

Although it would appear that the Twiss brothers purchased wooden movements for their shelf clocks and cased them in the factory, they were capable of making their own movements, as documented below. Hiram, the father of the five brothers, applied for two clock movement patents, which were granted on 13 May 1834. According to the patent office, patent numbers 8194X and 8195X were issued. However, the original documents were destroyed in a fire in 1836. Copies of the Twiss patents are available in Washington, D.C., where copies of some of the "reconstructed" patents are kept.

One patent was for the Hiram Twiss Patent Clock. He claimed, as his invention and improvement, the successful and advantageous application of the slow motion of the balance pendulum to the movement of any portable wooden clock or timepiece. It "consisted of a bar of wood or metal with weights…balanced on an axis almost three inches long crossing the bar in the center at right angles. The weight is placed on a single bar below the center of motion thereby producing a repelling action in the balance. By vibrating on knife edges or points, the friction is greatly reduced." He connected it with a regulating device by means of which the effects of heat and cold on metal are counteracted. The spring of the balance, pressing on the sides of the socket of the regulator assists the recoil of the balance when at the highest point of vibration and most disposed to remain at rest.

He claimed "as my original invention the regulator…whereby the motion of the pendulum is controlled by a spring and its expansion and contraction counteracted. I also claim the construction of the striking part…whereby the fly shaft is made to revolve backwards by means of the lever on the verge …checking and controlling the hammer more perfectly and with less loss of power…and so requiring less weight and less space for its descent."

According to Dr. Snowden Taylor, "in the 8 day Hoadley 'balance' clock, the trains appear to be exactly as in the patent, but with a compound pendulum." This would indicate a close association between the Twiss family and Silas Hoadley.

In recent years there have been rumours of the existence of a tower clock made by the Twiss brothers. These rumours are now substantiated; the movement and dial have been located and fortunately the history of the clock is well documented.

The clock was supplied by the Twiss brothers in Montreal around 1830 and was installed in the post office at Ste. Scholastique. As related by Daniel Piche of Ste. Monique, the clock was taken to his grandfather about 1875 for repairs. While the clock was absent from the tower, the post office burned. A new post office with a new clock was built. No one called for the old clock. It was still at the Piche place when it was sold to an antique dealer in 1969. It has changed owners twice since that time. The dimensions of the clock are as follows:

Width of movement	18 inches
Height of movement	22 inches
Plates	9¼ inches apart
Height of dial	46 inches
Width of dial	36½ inches
Total length of minute hand	23 inches
Total length of hour hand	17 inches

The Twiss men were natives of Meriden, Connecticut, and, as noted previously, owned a clock factory in that town. At one time there was a branch of the factory in Nashville, Tennessee. The Twiss family learned their trade of clockmaking from Silas Hoadley. The members of the family who did not apprentice to Hoadley learned their trade from their father, Hiram.

Very few birth and death dates are known for the members of the Twiss family. Early United States census records list Benjamin (1830 and 1840), Joseph (1830 and 1840) and Hiram Twiss (1830). Canadian records show that the Twiss brothers, with the exception of Russell, left Canada well before 1850. Benjamin, who returned to Connecticut, was included in the United States census of 1850. His age was given as fifty-two years, thus placing his birth date around 1798.

Austin Twiss came to Lower Canada in 1821. Austin is listed in the 1825 Montreal census as head of a household of fourteen. For the ensuing years, the Twiss brothers proceeded to travel back and forth to Montreal and to bring with them Connecticut clock movements in at least two parts in order to avoid the high duty on imported "complete" clocks. The movements were then cased in Canada. James A. Dwight of Montreal was associated with Austin Twiss in the initial period of establishing the clock factory. On 10 May 1823 Austin Twiss signed a document before Judge Doucet, renting for a period of nine years, a large building from a tanner named Pascal Persillier Lachapelle. The building was on what was then known as "Lachapelle Lane" on the westerly corner of old Côte des Neiges Road,[18] about 3 miles from the city of Montreal.

In August 1824 Benjamin married Almira Dewey of Champlain, Quebec. Almira died, and by 1850 Benjamin was married to Lucy (b. 1808), and had three children, Lucy aged eleven, Herbert aged six and Bruce, who was aged two. Although Benjamin did travel back and forth between Montreal and Connecticut, the Connecticut censuses of 1830 and 1840 list Benjamin Twiss as a resident of Connecticut.

On 28 September 1825 Austin and his brother Joseph bought, from Joseph Barbeau Boirore, an establishment at Laprairie. In the contract drawn by notary public Lanctôt, Joseph was referred to as an itinerant clock merchant.

FIG. 209

FIG. 210

FIG. 209 *A weight from a Twiss clock.*
FIG. 210 *J. & H. Twiss clock in Upper Canada Village, Morrisburg, Ontario.*

Early in 1826 Austin died, leaving a widow, Vincey Andrews, and two children. In 1828 Ira came to Montreal, married Austin's widow, and became joint guardian to Austin's children. Ira continued the association with his brother Joseph B. until 1830, when they sold the Laprairie shop and began to sell clocks bearing the name I. Twiss.

The clocks with Ira's name alone have some unusual features. Of the clocks examined, about a third use movements purchased from the Riley Whiting factory and are easily identified as such. Another anomaly found only in Ira Twiss clocks concerns the dials. Mr. Francillon noted that the chapter ring of several dials was smaller than that found on other tall case clocks. He believes that the explanation for this is that a tracing from a shelf clock was used. In fact, on one of the dials of an Ira Twiss clock, the seconds bit circle is partially covered by the figures.

On 17 December 1829 a son, Waldo Clinton Twiss, was born to Ira and his wife. Another son was born in Montreal in 1834 and was named Ira; he died in Meriden, Connecticut, on 18 April 1885.

The census of 1831 in Lower Canada records that Ira Twiss, head of the household, had the following occupation: "Fabrique d'horologes de cardes et de corps de chapeaux, mouvements, mus par une cheval et par l'eau alternativement." (Translation: Clock factory, also manufacturing carding machines and hat forms, the machinery driven alternately by horse and water power.)

Ira remained in Canada until 1836, when he returned to Meriden.

FIG. 211 *Movement of the clock shown in Fig. 203. Note repair on second wheel on the time side (right side of picture).*

FIG. 211

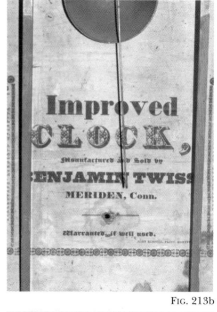

FIG. 212

FIG. 213a

FIG. 213b

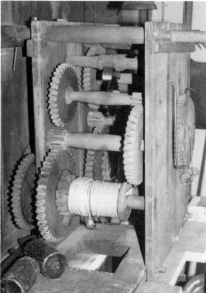

FIG. 214

FIG. 215

FIG. 216

FIG. 212 *Label from J.B. & R. Twiss clock.*
FIG. 213 *Clock and its label by Benjamin Twiss,
Meriden, Connecticut.*
FIG. 214 *Twiss shelf clock label. This clock was
fitted with a "groaner" movement.*
FIG. 215 *Twiss wooden tower clock. Hands
are not original.*
FIG. 216 *Twiss tower clock movement, side view.
The clock resembles a typical 30-hour wood
tall case movement, doubled in size.*
FIG. 217 *Twiss tower clock, rear view of movement.*

FIG. 217

FIG. 218

FIG. 219a

FIG. 218 *Ira Twiss dial.*
FIG. 219a *Dial of a J.B. & R. Twiss*
clock. – Courtesy Collections
d'ethnographie du Québec

He left the clockmaking business, and from 1839 to 1843 he operated a tavern at the corner of Broad and East Main streets. He acquired considerable land and erected sawmills at Twiss Pond. His son Waldo became manager of the mills, which prospered. According to G.E. Burrows, "Waldo and Ira were successful pioneers in bagging wheat flour of their own manufacture. Waldo went on to become engaged in the lumber business, the building of houses and moving of buildings. He also acquired extensive land holdings, including one purchase from William J. Ives, of 23 acres for which he paid $6,000 and by whom he was not required to give any security other than his word."

In 1832, the time that Ira was associated with the Twiss clockmaking establishment in Montreal, his brother Joseph returned to Montreal. Brother Russell must also have been in Montreal at this time, but no records were found to substantiate this. This supposition by the authors is based on the fact that records exist showing that in 1834 Russell was well established in Montreal, in partnership with Joseph. In addition, Montreal records show that Russell was married in Montreal on 5 November 1834 to a "minor and spinster" named Permella Hall from Connecticut. The marriage took place in the American Presbyterian

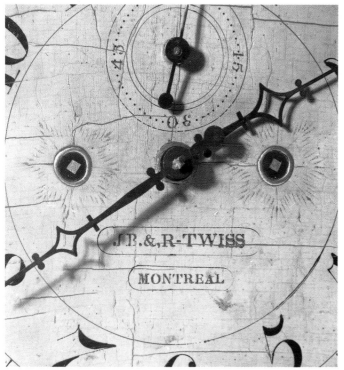

FIG. 219b

FIG. 220

FIG. 219b *Dial detail of a J.B. & R.*
Twiss clock. – Courtesy Collections
d'ethnographie du Québec
FIG. 220 *I. Twiss dial.*

Church in Montreal, with Rev. G.W. Perks officiating and brothers Ira and Joseph B. as witnesses. Their child, Joshua Austin, died on 26 October 1835 at the age of three months. Ira and Joseph B. were among those who attended the funeral. Another son of Russell's, Julius, eventually became secretary-treasurer of the National Savings Bank at New Haven, Connecticut. There was at least one more child in Russell's family – a parent of Carrie Twiss Burgess, who lived in Connecticut in the 1920s.

Because of the Rebellion of 1837 in Canada, the Twiss brothers were compelled to relocate their establishment. Joseph B. chose to dissolve the partnership with Russell and return to Connecticut. One reason for that decision may have been that Joseph B.'s wife, Rebecca G. Hall, died on 18 August of the previous year. Joseph B. and Ira were witnesses at the funeral. Joseph B. was in Quebec in 1850. He entered a fanning mill in the Industrial Fair at Montreal in that year and won second prize: £1.10s. There were two conditions to enter an article in the fair. The article must have been made by the entrant and it must have been made in Canada.

Russell remained in Lower Canada and established a shop in the Parish of Ligouri, County of Montcalm, and it is said (by Julius Twiss) that Russell continued to make clocks. Russell died in 1851 and is buried at Rawdon, Quebec.

It is not known if Robert Walpole Twiss, a lieutenant in the Royal Navy, was a member of the clockmaking family. He was given land in Lower Canada in 1837.

As stated previously, the Twiss clocks bore the initials of the Twiss men who were involved with their manufacture. The following initial or combination of initials are known to exist on clocks: (a) shelf clocks from Meriden, Connecticut: B. Twiss, B. & H. Twiss, J. & H. Twiss, J. & R. Twiss; (b) tall case clocks from Lower Canada: A. Twiss, J.B. & I. Twiss, J.B. & R. Twiss, I. Twiss, J.B. Twiss, J. & H. Twiss, J. & R. Twiss, H. & R. Twiss, R. Twiss.

Pictured in this section are clocks and dials showing several of the above combinations of initials and dial styles.

A question arises concerning the initials J. B. appearing on the Twiss clocks. According to church records, Joseph's middle name began with a B. There is, therefore, the possibility that clocks with the initials J.B. refer to Joseph only, and are not meant to indicate that Benjamin was also involved, in spite of the fact that clocks do exist with the initial J. alone on Twiss clocks.

In a survey of clocks carried out by J. Varkaris in 1978, fifty-seven Twiss tall case clocks were reported. Seventy percent were in private collections and 30 percent of the clocks were in museums and public buildings. Of these fifty-seven Twiss clocks:

– 39% had initials J.B. & R.
– 25% had initial I.
– 14% had initials J. & R.
– 12% were clocks with initials unknown

– 3.5 % had initials J.B.
– 3.5 % had initials J. & H.
– 1.5 % had initial A.
– 1.5 % had initials H. & R.

Very seldom does one know the exact history of a specific clock and when one is discovered, it makes for interesting reading. The story of one such clock was uncovered by the authors during a survey of clocks in 1978. The story is based on documents accompanying a clock that is now in the possession of the museum established by the Upper Thames River Conservation Authority in southwestern Ontario.

In the clock were several letters, one of which was written by Rev. G.A. MacLennan (d. 1935), a long-term resident of Montreal. The "line of ownership" as given by Rev. MacLennan follows:

This clock carries the initials of J.B. & R. Twiss and is said to have been purchased from the makers around 1829 or 1830. It was made by Twiss in a shop at St. Laurent near Montreal.

1. The first possessor was a Frenchman at St. Théreside, Blairville [sic], Quebec. (The Frenchman lived in an old log house where the floor had settled and the clock ran for years out of true.) On his death it was sold to

2. Mr. Osborne of Ste. Therese, Quebec (who relegated it to an old shed as it would not run). On his death, it was sold by his widow to

3. The Rev. G.A. MacLennan, Westmount, Quebec (who restored it to good running order) and gave it to

4. Muriel, daughter of the Rev. MacLennan, a doctor in London, Ontario. Dr. Muriel MacLennan died of cancer of the spine in 1934.

5. The next owner of the clock was Rev. R.J. Bowen of London, Ontario who received the clock in 1935 "in memoriam of his kindness to the late Dr. Muriel MacLennan and her sister, Dr. Helen MacLennan (Mrs. Wm. Nixon)."

6. The clock is now in the collection of the Upper Thames River Conservation Authority in southwestern Ontario.

The information accompanying the clock is in the form of correspondence between MacLennan and Rev. Bowen. Also included among the documents was an excerpt from a letter written in 1923 by Carrie Twiss Burgess, granddaughter of Russell Twiss. She wrote:

I will be only too glad to tell you what I know about the Twiss clocks. All I know is what I have found among Uncle Jule's papers. The Twiss Bros. learned the trade of clockmaking from Silas Hoadley at Bristol, Conn., who was the pioneer clockmaker of this country (U.S.A.) They went to Montreal and it is said "they were the first to manufacture clocks in Canada with the running parts from the United States." The Twiss Bros. carried on business from about 1822 to 1851. The clock business in Montreal lasted only until 1837. After that the business was carried on at St. Liguore [sic] and also at Joliette. The names of the Twiss brothers were Austin, Ira, Benjamin, Joseph and Russell. Russell was my grandfather. They had different names on the clocks. Some have A. Twiss on the face of the clocks. Some, I. Twiss and some, I. J. and B. Twiss and J.B. and R. Twiss and J. and R. Twiss and R. Twiss.

A second tall case clock with wood movement purchased by Rev. MacLennan was given to his daughter, Helen (Mrs. Wm. Nixon). She also had copies of the documents found at the museum, giving the history of the clocks.

FIG. 221

FIG. 222

FIG. 221 *J. & R. Twiss dial.*
FIG. 222 *J.B. Twiss dial.*

B.A. UPSON

B.A. Upson sold clocks in St. Johns (St. Jean), Lower Canada, for a brief period, probably around 1830. Three quite similar examples are known. Styling is typical New England transitional pillar-and-splat. The movements are wood, Terry-type, 30-hour time-and-strike. One of these clocks was reported in the *Bulletin*[19] and another clock has been examined by the authors. Movements in the clocks are designated as type 8-132, using the Taylor system. The maker was either Ephraim Downs or Atkins and Downs. The cases may well have come from the same source.

One interesting feature of the Upson clocks is that the directions on the label are bilingual – English on the left side and French on the right. These clocks have the distinction of possessing the only bilingual labels encountered in the study of clocks in Canada. They were printed by the Vindicator Press, but no address was given for the printer.

No information has been uncovered about Upson himself. Several persons with the name of Upson were active in Connecticut as clockmakers and it is possible that B.A. Upson was a relative.

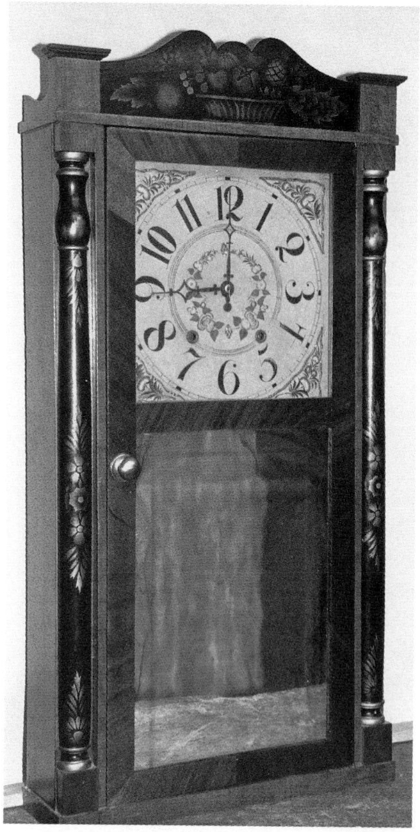

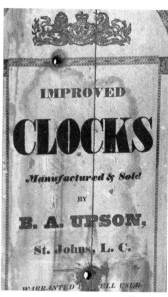

FIG. 223a

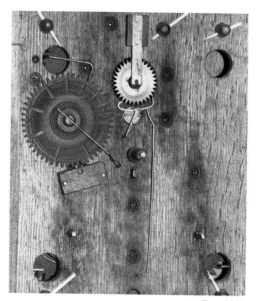

FIG. 223b

FIG. 224

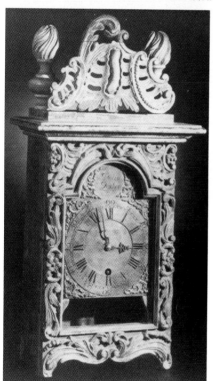

FIG. 225

AVERTISSEMENTS.
FRANÇOIS VALIN, *Horloger*,
FAIT ſavoir au Public, qu'il eſt à préſent réſident dans ſa maiſon à Québec, rue de La Fabrique, qu'il racommode Montres et Horloges de toutes les eſpèces, au plus juſte prix, et avec toute la diligence poſſible.
☞ *Il a auſſi le ſecret de bien faire les Trébuchets, et racommode ceux qui ne ſont pas juſtes.*

FRANCOIS VALIN, *Watch-maker*,
BEGS leave to acquaint the Public, that at preſent he has ſet up at his Houſe in the Upper-town, oppoſite the Jeſuit's College, where he continues his Buſineſs of Watch-making, and repairs Watches and Clocks, &c. at the moſt reaſonable Prices, and with the utmoſt Expedition.
☞ *He alſo underſtands making Money-ſcales and Weights, and rectifys ſuch as are not exact.*

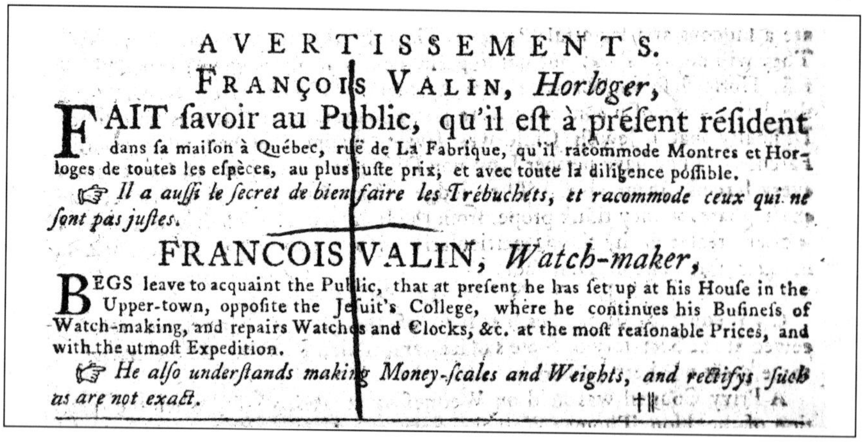

FIG. 226

FIG. 223 *A 30-hour pillar-and-splat clock and movement by B.A. Upson, St. Johns, Lower Canada. The wooden movement is Taylor type 8-132.* FIG. 224 *Label of Upson clock. This is the only bilingual label in this book. English on the left, French on the right.* FIG. 225 *Clock by François Valin in the collection of the Lake St.Louis Historical Society. – From the Canadian Collector, January/February 1977* FIG. 226 *Valin advertisement.*

FRANÇOIS VALIN

François Valin (b. 29 Nov. 1729 – d. about 1784) was a master locksmith, armourer, and clock- and watchmaker in Quebec City. At least two ornate mantel clocks survive that were sold by him about 1750. These clocks are mentioned in *The Early Furniture of French Canada* by Jean Palardy, who in 1965 reported one clock at Baie-St-Paul and another in Boston. It is fortunate that the clock thought to have been in Boston has now returned to Canada and is in the Macdonald Stewart Collection, Montreal. The clock case is pine with a carved front and is believed originally to have been painted with a blue-green paint. The dial is brass and the movement is steel. The numerals and signature on the dial have paint in the indented surface.

Valin's place of business was on Fabrique Street, Upper Town, Quebec City. One advertisement placed in the *Quebec Gazette* appeared in 1772. A second advertisement appeared on 8 June 1775.

The clocks described by Palardy appear to date from the time of New France. Valin continued to be active in Quebec after the arrival of the British in 1759. This is confirmed by advertisements illustrated in Fig. 226. Two later clocks are known and can be seen in Figs. 227 and 228.

Both clocks have typically British movements and, like other clockmakers in Quebec City and Montreal, Valin evidently imported clocks from England and sold them under his own name. The tall case clock in Fig. 227 has a simple pine case, which is English in style but could well have been made locally in Quebec City. The movement, hands, and dial are typical of those of British manufacture during the later period in which Valin worked. The dial is of brass, with an engraved brass chapter ring and matte centre. A strike/silent dial is situated in the centre of the arched top, surmounted by the words, "Valin, Quebec." There is a seconds dial immediately below the figure XII and a small square calendar aperture above figure VI. The dial has elaborate scroll work, spandrels, and decorations in the arch.

The bracket clock in Fig. 228 follows the English styling of the third and fourth quarters of the 18th century. The 8-day repeater movement has a verge escapement and a bob pendulum. The back plate, as illustrated in Fig. 229, is finely engraved, including an oriental pagoda motif. The case is mahogany, possessing brass feet and arched glass in the door. The top of the case is domed, equipped with a handle, and fitted with four brass finials. The brass dial has many similarities to the tall case dial described previously. It, too, has an arched top with strike/silent indicator centred in the arch. The dial has an engraved brass chapter ring and matte centre. No seconds dial is present; instead, under the figure XII, there is a small inset plate engraved, "Valin Quebec." There is also a calendar aperture above the figure VI. Again, there are elaborate spandrels and decorations in the arch.

François Valin was the son of François Valin (b. 1693) and Genevieve Trudel (b. 1707). He married Marie-Louise Bonhomme (b. 1732) on 11 January 1751. Their children were Marie-Louise (b. 18 Mar. 1753), François (b. 4 May 1755 – d. 10 Aug. 1760), Pierre (b. 18 Feb. 1758), and Marie-Félicité (b. 23 Dec. 1761). Apparently François Valin, the clockmaker, died in 1784, as his widow advertised that the house on Fabrique Street was to let.

FIG. 227 *Valin tall case clock.*

FIG. 227a

FIG. 227b

137

EDWARD WADE

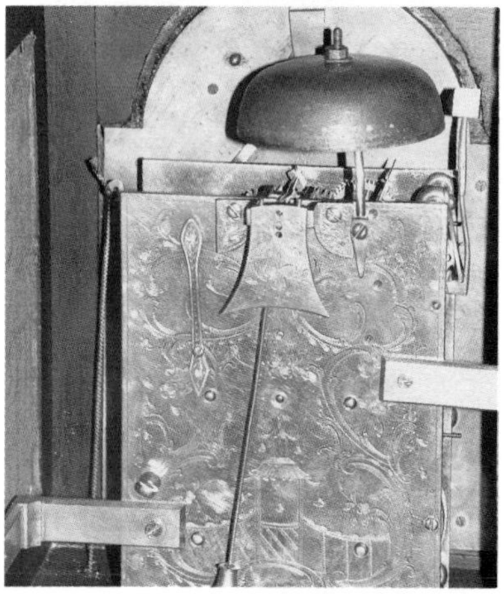

Edward Wade (b. 1800 – d. 28 March 1847) was a clock- and watch-maker who operated a shop in Upper Town, Quebec City. The Quebec City Directory first records a jewellery, clock, and watch establishment in his name in 1826 at 1 Buade Street. In 1844 the address of his shop was 14 Mountain Street. Several tall case clocks are known that display his name on their dials. One such clock is in a private collection in the Province of Quebec. The clock movement and dial appear to be English in origin.

Edward Wade's father, Francis (b. 1758 – d. 4 Feb. 1845), was a sergeant in His Majesty's 49th Regiment of Foot Guards. His mother's name was Mary. Edward had a brother, Francis, who died as a child and a sister Anne (b. 8 Mar. 1805 – d. after 1847). The family belonged to St. Andrew's Presbyterian Church.

Edward Wade married Jane Honstein, daughter of Marie and John Honstein, on 3 January 1835. There were five children: Mary Anne, who died as a baby, Francis Edward (b. 30 June 1837), Ellinor Henrietta (b. 2 Aug. 1840), John Augustus (b. 29 Aug. 1842), and William Brown (b. 5 Feb. 1845).

NELSON WALKER

Two clocks bearing the inscription, "N. Walker London and Montreal," have been seen by one of the authors and are being recorded here as a matter of interest. Unfortunately, little information and no illustrations are available.

The clocks were both mahogany bracket clocks, decorated with cast brass feet and pineapple-like finials. One clock had an engraved brass dial, while the other had a plain painted dial. The movements were typically English of the period, 8-day time-and-strike, with verge escapements. The back plates of both movements were copiously engraved, including the above inscription. The inscription was also on the dial.

Records exist for a Nelson Walker in the Montreal Directory of 1842–43, and his silversmith's mark is illustrated in *Canadian Silversmiths 1700–1900*, by John E. Langdon. The clocks mentioned above, with verge escapements and engraved back plates are typical of English clocks of an earlier period (up to about 1800). The authors have not determined whether Walker was selling "old stock" or if he was in business for a long time. It is possible also that two Nelson Walkers were active.

DAVID WEST

David West's name appears only in the 1844–45 and the 1847–48 Quebec City directories. He was listed as a watchmaker at 27 St. John Street. D. West won the prize of £2 for his skeleton clock at an industrial fair that was held in Montreal in 1850. In order to exhibit in the fair, two requirements had to be met: the article exhibited must have been made by the entrant and the article must have been made in Canada. Nothing further is known about this man.

FIG. 228 *François Valin bracket clock with typical British styling.*
FIG. 229 *Movement of Valin clock with crown wheel escapement and elegant engraving.*
FIG. 230 *"Edward Wade" clock in a private collection.*

FIG. 228

FIG. 229

FIG. 230

JOHN WOOD

John Wood (b. 1793 – d. 3 Feb. 1872) served Montreal as a watch- and clockmaker for over thirty years. A number of clocks exist with his name on the dial. One of these clocks, a round office clock circa 1850, is pictured on the cover of the book *The Days of John Wood: Watchmaker* by William A. Wood. Also in the book may be found pictures of John Wood and his family.

John was apprenticed in London, England, to William Isaac Hinton of Northampton Row. Following his apprenticeship, as required by the Clockmakers' Company, he became a journeyman for two years under John Barwise of St. Martin's Lane.

In 1832 he and his wife, Anna Wentworth (b. 1789 – d. 1864), and five children came to Montreal on the maiden voyage of *Africaine,* a barque of 316 tons. John leased space in a shop owned by a tinsmith on St. Lawrence Street. Suffering from smoke from the wood fires needed to keep the small shop warm, John Wood left Montreal to farm but, after discovering the hardships of clearing a farm, he and his family returned to Montreal in 1835.

John entered the employ of George Savage on St. Paul Street at the corner of St. Dizier Lane. In 1837, when George Savage opened a shop on Notre Dame Street two doors west of St. Gabriel Street, John Wood became manager. A further move to a larger store at the corner of St. Gabriel Street followed.

In 1839 John Wood went into business for himself at 130 St. Paul Street near St. Jean Baptiste Street and George Savage continued to use his services. At various times John's three sons helped in the business: Charles (b. 1817 – d. 1892), Peter (b. 1826 – d. 1907), and John (b. 1828 – d. 1905). Charles was the "son" in the name of the firm, "John Wood & Son."

In 1845 a branch of the firm was opened at 17 McGill Street near St. Ann's Market and Charles was put in charge while Peter worked with his father in the St. Paul Street store. John Jr. left the business to study for the ministry.

John Wood acquired David Savage's business at 137 Notre Dame Street[20] in March of 1848. David Savage was deeply in debt and declared bankruptcy. John Wood moved to the larger Notre Dame premises and Peter, his apprenticeship over, assumed more responsibility. In a slowing economy, depressed sales in both shops caused the McGill Street shop to be closed and Charles and his family left Montreal.

Business improved in 1850, and John made one of his frequent trips to England to purchase merchandise. However, England was not the only source of clocks. A letter from John to his son dated 4 August 1861 from New York stated that he "bought rather largely of clocks," revealing that U.S.A. was a source of clocks sold by John Wood and confirming the continuing prosperity of his business.

In the early 1860s the city of Montreal continued to appropriate and demolish old wooden buildings in order to modernize the city and reduce the chance of fire. John Wood was forced to relocate his business to 375 Notre Dame Street near St. John Street. At this time his family moved from above the shop to 44 St. Antoine Street.

In 1864 Peter left the business and John Wood took as partner Mr. Thomas Allan, giving him a third of the profit from the business. The name of the firm changed to "John Wood & Co." The following year it became Wood and Allan.

Among the clocks sold by the firm in 1866 was a clock for the new Molsons' Bank. John Wood also supplied a clock with illuminated dial for the Grand Trunk Railway Depot. At this time he was offered the job of keeping all the railway clocks in repair throughout the whole line to Detroit. However, because of his age, he felt that this undertaking would be too strenuous.

In 1869 Thomas Allan wished to buy out John Wood, who, on the advice of his sons, declined. Following the dissolution of the partnership, Thomas Allan established a business at 375 Notre Dame Street and "John Wood & Son" reappeared at the address of 325 Notre Dame Street. Peter's son, Wentworth, now helped in the shop allowing John to work fewer hours.

John Wood became ill in January 1872 and passed away on 3 February 1872.

FIG. 231a

FIG. 231b

TURNER L. ABEL

Turner Lillie Abel was a peddler of American clocks in the United Counties of Leeds and Grenville in the 19th century. Turner ordered a "lot" of clocks from the Seth Thomas Company of Connecticut, possibly at the suggestion of a close neighbour, S.J. Southworth, who was a seller of clocks for a period of twenty years. As only a few clocks are known to exist with the name T.L. Abel on the label, the authors have concluded that Turner sold clocks for a very short period of time. The OG clock pictured in Fig. 231 has a Seth Thomas movement type 1.241 as designated by Dr. Snowden Taylor and is stamped "S. Thomas, Plymouth, USA."

Several Empire-style column-and-cornice 8-day clocks are known by the authors. A clock and a label from an 8-day clock are shown.

Although no records exist that would indicate the exact period of time when Abel was peddling his clocks, from information available it would be reasonable to assume that it was between 1848 when he was eighteen years of age and 1851 when the census listed him as a shoemaker. With the exception of several short periods of absence between 1853 and 1855, a farm near Plumb Hollow, Bastard Township, Leeds County, continued to be his home.

Turner Abel was the fifth child of eight children of Lyman Abel (b. 1793 – d. 1869)[1] who came to Leeds County around 1802 from Connecticut. Lyman, of German ancestry, purchased all 200 acres of Lot 7, Concession VI, of Bastard Township in 1818, and in 1820 married Sally Miller (b. 1806 – d. 19 Apr. 1878).

In the mid-1850s Turner moved to Hallowell Township, Prince Edward County, where he married Adelaide (b. about 1837). Adelaide and Turner had two children: Peter, born in 1859, and Linsay (or Linley), born in 1861. In 1864 the family was living in Wellington, Prince Edward County, where Turner was listed as a sewing machine agent.

The census of 1871 does not have Turner and Adelaide listed in either Prince Edward County or Leeds County. However, their children were listed in Leeds County and were living with aunts and their grandmother. Therefore, it is assumed that Turner and his wife left Ontario for Riverside, California, at this time, to be joined later by their children.

In February 1885 a notice to the *Brockville Recorder,* sent by S.S. Southworth of Sacramento, California, announced the death of Turner Abel on 13 February 1885 as a result of typhoid fever.

The Abel family were active members of the Society of Friends (Quakers).

FIG. 231 *OG clock and label made for Turner L. Abel.*

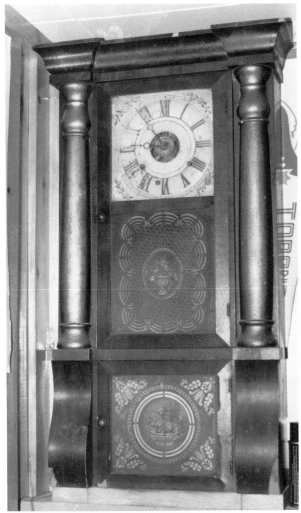

FIG. 232b

FIG. 232a

FIG. 233

AMERICAN WATCH CASE COMPANY
OF TORONTO

The American Watch Case Company of Toronto manufactured gold, gold-filled, and silver watch cases in Toronto. The company began business in 1885 on Adelaide Street West, moving in 1893 to occupy a large factory at 509–513 King Street West. The president of the company, John N. Lake, and the secretary treasurer, W.K. McNaught, were Canadians. The manager, R.J. Quigley, was born in New York. The basement of the building was used to melt gold and silver. On the ground floor were offices and machine shops. The top floor was for plating. The machinery was operated by electric motors and the factory employed 120 persons, making it the larger of the two watch case companies in Canada in the 1890s.

The building that housed the American Watch Case Company still exists, but is unoccupied at the time of writing.

FIG. 234

FIG. 232 *Column-and-cornice clock. Label from an 8 day clock by Turner Abel.*
FIG. 233 *The American Watch Case Company of Toronto. From* Toronto Illustrated.
FIG. 234 *This watch case label, printed in red, is slightly over one inch in diameter.*

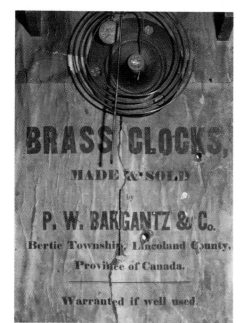

FIG. 235

P.W. BARGANTZ & CO.

One clock bearing the name of P.W. Bargantz and Co. has been reported to the authors. Unfortunately, no other information on the person has been found. The label declares an address, "Bertie Township, Lincoland County, Province of Canada." The clock can be dated to the 1841–1849 period. The term Province of Canada came into use in 1841. In 1849 Bertie Township became part of Welland County. The term "Lincoland" on the label appears to be an incorrect spelling of "Lincoln." Lincoln was one of the original counties of Upper Canada and, in turn, was part of the larger Home District in 1792 and the Niagara District after 1800. In 1849 the division into districts was discontinued and Bertie Township became part of the newly formed Welland County.

The clock itself has a typical veneered 30-hour OG case. The movement is of brass and, curiously enough, is spring-driven. The metal dial is attached to wooden side pieces, typical of a weight-driven clock. The label states that the clock was "made and sold" by Bargantz, but it has been pasted over another larger label. The clock remains somewhat of a mystery.

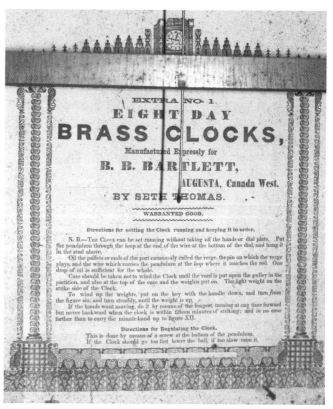

FIG. 236b

FIG. 235 *Label from a 30-hour OG, sold by P.W. Bargantz & Co. "Lincoland," which should read Lincoln County.*
FIG. 236 *Clock and label, sold by B.B. Bartlett.*

FIG. 236a

B.B. BARTLETT

Benjamin Buell Bartlett was a peddler of clocks in the United Counties of Leeds and Grenville in the middle of the 19th century. Clocks with his name on the label were made by the Seth Thomas Clock Company, Plymouth, Connecticut. Two styles of clocks have been found. One style is the half-column, brass-movement 30-hour weight clock. The other style is the Empire column-and-cornice 8-day brass weight clock. The movement in this latter clock is Taylor type 4.211 and is stamped "Seth Thomas Plymouth Hollow U.S.A." Because only a few clocks with the Bartlett name have been found, it has been assumed that Bartlett sold clocks for a relatively short period of time.

Benjamin B. Bartlett (b. 1833 – d. after 1893) was one of four children of Arvin Bartlett (b. 3 Nov. 1800 – d. after 1881), who was of Welsh extraction and came from Vermont about 1812 with his family. About 1830 he settled on Concession VIII on half of Lot 19, Augusta Township, Grenville County. The other half of the lot was owned by Arvin's father-in-law, Samuel Throop. The Bartletts were Wesleyan Methodist and the Throops were Baptist.

Benjamin's mother, Lucinda (Lucy Ann) Throop (b. 21 July 1804 – d. 7 Sept. 1876), was the second of seven children of Samuel Throop (b. 12 Aug. 1776 – d. 7 Aug. 1864) and Abigail Blakesly (b. 1779 – d. 3 Aug. 1843). Samuel, born in Litchfield, Connecticut, came to Canada about 1800. Through the Throop family, Lucinda was a direct descendant of Richard Warren (1580–1625), who came to America on the *Mayflower*. The Throop (Scroop) family can also trace their ancestors directly from William the Conqueror (r. 1066–1087). There are several publications that trace the genealogy of the Throop family and give incidents in the lives of the early family members.

Benjamin B. Bartlett appears to have sold clocks between 1848 and 1850. By 1851 he was listed as a farmer in the census records and lived with his parents and brothers in a log house in Augusta Township. By 1860 Benjamin and his wife, Mary (b. about 1833), had moved to Hastings County and were living on the east side of King Street, Shannonville, Tyendinaga Township. A child named Francis was born in 1860. Also living with them was a person called Almirin Bartlett (b. about 1840). The relationship of this man to B.B. Bartlett is unknown. The 1871 census records show that Almirin, a Wesleyan Methodist, had married Elizabeth (b. about 1845) and there were two children: Jessie, three years of age and Laura, nine months of age. Almirin was an employee of the Grand Trunk Railroad.

Benjamin Bartlett was the station agent for the Grand Trunk Railroad in the early 1860s. The company closed the station by 1868 and the building was turned into an inn run by Reuben Davis. It is not known where Benjamin went when the station was closed.

During the ten-year period between 1860 and 1870, Benjamin Bartlett was involved in a number of land transactions. The land records show him as a resident of both Belleville and Tyendinaga Township during this time. Several transactions involved Benjamin in partnership with a man called William H. Dake from Augusta Township. He also had dealings with his parents and grandfather.

FIG. 237

FIG. 238

FIG. 237 *Half-column clock sold by B.B. Bartlett with typical Seth Thomas 30-hour styling.*

FIG. 238 *The label from B.B. Bartlett clock in Fig. 237. There are two Bartlett label variants and from examination of other Seth Thomas clocks, this one is the later printing.*

EDWARD BEETON

Edward A. Beeton (b. about 1861 – d. 20 Oct. 1943) was the co-founder of the Canadian Horological Institute of Toronto, Ontario (see p. 156). In addition to his own watch repair business and, for a few years, his own watch company, he was on the staff of *The Trader* (now called *The Canadian Jeweller*), a journal that serves the jewellery and watch-repair trade. He was considered by his peers to be "one of Canada's greatest watchmakers."

About 1875 Edward Beeton was apprenticed to the firm of Fowler and Company, St. Catharines, where he probably trained under James B. Fowler. His skill as a watchmaker was honed through many years of study and intensive application to the science of horology.

Upon completion of his indenture, he went to Toronto to work as a watchmaker for P.W. Ellis and Company, manufacturing jewellers at 31 King Street East. The next year, he was employed by the Kent Brothers (Ambrose and Benjamin) at 168 Yonge Street, Toronto, where by 1886 he had risen to the head of the technical staff. From May 1886 to September 1887, H.R. Playtner, his future partner at the Institute, was employed by Kent Brothers and Company and worked under Edward Beeton. Playtner grew to respect Beeton and asked for the difficult work in order to learn as much as possible from this highly skilled man.

In 1887 Edward Beeton began his long association with *The Trader* and was placed in charge of the watch-repair section of the paper. By 1896 he had risen to technical editor, a post he held for a number of years.

Toward the end of 1889, Beeton left the employ of Kent Brothers and established his own business, first on Adelaide Street and then at 25 Leader Lane. The Leader Lane shop was the first place in Toronto to be lit with electric lights powered by a new system of storage batteries on the premises. By June 1890 Beeton took H.R. Playtner as his partner at Leader Lane.

In that same year Alexander Moffat, a well-known jeweller in Port Elgin, Ontario, was urging Edward Beeton to fulfill Moffat's dream of forming a horological school. Beeton was determined that this idea would become a reality. He was interested in improving the future of watch-repair men through establishing higher standards of performance. This would allow them to enjoy a higher standard of living through higher wages. To this end, Beeton had previously served as secretary of the Jewellers' Security Alliance.

Both Beeton and Playtner wrote letters to the editor of *The Trader* concerning the need for a horological school. The letters were published in the April 1890 issue of the periodical. Both men saw the establishment of a watchmakers' school as the solution to educating the many incompetent repair men. Playtner envisioned it as a place to "teach the pupil how to design and build a watch from one end to the other." Beeton was determined to teach repair men not to "botch." He was concerned with the large number of watches that were mutilated and damaged while undergoing repairs. In his long letter, he outlined in great detail the equipment, system of instruction, and curriculum. To him, the making of a watch was a learning experience. The student would then be able to make broken and missing parts when repairing watches. Theoretical lectures accompanying practical workshops would teach not only the "how" but the "why." A student would earn one of several levels of diplomas, depending on his level of accomplishment. Beeton hoped that wages for the graduate would rise from the level of $10 to $20 per week.

With Playtner as a partner, Beeton proceeded with his plans for the

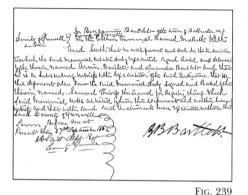

FIG. 239

FIG. 239 *Memorial signed by B.B. Bartlett.*

school, which opened as the Canadian Horological Institute at 133 King Street East in Toronto. In June 1890 the school had two pupils, Mr. J. Kincaid of Chicago and Mr. A. Zilliax of Listowel, Ontario.

However, on 11 August 1890, Beeton left the school, presumably due to ill health. Playtner continued to run the school in Canada until 1913, and it is of interest to note that the instruction that was offered continued to follow very closely the outline proposed in Beeton's letter in *The Trader.* It was also during the short time of Beeton's association with the school that the large models of escapements were built. They continued to be used for demonstration purposes throughout the duration of the school in Canada, and when Playtner taught at the Elgin Watchmakers' College, Chicago, he took them with him.

After leaving the school, Edward Beeton continued his business of watch and clock repairs at various locations, finally becoming established at 47 Adelaide Street. In 1903 he became the travelling representative "on a goodwill and technical mission for the firm" of the Elgin National Watch Company. In 1904 he opened the first Canadian office for that company to develop the Elgin business in Canada.

In 1912 Edward Beeton was manager of the Beeton Watch Company Limited. His sons Edward W. and Frank Ross were working for him in this company, which was first at 67 Yonge Street (Traders Bank Building) and then at 68 Yonge Street, ground floor of the Dominion Bank Building. In 1914 F. Ross Beeton went to work for Edmund Scheuer, moving by 1920 to Kent Brothers, where his father had worked nearly forty years before. Edward W. also left the Beeton Watch Company, which apparently closed around 1915.

After several years away from Toronto, Ed. Beeton opened a watch repair business at 25 Adelaide Street West, where he remained until he retired in 1933. He could then pursue his lifetime hobby of whist. He was considered to be one of Canada's leading whist players.

According to *The Trader and Canadian Jeweller,* "at the bench Edward Beeton had no peers. (He was) a great reader of horological matters and gave his best for the Canadian watch trade." However, it was also mentioned that "it required a great deal of understanding to reach the inner quality of the man. A great technician and horologist, he had little time for what he considered stupidity or carelessness in watch work. While a great craftsman, he lacked the common tact which wins friends and influences people. As it happens this never mattered to Mr. Beeton."

Edward Beeton was the son of Joseph E. Beeton (b. about 1828 – d. 1903), a druggist born in England, who came to the United States and married Mary Ann (b. about 1840). Edward A. Beeton was one of two children born in the United States. About 1868 Joseph and his family came to St. Catharines to join Joseph's brother William B. Beeton, and the brothers formed Beeton and Company, druggists. When William died about 1883, Joseph continued as a druggist with a new partner. Joseph lived with Edward and his family for the last ten years of his life.

By 1880 Edward had married Lily (b. 1860 – d. about 1925). Six children were born: Maud ("May," b. 1882), Edith (b. 1884), Nellie (b. 1886), Frank Ross (b. 1889), Edward W. (b. Nov. 1890), and John E. (in active service in 1943). During their years in Toronto, Edward A. Beeton and his family moved over seventeen times.

Edward Beeton died in Toronto on 20 October 1943 after a short illness and was buried in Orillia, Ontario. The Beeton family belonged to the Church of England.

FIG. 240

FIG. 240 *Edward Beeton. From* The Trader and Canadian Jeweller.
  – Courtesy the National Library of Canada

A.H. BROWN

Albert Henry Brown (b. 1826 – d. 31 Dec. 1894) was a clock peddler who lived all his life in Elizabethtown Township, Leeds County. He was born on a farm near Lyn, which was then known as Coleman's Corners. Members of the Brown family have lived in this area for over two hundred years.

A.H. Brown purchased clocks from the Seth Thomas Company, Plymouth, Connecticut, and his name appears on the labels of the clocks. Clocks with "A.H. Brown, Leeds County, Canada West" on the labels are relatively rare,[2] and it is probable that he sold clocks for a fairly short time. Elihu Geer, the printer of the label used on at least one clock, was at 16 State Street, Hartford, Connecticut, from 1856 to 1887. Therefore, A.H. Brown would have sold clocks sometime between 1856 and 1867, when Canada West became Ontario.

The clocks known to have been sold by Brown are weight clocks. Two styles have been identified. One 8-day clock is column-and-cornice style and is fitted with a lyre movement, Taylor type 4.211 (8-day brass weight). Several 30-hour half-column-style clocks are known. One of these clocks has an overpasted label and is fitted with an alarm.

Although Brown may not have sold many clocks, he must have had a varied inventory as he considered himself to be a peddler for at least twenty-five years. There were many periods when A.H. Brown was away from home, because his name appeared on the lists of names published by the *Brockville Recorder* of those people who had letters awaiting them at the post office. His name also appeared on lists of people granted peddlers' licences from at least 1854 until 1877. By that time, according to the 1876 *Ottawa and Central Canada Directory,* his family had left the family farm at Lyn and were living on Pearl Street in Brockville.

FIG. 241a

FIG. 241 *Column-and-cornice clock and label sold by A.H. Brown.*

FIG. 241b

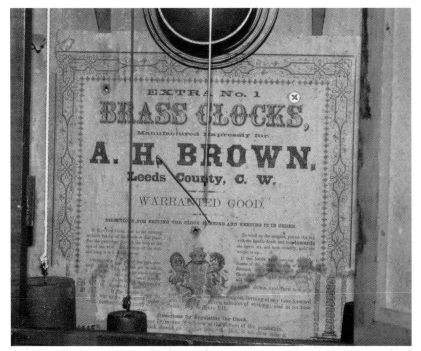

FIG. 242

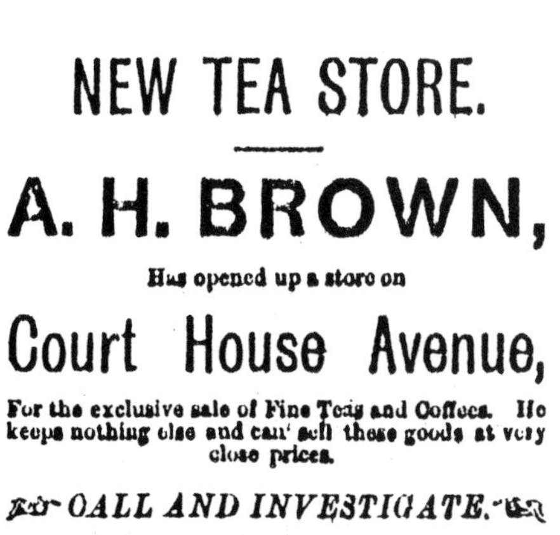

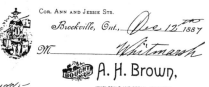

NEW TEA STORE.

A. H. BROWN,

Has opened up a store on

Court House Avenue,

For the exclusive sale of Fine Teas and Coffees. He keeps nothing else and can sell these goods at very close prices.

CALL AND INVESTIGATE.

FIG. 243

COR. ANN AND JESSIE STS.

Brockville, Ont., Dec 12th 1887

M Whitmarsh

A. H. Brown,

1885

P. O. Box 63.

WHOLESALE AND RETAIL DEALER IN

CHOICE TEAS.

FIG. 244

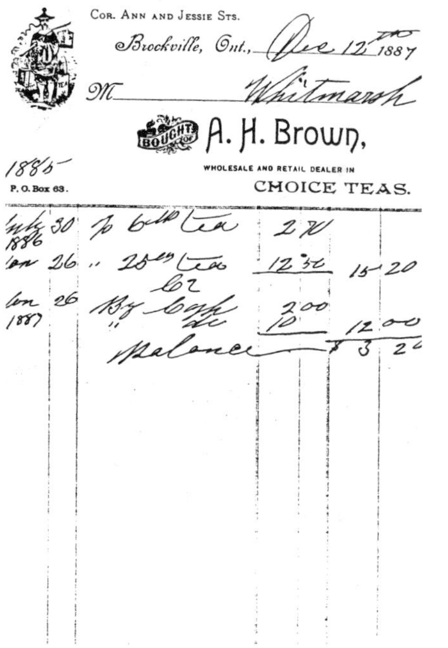

By 1881 A.H. Brown had begun a new venture in Brockville, opening a store on Court House Avenue for the exclusive sale of teas. A new store was opened on Court House Avenue in 1884. His shop was still in business in 1891. At that time, A.H. Brown and family lived on the corner of Ann and Jessie streets in Brockville. Pictured in Figs. 243 and 244 are documents from this period in the career of A.H. Brown.

Albert H. Brown was a descendant of United Empire Loyalists. His grandfather, Abraham, who fought in the Loyal Rangers under Major Jessop during the American Revolution, came to "Canada" with his wife and four children, landing at Chaleur Bay, Lower Canada. By 1792 the family had moved to Upper Canada where they were granted land. In 1798 one of Abraham's sons, Edward (b. 1777), received land, as the son of a Loyalist. In 1802 A.H. Brown's father, Benjamin VanAmber Brown (b. 1781 – d. Feb. 1862), was also given land. A brother of A.H. Brown, Benjamin Coleman Brown (b. 28 Jan. 1821 – d. 11 June 1904), had four children, the eldest of whom, Matthew Munsell (b. 1870 – d. July 1951), was a well-known attorney in Brockville.

A.H. Brown married Ellen West (b. 1836) of New Dublin, Upper Canada. There were seven children, five of whom reached adulthood: Ida Alicia, Mrs. Joseph Lane (b. 21 June 1858 – d. 14 Feb. 1944), Fred (b. 1863), Frank Albert (b. 1864 – d. 10 Mar. 1944), Jessie, Mrs. W.P. Burrows (b. 1872), and Charles L.H. (b. 1874 – d. 21 Nov. 1903).

Mr. Brown was a devout Methodist and often boarded the Methodist conference members at his home in Brockville. The night he died he suffered a heart attack on his way to attend the watch service at George Street Methodist Church. He was taken home and died in the arms of his son. He is buried in Lyn cemetery, the oldest Ontario cemetery still in use.

After the death of her husband, Ellen moved to the village of Rothsay in Wellington County.

FIG. 242 *Label of 30-hour half-column clock sold by A.H. Brown. Note extra weight for alarm mechanism.*
FIG. 243 *Advertisement for Brown's new tea store.* Brockville Recorder, *in spring of 1884.*
FIG. 244 *Bill of sale.*
– Courtesy Glenn Lockwood

FIG. 245

FIG. 246

FIG. 245 *H. & C. Burr clock.*
– Courtesy the Dundas
Historical Society Museum
FIG. 246 *Label detail from a clock*
by H. & C. Burr.

HORACE AND CHARLES BURR

Horace Burr and his brother, Charles, sold clocks at Dundas, Upper Canada in the 1830–1836 period. They seem to have enjoyed a good measure of success, since clocks bearing their names have survived in considerable numbers. Thanks to extensive research carried out about Dundas and its inhabitants by the Dundas Historical Society, a series of books entitled *History of the Town of Dundas* has been compiled by T. Roy Woodhouse. This reference confirms that Horace and Charles Burr arrived in Dundas, perhaps by way of nearby Ancaster, in 1830. Upon their arrival the Burrs opened Dundas's second jewellery store[3] and a clock-making establishment. Both Burrs had left Dundas by 1836 and their establishment was not among those listed in 1837.

Coincidentally, another clockmaker, Jonathan Burr, was active during the same period in Lexington, Massachusetts, alone and in partnership with one Austin Chittenden. The authors have been intrigued by the possibility that the three Burrs may have been related. Proving such a relationship has been made difficult because the same Christian names have been used in several Burr families.

Through the good offices of the Connecticut Historical Society, an extensive genealogy of the Burr families has been obtained. The Society has suggested that two sons of Captain George Burr of Hartford, Connecticut, namely Horace (b. 26 Mar. 1781 – d. 2 Oct. 1863) and Charles (b. 14 Feb. 1786 – d. 12 Aug. 1851), may have left Hartford to come to Dundas, U.C., in 1830. They were back in Hartford in 1836. Horace was for many years a cashier at a bank and Charles was a printer for the *Hartford Courant*.

A more probable identification has been pointed out in this genealogy by Dr. Snowden Taylor, research committee chairman of the NAWCC. He refers to three sons of Elijah Burr of Worthington, Massachusetts, namely Horace (b. 6 Dec. 1792 – d. 31 July 1853), Jonathan (b. 6 Mar. 1794 – d. about 1869), and Charles (b. 19 Nov. 1804). Horace moved to Michigan City, Indiana, in the 1830s where he married and raised a family. He became a dealer in western lands and left a large estate.

Jonathan moved to Chicago supposedly in 1834 and also amassed a large estate. Charles married in Worthington, Massachusetts, in 1830, raised a large family, and moved to Wameon, Ohio in the mid-1830s. The genealogy does not list clockmaking as a trade of any of these men from either family. This is not unusual, however, since genealogy is more concerned with lineage than occupation.

Both G.E. Burrows[4] and Dr. Taylor have noted that the clocks of Jonathan Burr are similar in certain respects to those of Horace and Charles. Both firms assembled clocks, but did not manufacture movements, although these often came from a common source. The printed instructions on the labels were identical and the two firms operated in the same time period.

Burr clocks in Canada fall into two distinct categories. Some are labelled "Made and Sold by H. & C. Burr Dundas U.C." and others, "Manufactured and Sold by Horace Burr Dundas U.C." The authors believe that the "H. & C." clocks were produced first and that Charles returned to the U.S.A. some years before Horace, who continued the business alone. This is confirmed by the activity periods of the printing firms that prepared the various labels. A majority of the surviving clocks bear the "Horace" labels.

148

Most known clocks by Horace and Charles Burr are 30-hour time-and-strike, wooden-movement weight clocks in typical pillar-and-splat cases. As has been noted elsewhere in this book, two styles exist for these cases. In one style, the door forms the complete front of the case and half columns are attached to it on either side of the glass; hinges may be placed on either side of the door, but are commonly on the left. In the other style, the columns are attached to the case itself on either side of a somewhat smaller door, which is usually hinged on the right. The clocks of H. & C. Burr are in cases of the first style and those of Horace are in the second. It is thought that the Burr clock cases were made locally, perhaps at one of the three local furniture shops that were in business in Dundas at the time. It has been observed that H. & C. Burr cases were made very substantially. The sides of the cases, for instance, are 1⅛ inches thick. On many of these clocks the strips of wood framing the door do not have mitred corners.

It has also been observed in a few clocks by H. & C. Burr that the wooden movements have simply been attached to the backboard of the case by four screw nails that pass through holes in the rear plate. This is a rather crude technique that has occasionally been used by a few other makers. The majority of Burr clocks have their movements mounted conventionally between vertical mounting boards, as can be seen in Fig. 247.

It has been noted by Dr. Snowden Taylor and by G. Edmond Burrows that identical movements (Taylor type 8.136) can be found in clocks by Jonathan Burr of Lexington, Massachusetts, as well as in H. & C. Burr clocks of Dundas. These movements were used by a large number of clockmakers and the maker has not been identified. A similar movement has been found in a clock by Horace Burr. A second movement (Taylor type 9.223) produced by one of the Chauncey Boardman firms was also used by Horace Burr.

Two different label printings are found on clocks of Horace Burr (see Fig. 250). Both bear the name of the printer, G.H. Hackstaff. The activities of George Hackstaff are well documented, as noted below, and have been useful in clarifying the story of the Burr brothers.

Almost all of the Horace Burr clocks have 30-hour wooden movements housed in pillar-and-splat cases. However, at least one 8-day clock exists. This clock is fitted with a strap-brass movement made by one of the Bartholomew firms – probably Barnes, Bartholomew & Company, which was in business from 1833 to 1836. According to Dr. Snowden Taylor, "the verge pin area has been replaced by one from a Seth Thomas clock, but the 'shadow' of the original is clearly seen beneath it" (see Fig. 253).

The case of the clock is made of maple and it stands 40 inches high on four feet. The clock is 21½ inches wide. It has two doors on the front and, unlike other Burr clocks, the top of the clock has a cornice. Half columns are attached to the case on either side of the doors. It is believed that the case was made locally. The label of the clock is identical to Fig. 250a.

Labels in the H. & C. Burr clocks were printed by "Smith's Print" in Hamilton. Those in the clocks of Horace Burr state "Printed at the Office

FIG. 247

FIG. 248

FIG. 247 *Terry-type movement of an H. & C. Burr clock.*
FIG. 248 *Clock sold by Horace Burr.*
  – Courtesy the Dundas Historical Society Museum

FIG. 249

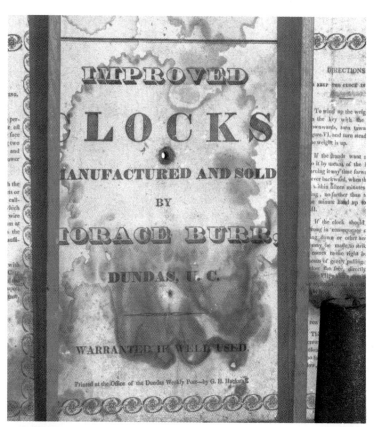

FIG. 250b

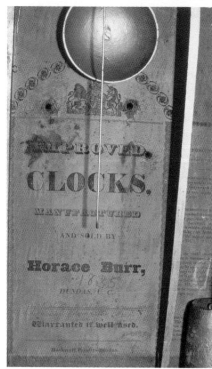

FIG. 250a

FIG. 249 *Horace Burr clock.*
FIG. 250 *Two styles of labels found in*
Horace Burr clocks. Both labels were printed
by G.H. Hackstaff, Dundas, Ontario.
FIG. 251 *Movement of a*
Horace Burr clock.

FIG. 251

of the Dundas Weekly Post – by G.H. Hackstaff." Smith was in business alone at first, then Hackstaff, who came to Canada in 1828, became a partner (in 1832). Hackstaff moved the business to Dundas in 1834. This provides confirmation that the H. & C. Burr clocks precede those of Horace Burr.

The authors have uncovered several interesting incidents in the life of George Hackstaff during this period. In Hamilton, he printed one of Hamilton's earliest books, entitled *New Guide to Health or Botana Family Physician*. In Dundas on 22 April 1834 the first edition of the *Dundas Weekly Post* was published. In August 1835 Hackstaff apologized for the lack of an editorial in his newspaper because of sickness in the family. The following week his daughter, Charlotte, died at the age of one year and twelve days. In December 1835 a paper shortage caused Hackstaff to reduce the size of his newspaper. He made no charge for the smaller paper. In July 1836 the *Dundas Weekly Post* published its last known copy. The paper on which the newspaper was printed (and no doubt the clock labels as well) was a product of the pioneer mill owned by James Crooks.

During his stay in Dundas, George Hackstaff had the distinction of putting into print the only hardcover book ever published in Dundas. The book, written by Dr. Thomas Rolph, bears the incredibly long title: *A Brief Account Together with Observations Made During a Visit to the West Indies and a Tour Through the United States in Parts of the Years 1832–33, Together with a Statistical Account of Upper Canada.*

FIG. 252

The book was printed on paper made in Crooks Hollow near Greensville. There are copies of the book in both the Dundas Museum and the Dundas Public Library.

In 1836 Hackstaff went to Toronto and started the *Toronto Herald*. He was also associated with newspapers in London, Ontario, and Buffalo, New York. In 1846 he went to Niagara, Ontario. George Hackstaff died on 25 March 1858 in his forty-second year.

Dundas prospered during the 1830s. The population rose from 580 in 1830 to 1,200 in 1839 and included a number of blacksmiths, who were the "tooth pullers" of the town. The Desjardins Canal opened in 1837 and was an excellent outlet for products of Dundas and other towns along its banks.

In addition to the Burr brothers, Dundas history includes stories of several well-known Canadian persons. The rebel leader, William Lyon Mackenzie, was a druggist there from 1820 to 1823. Also, in 1826, Dundas history included an episode of "tar-and-feathering" which involved Allan N. MacNab, a future leader of the Tory party.

FIG. 252 *Horace Burr clock with 8-day strap-brass movement.*
FIG. 253 *Strap-brass movement in the Horace Burr clock in Fig. 252.*

FIG. 253

151

REUBEN BURR

Reuben Burr (b. 14 Mar. 1766 – d. 21 Sept. 1842) came to Upper Canada from Pennsylvania. He was a carpenter and joiner and is one of the few men known to have made clock cases in Canada. The wooden movements and the dials found in these clocks were supplied by John Kline, also from Pennsylvania, who lived in Vaughan Township, York County, not far from Reuben Burr.

The clock illustrated in Fig. 254 was given to Priscilla Taylor, granddaughter of Reuben Burr, on the occasion of her marriage in 1839. It was for many years in the possession of J.B. Tyrrell, his great-grandson. It is now a part of the collection of the Ontario Archives and was on display (1991) in the Royal Ontario Museum.

The movement of the clock is of typical New England wood construction, 30-hour time-and-strike, with seconds and calendar dials and pull-up weights. The dial is painted wood. The hands currently on the clock are English-type brass. It is not clear whether they are original. The clock's hood has a door to allow access to the hands.

The cherrywood case is well made and is dark brown in colour. The styling is typical of a country carpenter. The most unusual feature is the shape of the finials, which have a smoothly hollowed interior resembling three egg cups.

Reuben Burr was a descendant of Henry Burr (d. 1743), a Quaker from England and a friend of William Penn. Henry and his wife, Elizabeth, arrived in New Jersey in 1682 and settled at Mount Holly, about 20 miles east of Philadelphia. Reuben's father, William (b. 4 May 1740 – d. 15 Sept. 1833), was the ninth child of Henry's son Joseph (b. 1604), a wealthy landowner and slaveholder.

Reuben was born in Bucks County, Pennsylvania, while the family was en route to Columbia County. He trained as a carpenter, and in 1787 went to Lincoln County in the Niagara Peninsula, where he worked for three years in a sawmill. Wishing to return home to Pennsylvania in 1790, he found his way barred due to recent hostilities and an unsettled peace. He finally obtained a passport from a British lieutenant-colonel at Niagara to travel by a roundabout way.

FIG. 254

Fig. 254 *Clock case and chairs made by Reuben Burr. Ontario Historical Society Report.* – *Courtesy the National Library of Canada*

Once home, he married Elizabeth Cleaver (d. 17 Mar. 1839) and settled in Union County, Pennsylvania, where he worked as a carpenter and joiner. His specialty was fanning mills, used by farmers to remove the chaff and seeds of noxious weeds from their grain. He also made household furniture. In 1805 Burr returned to Upper Canada with his wife and five children, encouraged by his sister Janie and his brother-in-law Benjamin Pearson, who farmed north of York and wrote of the abundance of deer and wild game. Reuben's brother, Henry, accompanied them on the five-week trip. Reuben purchased 25 acres northeast of present-day Aurora and built his carpentry shop. He remained in business ten years, supplying the occupants of the countryside with household furniture and fanning mills. He was also appointed "path-master" to look after the upkeep of the roads in that part of the township.

Reuben also took apprentices and, beginning in 1808, his son Rowland (b. 1798) learned the trade. During the War of 1812 and the "unsettled peace," the exchange of letters between Reuben and his father in Pennsylvania ceased for more than three and a half years. Their resumption was happily welcomed. In the first letter Reuben received, his father referred to the war as "the late unfortunate contest."

In 1814, following his wife's wishes to farm and leaving Rowland to carry on the carpentry business, Reuben sold his land and purchased 285 acres 1 mile north of Richmond Hill. He traded this land for a farm 5 miles north of Markham and also purchased land in the Township of Uxbridge. Reuben Burr was decidedly unhappy as a farmer and in 1827 he rented his farm and went back to carpentry, this time in the Township of King, west of Yonge Street. He was joined in this venture by his son Nathaniel.

When they lived in the Township of Whitchurch there were few Quakers and the family joined the Methodist Church. In addition to Rowland and Nathaniel, the family included John and four daughters. When he moved to King Township, he donated 1 acre of land, part of Lot 92, Concession 1, to the Society of Friends, who had been active in the area since 1804.

In 1832 Reuben and his wife went to live with Rowland, and he worked in his carpentry shop, making furniture and fanning mills. In 1838 Reuben sold all his lands, and the next year when his grandson William Burr Terry married, the Burrs moved in with him and his wife near the banks of the Humber River in the area of present-day Woodbridge. It was during this period that Reuben made tall case clocks using the movements of Mr. Kline. They were made in the shop of his grandson, William, who was also a carpenter. One of these clocks, given to William and his wife as a wedding present, remained in the family for generations.

Reuben Burr died in the home of William B. Terry and was buried in the Methodist Cemetery at Woodbridge.

FIG. 255 *Burr House, now a museum at Richmond Hill, Ontario was built by Rowland Burr, son of Reuben. Reuben Burr and his wife lived here for several years after 1832.*

FIG. 255

FIG. 256

THE CANADA CLOCK COMPANY LIMITED, HAMILTON, ONTARIO

This clock manufacturing company was the third in a series. It succeeded the Hamilton Clock Company, which in turn had followed the original Canada Clock Company of Whitby, Ontario. Details of all of these companies and their clocks can be found in the book *The Canada and Hamilton Clock Companies* by Jane Varkaris and James E. Connell.

The Canada Clock Company Limited was organized in Hamilton by its president, James Simpson, and other investors and was incorporated in March 1881. The company continued operation in a factory formerly used by the Hamilton Clock Company. Upwards of fifty people were employed and a variety of OG, mantel, and wall clocks were manufactured. At the time it was claimed that the company produced seven movement styles, and it is known that the cases were made there as well. After an initial period of prosperity, the company fell on hard times and suffered a disastrous bankruptcy in December 1884. The assets were dispersed to creditors in a welter of lawsuits. Production of clocks by Canadian factories ceased for twenty years, until the Arthur Pequegnat Clock Company began operations in 1904.

THE CANADA CLOCK COMPANY, WHITBY, ONTARIO

A detailed discussion of industrial clock manufacturing is beyond the scope of this book. However, the Canada Clock Company, Whitby, was active during the time frame the authors have selected, so a brief summary of the company's activities is included below. A detailed description of the company and its clocks can be found in the book *The Canada and Hamilton Clock Companies* by Jane Varkaris and James E. Connell.

The Canada Clock Company, Canada's first clock factory, was established in Whitby, Ontario, in 1872 by William, John, and Edward Collins. Initial production consisted of 30-hour weight-driven OG clocks. Both cases and movements were manufactured at the plant. In 1873 William Collins withdrew and the Whitby mayor, John Hamer Greenwood, became president and principal investor. John Collins remained as manager. Greenwood declared bankruptcy in 1875 and ownership of the company passed to James Wallace, who called himself "Proprietor." Wallace unsuccessfully attempted to find additional capital during 1875, but in December of that year a fire at the factory put a stop to production. In 1876 Wallace sold the production machinery to James Simpson and George Lee of Hamilton, Ontario. They moved the equipment to Hamilton and founded a new company, the Hamilton Clock Company. Wallace sold off his remaining stock of clocks, which consisted primarily of OGs, although some other case styles were listed in the sales bills. John Collins was re-employed at the new clock factory in Hamilton as mechanical superintendent.

FIG. 257

FIG. 256 *This clock, "The Dominion," by the Canada Clock Co. Ltd, Hamilton, Ontario, has only recently been documented.*
FIG. 257 *A clock made by the Canada Clock Company, Whitby.*

154

THE CANADA WATCH COMPANY

A stem-wind watch marked "Canada Watch Company" on both dial and movement is shown in Figs. 258 and 259. The movement is 15 jewels and bears no serial number or other identifying mark. It has been identified as being of Swiss origin, at a period when certain Swiss makers offered movements that imitated American production.

These Swiss movements were of good quality and could be ordered with customized inscriptions on dials and movements. It was a common practice to import them without cases, since cases could easily be obtained from Canadian or American factories. The watch illustrated here has a case marked "Dueber Silverine Canton O."

Similar watches bearing names of Canadian jewellers are frequently seen and generally date from the late 1800s to the early 1900s. The names of the proprietors of the "Canada Watch Company" have not been identified. They do not appear to have sold many such watches, since they are comparatively rare. The watch has been included in this book because of the inscription, but it is Canadian in name only.

FIG. 258 *Watch showing dial marked "Canada Watch Company."*
FIG. 259 *Movement of watch in Fig. 258.*

FIGS. 258 & 259

THE CANADIAN HOROLOGICAL INSTITUTE
AND HENRY RICHARD PLAYTNER

Henry R. Playtner (b. 18 Dec. 1864 – d. 20 Sept. 1943), a watchmaker, was born in Preston (now Cambridge), Ontario. His place in horological history was assured when, in 1890, he opened a "manual and technical institute" bearing the name Canadian Horological Institute in Toronto. The institute was considered to be unique. It was the first and only school where students could learn all aspects of the watchmaker's trade in Canada. Concerning the school, Playtner wrote, "My main aim in establishing it was to make it easier and less costly to others to gain such skills as I possess which required many years of persevering work and study.... The school is conducted on honour. I have set the highest standards for it and neither can nor will recede from that position."

The duration of the course was two years, with a further six months' instruction without tuition. Playtner claimed, "no two-week diploma nonsense here!" The "improvers'" course was at least six weeks in duration. The students followed an intensive and comprehensive training program, including theory, drafting, and practical bench work.

During this time, it was a requirement that each student make a "masterpiece" watch from raw materials in order to earn the highest level of diploma, the "Grade A 1." The exacting demands of this Grade A 1 diploma course are attested to by the fact that from 1890 to 1913 only twenty Grade A 1 watches were made and almost all of them are documented.[5] These watches are the only watches made in their entirety in Canada.

On 15 June 1880 H.R. Playtner was apprenticed to Edward Fox, a watchmaker in Kincardine, Ontario. He augmented his training in watchmaking by studying mathematics, mechanical drawing, and model-making. Leaving Mr. Fox on 10 February 1886, he moved to Toronto, where he found employment in the following shops:

14 April 1886 – 11 May 1886, employed by Robert Cuthbert;

12 May 1886 – 24 September 1887, worked for Kent Brothers at 168 Yonge Street;

25 September 1887 – 17 May 1890, worked for John P. Mill at 445½ Yonge Street.

While working at the Kent Brothers' shop, Playtner had the opportunity to work for Edward Beeton, a first-class watchmaker. When Beeton established his own business as a watch specialist in 1890 at 25 Leader Lane, he asked Playtner to become his partner. The partnership was announced in *The Trader*, now *The Canadian Jeweller*, a trade journal for which Beeton was in charge of the watch repair department. In 1890 Playtner was described as being "one of the most skillful journeyman watchmakers in this city."

It was in *The Trader* that Playtner first pointed out the need for a Canadian horological school and outlined a possible curriculum of study. Both Playtner and Beeton wrote letters on the subject. They were printed side by side in the April 1890 issue. These letters psychologically prepared the readers of the periodical for the opening by Playtner and Beeton of the Canadian Horological Institute in June 1890.

The first location of the Institute was the second floor of a four-storey building at 133 King Street East in Toronto. The 1,000 square feet of space was divided into the "Practical Department" and "Technical Theoretical Department." Workshops were supplied with the finest equipment, tools, and appliances. The premises were heated by steam and the

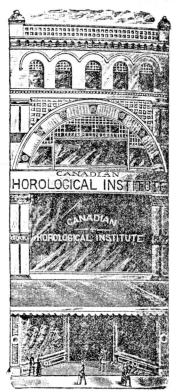

FIG. 260

FIG. 260 *Henry R. Playtner.*
– Courtesy Eugene Fuller
FIG. 261 *First location of the Canadian Horological Institute.*
– Courtesy the Thomas Fisher Rare Book Library, University of Toronto

FIG. 261

building was one of the first in Toronto to be lit with electric lights, powered by a system of storage batteries on the premises.

Shortly after the school opened, a reporter paid it a visit. An article appeared in the July 1890 issue of *The Trader*, describing the excellent facilities at the school. The writer was impressed with how quickly the students had progressed in two weeks in the well-equipped workshop. The writer also noted the excellent natural light entering the workshop through a huge plate-glass window. Through this window the students were able to view St. James Cathedral, which had the highest steeple in North America and the second largest clock in the world.

In June 1890 the school had opened with two students, Mr. J. Kincaid of Chicago and Mr. A. Zilliax of Listowel, Ontario. During its twenty-three-year history in Toronto, the school attracted students from across Canada and the United States and some students came from as far away as Bermuda and England. The attendance was divided fairly equally between Canadians and Americans.

In August 1890 Mr. Beeton resigned from the Institute, presumably because of ill health. However, he continued his shop on Leader Lane and also his responsibilities with *The Trader*, having risen to technical editor of that journal. Henry Playtner became proprietor and manager of the Institute and its instructor, teaching in both English and German. He continued to advertise the Institute in *The Trader*.

Through lectures, each student became proficient in calculating all actions of a watch and clock. During the first year of study, each student made a few specialized tools and learned clock and watch repair, the cutting of wheels, and repairing of escapements. He could calculate new mechanisms or, if confronted with a watch with a lost or broken part, could proceed to make the part for replacement. Half the profit from the trade work was received by the student completing the job and the amount could be applied against the tuition fee of $165 per year (in 1900). In the second year of the course, each student completed his "masterpiece." Each student was required to "compute" at least fifty drawings, containing over a hundred actions during the two-year program. One such drawing is shown in Fig. 263.

The students at the Institute were divided into three groups and the grouping determined the type of diploma that the student would receive. Group 1 students were, according to Playtner, "born mechanics, young men who are artist mechanics by nature and endowed for the highest positions...accorded the special privilege of calculating, designing and making outright from the raw material, a "masterpiece"...a watch of the highest class." Every component was made, with the exception of the dial, mainspring and jewels. Playtner also insisted that no two "Grade A 1" masterpieces could be made of the same design.

Once completed, the watch was adjusted for temperature isochronism and for six positions. The period of adjustment usually lasted about three months, and during this period, the student made a model of his choice of escapement mechanism. Before receiving the diploma, the student completed both oral and written examinations before a board of examiners chosen from representatives of the trade.

Group 2 students were each required to make a draft of a movement that was purchased in the "rough punchings" and to finish the watch in a first-class style. The student then designed and constructed, from raw material, his choice of an escapement model. The student could then be examined for a Grade A diploma.

FIG. 262

FIG. 262 *Practical workshop at 133 King Street East.*
– Courtesy the National Archives of Canada

Group 3 students spent up to two years repairing watches to gain experience and speed. They did not receive a diploma and were referred to as "improvers."

The first masterpiece watches were not signed or numbered. However, the watch made by J. Kincaid, one of the first students, is located in the Rockford Time Museum collection; the location of the watch of his colleague Mr. Zilliax is not known. Although each watch made to qualify a student for a Grade A 1 diploma was numbered, the numbers were not engraved on the watches. The signing of watches by students did not begin until 1894, and numbering and signing of each watch was not introduced until 1896. From an advertisement, however, it is known that one of the earliest Grade A 1 diploma masterpieces, a precision lever watch, was made in 1891 by L.A. Perret, and it is believed to be Number 3. Because of the rigid requirements for the achievement of the Grade A 1 diploma, it appears that only twenty such diplomas were earned. After L.A. Perret produced his Grade A 1 masterpiece, the next two Grade A 1 masterpieces were produced in 1894 by J.T. Whelan and J.L. Thuman. These watches are signed and dated and are believed to be masterpiece numbers 4 and 5.

On 12 September 1893 in Toronto, the Canadian Horological Institute played host to the newly formed Canadian Watchmakers' and Retail Jewellers' Association of Canada. George Klinck from Elmira, Ontario, was president and Edward Fox from Kincardine was vice-president. H. R. Playtner was secretary treasurer. Sometime in 1894 Playtner delivered a lecture or a series of lectures to the Association. The lectures, entitled "An Analysis of the Lever Escapement," were published in nine parts and appeared from December 1894 to August 1895 in the *American Jeweler*. Late in 1895 the series was reprinted as an eighty-eight-page hard-cover book, illustrated with twenty-eight line figures. It was distributed at a price of fifty cents. This comprehensive study did much to establish an international reputation for H.R. Playtner and the Canadian Horological Institute. It received praise from as far away as England and Germany. The booklet was reprinted with the permission of Playtner in the late 1920s or early 1930s for the E. & A. Gunther Company Ltd. of Brantford, who distributed the book to their customers without cost.

The publicity generated by the lecture series caused increased enrolment in the Institute in 1895. In that year as well, the students' work was exhibited at the Toronto Industrial Exhibition and received a silver medal, the highest award for skilled mechanical work.

In order to accommodate the increased number of students, H.R. Playtner moved the Institute to the Oak Hall Building, a short distance away at 115–121 King Street East.

The second floor of the building gave the Institute 2,000 square feet. A Riefler astronomical regulator was obtained for the work of adjustment. There were also two marine chronometers for comparison, and use was made of the time signals sent out by the Toronto observatory. According to an advertisement appearing in the *American Jeweler*, in December 1895 there were twenty-five students and eighty-three improvers. The masterpieces became more complex, and tourbillons, chronometers, and karrusels were being made.

In late 1895 the publisher of the *American Jeweler* offered a $10 cash prize for each of the four best essays on a subject in the field of the watch or jewellery business. Using code names, students of the Canadian Horological Institute met this challenge and won all four prizes. John A. Campbell of Guelph, Ontario, won two of the prizes for essays entitled "Making a Com-

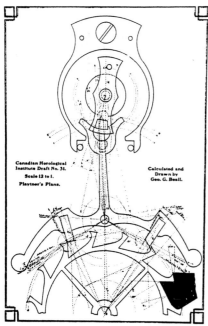

FIG. 263 *Drawing made by George G. Beall as part of his course.*

pensating Balance" and "Faults of the American Watch." Both essays were printed in subsequent issues of the journal. Immediate discontinuation of the prize essay contest may have been the result of the fact that the students from the Institute had so easily won all the prizes. Also, it may be suspected that the American watch companies who were advertisers in the journal were less than pleased to read an essay on their faults.

However, in 1897 Mr. Edward Rivett, the famous lathe-maker and president of the Faneuil Watch Tool Company of Boston, announced a prize for horological school students in each of three levels of experience. The judges were from the Philadelphia Horological Society. Again, code names were used by the contestants and once again students from the Canadian Horological Institute received all three prizes offered. Fred H. Spriggs received as first prize a $50 Rivett lathe for his masterpiece. The second prize, a $35 Rivett slide rest, went to W.L. Smith for a partly completed movement. Third-prize winner Lorne Totton received a $15 Rivett staking tool for part of a movement. H. R. Playtner was sent a letter from William T. Lewis, President of the Philadelphia Horological Society, complimenting him on his "highly creditable" instruction in the science and art of horology. Subsequently, the letter and pictures of the students appeared in many Canadian Horological Institute advertisements.

Shortly after the prizes were awarded, Mr. Rivett visited the Institute. In an address, he gave high praise to H.R. Playtner and his students. "Mr. Playtner is held in high esteem as a scientific mechanic and thoroughly practical educationalist. I am glad to see that this school is such a success, no matter from which standpoint it is considered."

Playtner himself offered a Playtner Prize early in 1898 to stimulate the desire to excel in technical drawings. This competition was conducted on an annual basis.

In the years 1896 and 1897, nine masterpieces were completed to achieve the Grade A 1 diploma (see Table 1). Although the Institute continued to operate until 1913, it may be noted that no Grade A 1 diploma masterpieces are listed after masterpiece number 20, completed in 1904. Students in the first decade of this century may have elected to enter the Grade A diploma course. Advertisements were the major source of information on the students and their accomplishments, but they became fewer and smaller in this century. Readers can only hope that an as yet undiscovered advertisement will come to light, featuring more masterpieces – if they exist.

Six of the Grade A 1 diploma watches have been located: unnumbered masterpiece by J. Kincaid, masterpiece number 6 by C.K. Weaver, masterpiece number 16 by W.L. Smith, masterpiece number 17 by Lorne Totton, masterpiece number 18 by Jerry Smith, and masterpiece number 20 by W. Douglas Smith. The tourbillon chronometer, masterpiece number 16, made by W.L. Smith is pictured in Fig. 266. This watch was illustrated in an advertisement in the April 1898 issue of *The Keystone,* where it was described as "the finest, most delicate, complicated and difficult piece of mechanism yet undertaken by the pupils." A detailed description of the watch mentioned the "up and down mechanism by the instructor.... In design and execution it is a poem and the pride of the school." After obtaining his diploma, W.L. Smith, a native of Toronto, moved to Nelson, British Columbia, where he was employed by the Patenaude brothers.

In September 1898 *The Keystone* ran Playtner's only two-page advertisement. It illustrated twelve different watches made by the Institute students. He also made clear his views that no school could train a watchmaker in

FIG. 264

FIG. 265

FIG. 264 *Silver medal, highest award for skilled mechanical work awarded to the students of the Canadian Horological Institute at the Toronto Industrial Exhibition, 1895.*
FIG. 265 *Oak Hall Building.*

FIG. 266

FIG. 266 *Grade A 1 masterpieces by W.L. Smith and Lorne Totton. Prize winners, 1897 Faneuil competition.* – *Courtesy Eugene Fuller*

less than two years. Those who indicated that they could were "disgracing the field of horology." He challenged any other school of horology in North America to produce watches equal to those being made by the students at the Canadian Horological Institute.

TABLE 1
GRADE A 1 MASTERPIECES

Kincaid?	Precision Lever	1	?
?	Precision Lever	2	?
L.A. Perret	Precision Lever	3	1891
J.T. Whelan	Precision Lever	4	1894
J.L. Thuman	Precision Lever	5	1894
C.K. Weaver	Tourbillon Lever	6	1899
A.C. Liphardt	Chronometer	7	1896
L.W. Hodgins	Chronometer	8	1896
W.E. Hagel	Chronometer	9	1896
Arnold Jansen	Chronometer	10	1896
Geo. T. Gilpen	Chronometer	11	1897
Fred H. Spriggs	Chronometer	12	1897
H.G. O'Dell	Chronometer	13	1897
Geo. E. Gendron	Chronometer	14	1897
E. Roy Butler	Chronometer	15	1897
W. L. Smith	Tourbillon Chronometer	16	1898
Lorne Totton	Karrusel Chronometer	17	1898
Jerry Smith	Precision Lever	18	1899
C.W. Parker	Chronometer	19	1900
W.D. Smith	Precision Lever	20	1904

It is possible that the watch made by Kincaid was masterpiece number 1 or 2. Moreover, missing watches, though assigned, may never have been completed. Masterpiece number 6 was not completed by C.K. Weaver until 1899, though assigned about 1895.

160

A precision lever watch earned Jerry Smith his Grade A 1 diploma in 1899. This masterpiece was assigned number 18. An article in the *Toronto Evening Telegram* of 10 December 1927 gives much information about the influence of the Canadian Horological Institute on the life of Jerry Smith. Smith's activities are outlined later in this book.

One Grade A 1 diploma watch, a precision lever, was assigned number 20 and was completed in 1904 by W.D. Smith.

Although the watch resembles an 18 size Elgin Veritas, it was not constructed from rough punchings from the company, but made from raw metal in accordance with the requirements of a Grade A 1 diploma. There are important differences between the Smith watch and the watches made in the Elgin factory. The prize won by W.D. Smith was the case for the watch, made by the American Watch Case Company. The details of the presentation are engraved on the case.

Grade A diplomas were given to students who completed watches from rough punchings instead of from raw metal. This approach was followed by most of the students at the Institute and apparently by all but two students between 1900 and 1913. Much less publicity was given by Playtner to these students, and therefore, few of these watches can be documented. From illustrations appearing in Canadian Horological Institute advertisements, it is apparent that in numbering the watches, a number followed by the letter "A" appeared on the watch.

Two Grade A watches were illustrated in the 1902 catalogue. George G. Koeberle completed Watch 3A in 1900 and George F. Caven completed watch 4A in 1901. Both were size 16 and probably lever set with at least 17 jewels.

George Koeberle of Sumner, Iowa was in charge of an exhibit consisting of various Institute items at the Watch and Clock Trades Exhibition in Chicago during October 1901. His main task was to persuade visiting watchmakers that the excellent exhibit of drawings, precision movements, and models had, in fact, been made by students themselves at the Canadian Horological Institute.

The watch completed by George F. Caven and featured in at least two school catalogues was adjusted to very high standards. The 1902 catalogue, the last catalogue known, recorded the accuracy of this watch, which "showed a variation of 1 second in six positions after a full day's run in each. One with an error within 5 seconds per day is considered a first class watch."

Other students known to have received Grade A diplomas from the Institute were: R.J. Agnew, 1892; Joseph J. Schuster, 1900; George G. Beall, 1902; N.S. Porter; A.D. Savage; Edward Teffein; and E.S. Walker. E.S. Walker received his diploma by completing, from simple punchings, a five-minute repeater with glass dial showing the mechanism below. This is the only documented repeater watch produced at the Institute.

A total of thirty-six graduates have been identified. Six graduates and their locations after graduation are known, but the type of their diplomas is unknown:

J.H. McEwen	Hublersburg, PA
J.O. Patenaude	Nelson, B.C.
A.K. Chattaway	Philadelphia, PA
L.G. Bolt	New York, NY
John A. Campbell	Guelph, Ont.
Hubert Langevin	Montreal, Que.

Henry R. Playtner greatly appreciated hearing from and about his students, and when he received a letter from the father of George Beall, mentioning the excellence of the finished watch and the great value of a diploma signed

FIG. 267

FIG. 267 *This advertisement appeared in* The Globe, *Toronto, in the Christmas issue, 1898.*

FIG. 268

FIG. 268 *Masterpiece No. 6, a tourbillon lever, by C.K. Weaver.* – *Courtesy Ira Leonard*

by Playtner, he devoted an entire advertisement to reproducing this letter. Also in the early 1900s, Playtner published a little booklet entitled *Appreciative Words,* which he distributed to prospective students.

By 1900 the international reputation of H.R. Playtner attracted more students than could be accommodated in the Oak Hall Building on King Street, and there was therefore little need to attract students by elaborate advertisements. For the large enrolment, Playtner planned a new facility. An advertisement in *The Keystone* in July and August 1907 announced, "We are now erecting a modern school building planned especially for our use in an advantageous corner location at Church and Wellesley Streets, Toronto...our new building is growing apace. It will be brick, 3 storeys, concrete foundation...steel beams will rest on concrete piers and support a double floor of hardwood on which benches will be placed...8 large windows will admit north light...we hope to occupy in October."

The move was completed on 21 November 1907. A photograph of the new building at 330 Church Street appeared in several advertisements in *The Keystone* during 1908.

Advertisements continued until 1911, showing illustrations of Grade A 1 diploma masterpieces and extolling the values of the school. However, few advertisements for the school appeared in 1912, and in February 1913 the building was sold to the Royal Bank. Classes were discontinued as of 13 July 1913. Henry R. Playtner had closed his successful institute and returned to his home town of Preston, Ontario.

Early in this century, there had been contact between the Canadian

Horological Institute and the Elgin National Watch Company of the United States through their Canadian representative. The details of a "proposition" made at a Canadian Horological Institute banquet by this representative and mentioned in an advertisement are unfortunately not known. However, it is highly possible that this representative was Ed Beeton, former partner of H.R. Playtner and co-founder of the Institute. Beeton represented the Elgin National Watch Company in Canada from 1903 to 1912. Therefore, it is believed that Playtner did not "close the doors of the Institute due to ill health" as suggested by his obituary thirty years later. Judging from his association with the Elgin National Watch Company early in the century and from the fact that Playtner eventually went to Elgin, it is apparent that his old friend Ed Beeton had finally persuaded Playtner to launch a new school under the sponsorship of the Elgin National Watch Company at Elgin, Illinois.

Unfortunately, the war of 1914–1918 intervened. It is not known how Playtner spent the war years, but he would have been very conscious of his German heritage during that period.

However, once the war was over, Playtner reached an agreement with the Elgin company and was appointed president of the proposed Elgin Watchmakers' College. The three-storey brick building erected in 1920 across the street from the watch factory, was similar in style to the building that had housed the Institute on Church Street, Toronto. It was equipped with the best and most modern equipment. The College officially opened in 1921.

One of the superiors appointed over Playtner was De Forest Hulburt, assistant to the president of the watch factory at the company's executive offices in Chicago and son of Charles Hulburt, president of the Elgin National Watch Company. The other man superior to Playtner was E.N. Herbster who was on staff at the watch company. Playtner accepted two superiors because he recognized that Elgin dominated the domestic watch industry. They produced an average of one million movements annually from 1920 through 1928, more than half of the U.S. domestic watch production. Also, the Elgin Watchmakers' College was liberally endowed by the watch company.

In 1921 when the college opened, W. H. Samelius, who had previously taught at Washburn Technical School in Chicago, came to Elgin as an instructor under H.R. Playtner. Both men were considered to be top horologists and their presence assured the new school of the finest instruction of any school on the continent.

Also in 1921, the Horological Institute of America was organized "to advance the interests of horological science and practice." On 22 January 1922 Playtner addressed the H.I.A. at its first banquet at the Hotel Astor in New York City. The fact that Henry R. Playtner was invited to give the principal address at the very first H.I.A. banquet speaks highly of the esteem in which he was held in the field of horology. In printed form, his speech was entitled, "The Future of Watchmaking in America." It was apparent from his speech that he now believed that three years of training were necessary to produce a competent watchmaker. He reiterated the necessity for a student to produce a masterpiece made from his own plans and calculations, out of sheet metal strips and wire rods, using such tools as were to be found in a repair shop. He also advocated that uniform curriculum and standards be established in all horological schools and implied that his school should be the model. He believed that no student should be prevented from entering a school for lack of funds, and he proposed that a fund be established from which a student could borrow at low interest rates.

FIG. 269

FIG. 269 *Masterpiece No. 20, a precision lever, made by W. D. Smith.*
– Courtesy Eugene Fuller

FIG. 270

FIG. 270 *The prize case won by W.D. Smith.*
– Courtesy Eugene Fuller

The standards for a horological school's curriculum as outlined in Playtner's speech were found in few, if any, horological schools in the United States in the early 1920s. However, perhaps as a result of the speech, the Horological Institute of America did take as its primary goals, first, the standardization of the curriculum and, second, the certification of watchmakers.

Under Playtner's guidance, the Elgin Watchmakers' College offered, as one of its options, a three-year course with masterpiece requirements. However, it became apparent to his superiors at the school that Playtner's ideal watchmakers' school requirements were not really what they considered appropriate. According to *Through the Ages with Father Time* by Roy Bailey (1922), "The purpose of the Elgin Watchmakers' College was for training competent, all-round workmen for the watch repairing trade." Because the Elgin National Watch Company produced the watches, the superiors at the school no doubt considered that the training should produce graduates with the ability to repair those watches.

Playtner would not lower his educational standards. Therefore, by mid-1923, Playtner was not happy with the college, and the college was not happy with him. Correspondence in June 1923 between De Forest Hulburt and Playtner gives the "official" reason for Playtner's subsequent resignation: "I believe you would be doing the greatest good to the Elgin Watchmakers' College if you were to retire, returning to your former home and there commence work on a complete treatise for the instruction of horology in American watchmaking schools. I would suggest that you retain the title of President Emeritus…then for the time being there will be no President…but Mr. Herbster and I will continue its management as Vice-presidents and will appoint Mr. Samelius as Director."

Playtner's salary was paid until the end of the year. His suggestion that one of the talented students should be hired as a drafting instructor was carried out. Also, prizes offered by him were to continue to be offered until all students at the school during 1923 should graduate. Playtner promised to send copies of his textbook to the college before it was printed.

Playtner left the college on 1 August 1923 and W.H. Samelius was appointed as director, a position he held until he retired on 1 July 1954. There was never another president appointed at the Elgin Watchmakers' College.

A large model of the tourbillon escapement made for teaching purposes while Playtner was at the Canadian Horological Institute was left at the school and was purchased by Director W. Samelius, who dreamed of forming a horological museum. The dream of the museum was Samelius's, but "he passed on to a younger man all his hopes and aspirations." In 1960 the museum was established by Mr. Orville Hagans and was named "Hagans Clock Manor." It was here that Playtner's model was displayed. Unfortunately, the museum is no longer in existence and the present location of the model is not known.

H.R. Playtner was born in Preston, Ontario (now part of Cambridge), to a Lutheran couple from Germany, August and Martha (nee Heise) Playtner. Although the name, originally Ploethner, was simplified to reflect its phonetic pronunciation, Martha, still alive in 1914, retained the German spelling of her name.

After leaving the college, Playtner returned to Canada and purchased a house in Kitchener, Ontario, at 16 Peter Street, where he lived out the remainder of his life. He became a recluse in his later years and there is no evidence that he completed the planned textbook.

Playtner's brother Oscar predeceased him. His sister Mathilde married John Sohrt. Playtner died at his home in Kitchener on 20 September 1943, in his seventy-ninth year. He is buried in the Preston Cemetery. His nephew Herbert J. Sohrt and his nieces Margaret and Helen were beneficiaries of Henry R. Playtner's will. His tools were loaned to a Preston jeweller and their location was not known in the 1960s.

Playtner was exacting in his relationships with people and, except for the period he spent as an apprentice, he went quickly from job to job. He was unable to work with his first partner at the Canadian Horological Institute and he most probably left the Elgin Watchmakers' College because of a clash of personalities. His exacting nature was even evident in his will. He cut his housekeeper out of his will because "she left me in the lurch...receives nothing." He cancelled a bequest of his Rivett Number 4 lathe because it had not been cleaned as promised. "As he did not clean it he has no claim to it." At the same time, the will cancelled a 1936 debt and included a small bequest to an old friend of his brother at the House of Refuge.

His obituary, in December 1943, described him as a brilliant, internationally known Canadian horologist, whose penchant for long hours and little recreation had resulted in an exemplary career.

Playtner, a pioneer in horological education, was as demanding of himself as he was of his associates and students. He was a perfectionist who demanded excellence. Consequently, his students became very proficient, as evidenced by examples of their work. Their gratitude is evident, too, in their testimonials published years ago. In an article in the *Toronto Telegram*, Jerry Smith was quoted as saying about his experience as a student at the Institute, "There I found myself under the guidance of a master mechanic in the person of H.R. Playtner, the greatest horologist of all time."

FIG. 271

FIG. 272

FIG. 271 *Watches made to earn Grade A diplomas.*
Above: Hunting case lever watch No. 3A, completed in 1900 by Geo. G. Koeberle.
Below: Open-face lever watch No. 4A, completed in 1901 by Geo. F. Caven.
FIG. 272 *Third location of the Canadian Horological Institute at 330 Church Street, Toronto.*

CHARLES CLINKUNBROOMER

The story of Charles Clinkunbroomer (b. 1799 – d. 12 Jan. 1881), clock-maker, watchmaker, and silversmith of York and Toronto, has consider-able historical interest. The family surname itself certainly attracts atten-tion. According to descendants still living in the area, at least ten differ-ent spellings have been used over the last two hundred years. The form used here was favoured by the clockmaker himself.

The original settler was Nicholas Klingenbrumer (or Klingenbrenner), who arrived in York, Upper Canada, in 1795. He was a native of Germany and a career soldier in the British army. It has been recorded that on sev-eral occasions British recruiting teams were sent to Germany to enlist mercenary soldiers at times when volunteers were hard to find in the British Isles. Family tradition maintains that Nicholas was actually "shang-haied" by a press gang, but in any event, he spent much of his adult life fighting for Britain. Family tradition also has it that he fought at the cap-ture of Quebec on the Plains of Abraham with General Wolfe in 1759. He later fought with the British forces during the American War of Inde-pendence. He was said to have taken part in the Battle of Bunker Hill, Boston, in 1775 and to have served under General Burgoyne at Saratoga in 1777; he was briefly a prisoner of war. In 1783 Klingenbrumer left the army. He worked for a short time as a fur trader for the Northwest Com-pany. As an army veteran, he was entitled to a grant of free land and was deeded Town Lot 240 in Newark (now Niagara-on-the-Lake), where he worked for William Jones, an army tailor. For a period of time, he also owned a 100-acre lot at Dundas, Upper Canada.

In 1795 Nicholas Klingenbrumer moved to the newly founded settlement of York.[6] He became the first tailor of York and opened a shop at the corner of present-day Adelaide and Jarvis streets. The following year, this set-tlement of fewer than two hundred people became the capital of Upper Canada. On 4 January 1799 Nicholas was married to Sarah White. Nicholas's date of birth is not known, but by 1799 he would probably have been between fifty and sixty years old. Klingenbrumer served as Constable of York and as such presided over the pub-lic meeting held on 4 March 1799, at which a number of parish and town officers were first appointed.

Charles, the eldest son of Nicholas and Sarah Klin-genbrumer, was born in 1799 on Nelson Street (later Jarvis Street). The census record of 1805 shows that the Klingenbrumers had at that time a family of four, but later records only exist for three children. A second son, Joseph (b. 10 Mar. 1801 – d. 22 May 1884) became well known in York and Toronto in later years as a tailor. The third son, Exaveras (sometimes called Xavier), became a stonemason and his name appears in the records of York, Toronto, and North York. Nicholas ap-plied for and in 1802 received a grant of land in York. He was therefore one of the first patentees of town lots in York. The lot was on the north side of Hospital Street (now Richmond Street). The assessment roll of 1802 lists the family at 10 Hospital Street. Additional information about the descendants of the three brothers appears in Appendix 1. Nicholas Klingenbrumer's name can be found in assessment

FIG. 273

FIG. 274

FIG. 273 *Tall case clock by Charles Clinkunbroomer, York, Upper Canada, the only known example.*
FIG. 274 *Heavy copper dial from Charles Clinkunbroomer clock showing "Fraktur" type ornamentation.*

166

rolls until 1812. In 1810 he was living with three children but no wife, and in 1812 he was living alone. No record of his death has been found.

At some later date, his family agreed that their name should henceforth be spelled Clinkunbroomer. Descendants living in the Toronto area are proud of the fact that seven generations have been born there. They are undoubtedly one of the very oldest Toronto families.

Some time after his father's death, Charles apprenticed to Jordan Post, the first clockmaker in Upper Canada. By then, Post was an established landowner and maker of clocks, silverware, etc. Charles completed his apprenticeship with Post and is believed to have continued in his employ until 1833. His two brothers also apprenticed in their chosen trades.

Shortly after 1833 Charles established his own business as a watch- and clockmaker and jeweller His silversmith's mark is recorded in the book, *Canadian Silversmiths and their Marks 1667–1867* by J.E Langdon. He opened his first shop at 119 King Street East, between Yonge and Church Streets. The building was assessed at £83 in 1834. He conducted business at 73–75 Richmond Street East from about 1846 to the early 1860s. (The *Toronto Directory* of 1859 lists his shop at 96 Richmond Street East.) During the 1860s, he moved to a three-storey building that he had built at 156 Queen Street West and remained there until he retired in 1870.

Charles Clinkunbroomer married Hannah Anderson (b. 1811 – d. 1872) in the early 1830s and they had a family of nine children.[7] After Charles's wife died in 1872, he spent his last years with his daughter Eliza (Mrs. John Alexander) at 24 Baldwin Street. At the time of his death in 1881 he was believed to be the oldest continuous resident of Toronto. His obituary notice states that "Charles was of a most unobtrusive and unassuming disposition [playing] no active part in politics and always walking to the polls instead of accepting the services of a cab." In politics, he was a Reformer and by religion, a Congregationalist. In spite of the obituary's claim about politics, the family believes that as a younger man, Charles, actively supported William Lyon Mackenzie in the Rebellion of 1837.

As a clockmaker, Charles Clinkunbroomer belongs to a very small group of true clockmakers who were both born and trained entirely in Canada. Most early clockmakers working in this country were trained in Great Britain, the United States, or elsewhere. Clinkunbroomer would in all likelihood have completed his apprenticeship by his twenty-first birthday in 1820. At that time the local market for clocks would have been very small. The population of York was by then only a few thousand, most of whom lacked the resources to buy a clock of any sort. In addition, the activities of Eli Terry and his contemporaries in Connecticut were already heralding the end of hand-crafted clocks. The output of clocks by Charles Clinkunbroomer, therefore, was undoubtedly very small. Although more clocks may exist, this discussion of his work is restricted to one known specimen. The existing clock, however, has enough unusual features to make it an interesting study. The commentary can be broken down into several aspects:

THE CASE

In its present form, the case of the clock stands 7 feet 5 inches high. From marks remaining on the base it is clear that the clock originally stood on turned or bun feet. These feet are lost and the present feet as pictured are modern. The style of the case could be described as "Coun-

FIG. 275

FIG. 275 *Movement from Clinkunbroomer clock showing skeletonized plates. (Front view above)*

167

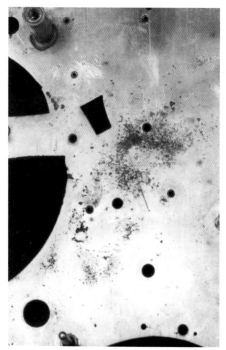

FIG. 276

FIG. 277

FIG. 276 *Plate detail from Clinkunbroomer clock. Pitted surface came from casting mould. Post at upper left supports the dial, receiving a screw that passes through the dial.*

FIG. 277 *Great wheel, strike side, from Clinkunbroomer clock, showing irregular hand-filed teeth and casting flaw in the spoke.*

try Chippendale." The case is constructed mostly of cherry with pine as a secondary wood. There are two "horns" on the hood, but there is no evidence that finials were originally present. The front of the case was originally decorated with an unusual variety of veneers, inlays, and bandings. The case has an overall appearance of elegance, which is out of character for the primitive, early days of York. The authors are of the opinion that the case is of a later vintage, dating from a period when Clinkunbroomer might have been more affluent. In the not too distant past, the clock underwent an unfortunate attempt at restoration. A workman attempted to obliterate all the original decoration by using a combination of modern trim and dark stain. These additions have been removed, requiring considerable restoration in two or three areas. Fortunately, most of the decoration is now as it was when the clock was made, restoration in some areas being aided by scraps of original material that had survived the disfigurement. It is always unfortunate when clocks of this rarity fall into unsympathetic hands.

THE WEIGHTS

One of the original weights has survived. It is similar to those used by Clinkunbroomer's master, Jordan Post. The weight is made of thin sheet iron, probably filled with sand, and it is plugged at the top with lead. It is 3 inches in diameter and 14 inches long. It weighs 14 pounds.

THE DIAL

As can be seen in Fig. 274, the dial is quite unusual. The material used in its construction is solid copper, measuring 12¼ inches wide, 17½ inches high, and over 1⁄16 of an inch thick. The use of copper for a dial is uncommon and the only other clock with a copper dial (much thinner copper over iron) known to the authors bears the name of Jordan Post.

Although the reverse side of the Clinkunbroomer dial has original hammer marks, the front is smoothed and polished. The dial is extensively decorated with birds, branches, leaves, and flowers. This decoration closely resembles "Fraktur Art," which is usually associated with "Pennsylvania Dutch" or Mennonite, and other German ethnic groups. A bell-shaped opening at the top of the dial was provided to display an oscillating panel attached to the escapement anchor. This feature is also a form of decoration favoured by Jordan Post. Post, however, used objects such as rocking ships, which moved in front of the dial. In the Clinkunbroomer clock the moving object was immediately behind the dial. One of the tragedies of this one known clock is that the original panel has been lost. A replacement has been fashioned from a sheet of brass to fill the gap and features a flower that swings back and forth. Probably the style of the original panel will always remain unknown.

Below the central arbor, another semi-circular opening displays the original brass calendar dial on which the dates of the month have been rather crudely engraved. The seconds dial is decorated with a six-pointed "compass rose," which is often associated with Mennonite hex signs.

A fragment of the original minute hand has survived. It was filed from steel and is black in colour. The hands now on the clock are English hands of a later vintage.

The maker's name, "Charles Clinkunbroomer, York," was engraved in a simple, almost childlike script, which tends to slant downward. The numerals, too, have a naive character. The engraver, obviously engrossed in his work, allowed some of the numerals to slant in an almost comic fashion. The overall result is a dial of great charm.

THE MOVEMENT

The influence of Jordan Post is very evident in the mechanism. Comparison with an original Post movement (see Figs. 364 and 365) shows the same economy in the use of brass. The plates have been cut out deeply. The general layout of the movement follows the usual pattern of English 8-day weight-driven clocks, with rack and snail striking. The resemblance ceases there, however. One distinction that is immediately obvious is the fact that the dial was affixed directly to four posts on the front of the movement plate. There is no false plate and the screws actually pass through the dial. This is a rather crude technique used by few makers but, again, favoured by Jordan Post.

In fact, a somewhat "rough and ready" approach to the finer points of clockmaking was apparently a characteristic of clocks made in York at Post's shop. Because of its similarity in many aspects to the Post movement, a more complete discussion of the Clinkunbroomer movement is found in the section on Jordan Post. A brief summary is given below:

1. The plates are cast brass and contain numerous rough spots and blow holes. No great effort has been made to file them smooth. Plate thickness varies considerably.
2. Viewed from the side, the arbors are not perfectly parallel. It would seem that the maker laid out one plate at a time and did not drill plates as matched pairs. Small cumulative errors led to a rather haphazard result.
3. Under magnification it is evident that the gear teeth have been filed by hand. File marks are seen at the root of each tooth and the tops have been roughly rounded, so that no two individual teeth have the same profile. Since the name York was changed to Toronto and Jordan Post gave up clockmaking in 1834, it is obvious that Clinkunbroomer made the clock during the years when he worked for Post. One can surmise, therefore, that Post did not possess a proper gear-cutting engine. However, he must have had some method for indexing the teeth evenly because, in spite of the occasional crudity of the workmanship, the clocks do run.

The Clinkunbroomer movement has one unique feature which has been only seen in clocks from Post's shop. The escapement anchor is of cast brass or bronze. The contact points of the pallets consist of tiny pieces of steel set into the softer metal. This achieved the desirable arrangements of steel pallets in contact with the escape wheel. The top of the anchor, incidently, has a tapped hole to provide a mount for the moving panel in the dial.

The bell hammer of the Clinkunbroomer clock is also made of brass and is fitted with a steel insert at the tip. The use of steel inserts is a novel way to avoid special tempering of larger steel components.

To summarize: "Muddy York" (the derogatory name given to the frontier settlement in the 1820s) was little more than a village. The population was not affluent, supplies were scarce, and life was generally difficult. The trade of clockmaking was a luxury. The clocks of Charles Clinkunbroomer and his master can be criticized as crude and primitive by more sophisticated standards, but both must be given credit for having tried to make clocks at all.

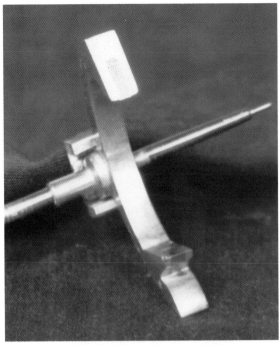

FIG. 278

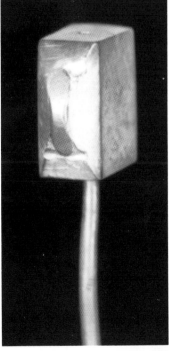

FIG. 279

FIG. 278 *Brass escapement anchor from Clinkunbroomer clock, with steel inserts.*
FIG. 279 *Bell hammer of brass with steel inserts, from Clinkunbroomer clock.*

169

JOHN DUNCAN COLQUEHOUN

John D. Colquehoun (b. 1849 – d. 6 June 1946) was a watchmaker in Wales, Ontario, a village located on Lot 7 Concession I of Osnabruck Township, Stormont County. John was one of the few Canadians who held patents for improvements of clocks and watches. At least two clocks exhibiting his innovations are in existence.

The village of Wales, christened by the Prince of Wales in the last century, cannot be found on the current map of Ontario. The four hundred villagers of Wales, along with the population of six other villages along the St. Lawrence River, were forced to move when the St. Lawrence Seaway was constructed in 1958. The village of Wales is now several fathoms beneath the waters of the St. Lawrence River. Many of the village residents now live in Ingleside, Ontario.

Some of the buildings of the village were moved. According to newspaper reports, the hydraulic lifts raised the houses so gently that the dishes were safe in the cupboards and a glass of water on the window sill was not spilled.

Upper Canada Village at Morrisburg is a memorial to the villages and is composed of buildings removed from them. However, many of the buildings in the villages were burned, and it is reported that there was much looting. The building in Wales that housed the J.D. Colquehoun shop was not moved.

It was while operating his watch, clock, and jewellery business in this building that John D. Colquehoun applied for and received several Canadian patents. One such patent, number 24125, was granted in 1886. The patent was "sworn" before John Wanless, justice of the peace in Toronto, who was also a watchmaker. It read in part: "The object of the invention is to adapt the dial of a watch or clock for the twenty-four hour system without changing the numerals from the positions they occupy in the twelve-hour system and it consists essentially, of a dial having twelve openings at the points usually occupied by the numeral, and placing behind the dial a dial-ring having printed on its surface two sets of figures, one set indicating from one to twelve, the other set from thirteen to twenty-four" and shifting from one set to the other automatically. Further improvements were patented in 1888 in patent 28877.

Patent number 24125 is reproduced in *Canadian Clocks and Clockmakers* by G.E. Burrows. Copies of the patents may be obtained from the Government of Canada. The clocks using the innovations of these patents are octagonal "school" clocks.

Much of the information concerning J.D. Colquehoun was obtained from the Tweedsmuir History of the area compiled by the Ingleside Women's Institute. The Tweedsmuir Histories were written by local residents of small communities to preserve the history of rural Ontario.

According to the record, the Colquehoun family originated in Scotland in the reign of Alexander II, when Humphrey Kilpatrick was granted the barony of Colquehoun for military service. It was the custom to assume the name of the granted lands.

As with all clans, the Colquehoun clan had a tartan, a gathering tune,

FIG. 280

FIG. 281

Fig. 280 *J.D. Colquehoun.
– Courtesy Tweedsmuir Histories, Women's Institute*
Fig. 281 *J.D. Colquehoun's shop and post office.*

a march, a motto, and a war cry. It also had a feud with the Macgregors which lasted from 1600 to 1800. Sir James Colquehoun was responsible for resolving the difficulty. He danced the Highland fling with Sir John Atholl Macgregor on the summit of Ben Lomond, shaking hands at the end of the dance. Sir James also "sowed the shore of the lake (part of Loch Lomond) with daffodils which flutter and dance in the thousands there, scenting the western breezes."

The history of the clan was important to John D. Colquehoun. When the Clan Colquehoun Society was instituted in 1892, he became one of its members.

John D. Colquehoun's great-great-grandfather Peter was a cavalry officer who lived near Loch Lond. His sword was brought to Canada by John's grandfather. Peter's son, John Sr. (b. 1742 – d. 4 Jan. 1804), lived in Scotland in the parish of Luss. John Sr. married Marion McCutel. Five of their seven children came to Canada. Marian also came to Canada to live with her elder son, Alexander J. (b. 1779 – d. 1871), and his large family.

John D. Colquehoun's grandfather John Jr. was the second son of John Sr.; John Jr. married Agnes McKellar (b. 1791) and came to Canada in 1822 to live with his brother, Alexander. Agnes and their three sons followed them to Canada. In 1828 John Jr. bought land and founded Colquehoun Village. A school was built on the corner of the property. Anglican church services were held in John's home until 1831, when a Presbyterian church was built at N. Williamsburg. Although the church was 12 miles away, John Jr. and his family attended it.

John D. Colquehoun's father, Alexander (b. 1821), married Harriet Munro (b. 1828) in 1846. John D. was the second of Alexander's eleven children. He married Elizabeth Vance (d. 1934) of Colquehoun, Ontario. There were three children: Kathleen (Mrs. Garnet Whitney); George (b. 1884 – d. 29 June 1949), who married Mabel May and had a son, Donald; and a third child, Alexander J.

At the close of the American Civil War, John D. was working as a foreman at a broom factory in Shelleyville, Kentucky. He came to Wales in the 1870s to open a watch repair and jewellery shop and was listed as a watchmaker and jeweller in the Province of Ontario directory of 1884. In addition, John became postmaster of Wales in 1896 a position he held for the next forty years. He was honoured with a silver medal in 1935, on the occasion of the silver jubilee of King George V and Queen Mary, as one of the postmasters appointed prior to 1900.

An advertisement appeared in 1898 in the Wales newspaper, indicating that, in addition to being a postmaster and having a watch repair and jewellery shop, John D. had added eyeglasses and spectacles to his inventory.

He was also active in the Wales community where, according to the Tweedsmuir History, he took "deep interest in everything pertaining to the advancement and welfare of the village and was a kind neighbour and friend to all."

Both John D. and his wife were active in church work. They held many meetings of the British and Foreign Bible Society in their home. John D. acted as secretary for more than sixty years and was leader of the Bible study groups. In 1935 he was honoured by being made a life member of the society. John D. was the founder and elder of the Wales Presbyterian church, which after church union became the United Church. He and his wife also hosted meetings of the Temperance Union.

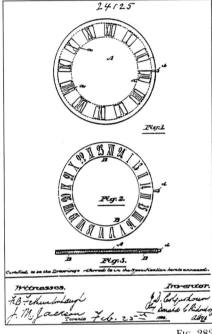

FIG. 282 *Illustration accompanying the patent application.*

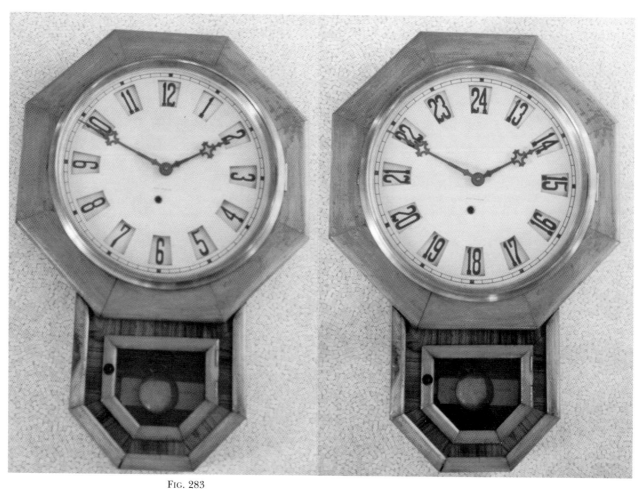

FIG. 283

FIG. 283 *Clock using the Colquehoun patents.*

A former resident of Wales, "Fran" La Flamme, recalls that John " was usually called Jock. In the late 1940s, he was a short, roly-poly, silver-haired man with ruddy cheeks. He wore a plaid tam-o-shanter and had a puckish smile. All in all, he looked like a beardless Santa Claus."

John D. Colquehoun died on 6 June 1946 in his ninety-eighth year. He was buried in the United Church cemetery in Wales, where he was given full Masonic honours at his funeral. When Wales was flooded, the stones from Wales cemetery were moved to the new cemetery that was established on the banks of the St. Lawrence River. However, only the bodies of those dead less than ten years were moved. The rest of the bodies, including that of John D. Colquehoun are now beneath the water.

One of John D. Colquehoun's children, George, succeeded his father as watchmaker and jeweller in Wales. George had taken a course in Toronto and was in charge of the railway watches and clocks for the area. He was also connected with the post office for forty-five years, at first as an assistant to his father. When his father retired in 1936, George became postmaster until his death of a heart attack in 1949. His wife, Mabel, then became postmistress until the post office of Wales was closed on 22 December 1957.

GEORGE L. DARLING

George Lacey Darling (b. 24 Nov. 1823 – d. 2 July 1899) was born in Cook's County, Pennsylvania. He was a descendant of George Darling, one of the Scots prisoners taken by Cromwell and transported in 1653 to New England, where he was required to labour for eight years. He settled in Salem, Massachusetts. George L. Darling's grandfather Samuel (b. 1763) lived in Cherry Ridge, Pennsylvania. His father, J.B., married Nancy Ann Lacey. Shortly thereafter, his family brought him to Canada West, his mother carrying him in her arms on horseback.

He was apprenticed to a Mr. Cook of London, Ontario, and became a clock- and watchmaker. He opened a business in Whitby, Ontario, but ill health forced him to give up his business there, and on his recovery he worked in Hamilton. In 1847 he settled in Simcoe, Norfolk County, where he worked for the next fifty-two years. The G.L. Darling block in Simcoe was completed in 1849 and Darling established his business on the west side of Norfolk Street between Robinson and Peel streets.

According to the 1865–1866 directory for Elgin and Norfolk counties, Darling was a "dealer in watches, clocks, gold and silverware, plate, jewellery, fancy ornaments and useful articles." Darling had his own silver stamp, and six marked spoons may be seen at the Eva Brook Donly Museum in Simcoe, Ontario.

In addition to looking after the business of his shop, George L. Darling made at least three regulators during his years in Simcoe. They are very high quality. One clock is in the possession of the Canadian Museum of Civilization, Hull, Quebec. It was made between 1857 and 1864. The brass-and-steel movement of the clock is hand-forged, and stamped with George L. Darling's silver stamp mark. The dial is silver plated and bears the names Newbury and Birley, a jewellery firm in Hamilton, Canada West. The free-blown glass weight receptacle holds 26 pounds of mercury. Crotch mahogany veneer was used on the case and under the dial is a border of roses, petunias, and leaves, hand-carved in walnut. Behind the long pendulum is a mirror. For many years the clock was in Petrolia, Ontario, brought there by George Darling's son, Arthur.

Darling retired around 1890 and his son, Clarence (b. 1869), took over the business. However, after the death of G.L. Darling in 1899, the business closed and Clarence is thought to have gone to Hamilton.

According to *The History of Simcoe: 1829–1929* by Lewis Brown, it is possible that this quiet man, George L. Darling, would have been all but forgotten in Woodhouse Township but for an incident in which he was the victim. It has gone down in the history of Simcoe as the "Ten Thousand Dollar Jewellery Robbery."

The robbery occurred on 6 November 1883. Details may be found in the *Simcoe Reformer* and in the book by Lewis Brown. Only the very highest-quality watches and jewellery were stolen. Later, in a deal with the lawyer, the details of which were never revealed, all but $400 of the loss was returned.

Although Daniel Almond, a popular citizen, and Joseph Adams were arrested, two other men with previous criminal records were brought to trial. At the trial Bruce Jackson, the lawyer who had

FIG. 284a

FIG. 284a *George L. Darling clock.*
– Courtesy the Canadian Museum of
Civilization, Hull, Quebec (78-3192)

made the deal with the thief, was forced under oath to reveal the name of his client, Daniel Almond. Almond rose from his seat, "a smile on his face and hat in hand, strolled leisurely to the county house doors... and outside." He was followed in great haste by the spectators and the members of the court, but he had completely disappeared. An exciting manhunt proved fruitless. Almond was never seen again.

The late William Z. Nixon, a Simcoe district historian, has commented that "long afterwards it was discovered that there was a secret room beneath the courtroom where Almond had concealed himself. It was assumed that, being a tinsmith, he had discovered this hiding place when doing some work on the furnace pipes."

FIG. 284b

FIG. 284b *Dial detail of the clock in Fig. 284a.*

According to Lewis Brown, Darling was a well-liked man who took no part in public affairs. During his half century in business, he was held in the highest esteem as a man of high moral and religious principles. He was a devoted member of the Methodist Church and for many years was a teacher in the Sunday school. When he first came to Simcoe, his residence was on the corner of Evergreen Drive and Culver Street. He later purchased about 4 acres of land on the corner of Evergreen Drive and Norfolk Street and built a large brick house. He was remembered for the proud manner in which he arrived to work each day with his horse and carriage. In fact, the census of 1861 makes note of his having in his possession a horse and two carriages, part of his total worth of over $8,000.

In Simcoe, in 1847, George L. Darling married Rebecca Walker of Woodhouse, but she lived for only a few years. In 1855 he married Mary R. Walker (b. about 1825). Of their four children, only James Leland (b. 1858) lived to adulthood. George L. Darling remarried in 1867. His third wife, Phebe Ann Kennedy (b. about 1843 – d. after 1926), was Irish, born in Smithville. Four children were born: Clarence K. (b. about 1869), Arthur C. (b. about 1871), Ernest H. (b. about 1878), and Victor S. (b. about 1879). After the death of her husband, Phebe left Simcoe for Hamilton, where she lived with her son Ernest.

George L. Darling is buried in the old Methodist Church cemetery at Woodhouse. At the time of his death, his estate was worth nearly $12,000.

THE DOMINION CLOCK COMPANY

Three simple 30-hour time-and-strike OG clocks with overpasted labels of the Dominion Clock Company have been found over the years in Ontario. These have been regarded as somewhat of a mystery since no address or printer's name is shown. The label is pasted over a typical Waterbury Clock Company label and the movement is an earlier Waterbury type (Taylor type 2.411). The plates are held together with steel pins rather than the later blued steel screws. The instructions on the Dominion label appear to be copied word for word from Waterbury.

The authors have not been able to find any trace of such a company. One possible explanation for the existence of these clocks is suggested here in the absence of any hard evidence. In 1872 the Canada Clock Company of Whitby, Ontario, began to produce 30-hour OG clocks which were closely copied from Waterbury in both movement and case design. The earliest labels simply stated "Canada Clock Company" with no address. It is tempting to think that perhaps Waterbury or some Canadian agent decided to retaliate with similar clocks labelled "Dominion Clock Company." There would have been support for "Made in Canada" clocks and someone may have seen this as an opportunity. See clocks in Fig. 285.

NATHAN FELLOWES DUPUIS

Nathan Fellowes Dupuis, M.A., L.L.D., F.R.S.E., F.R.S.C. (b. 13 Apr. 1836 – d. 20 July 1917) was a clockmaker and a professor at Queen's University, Kingston, Ontario. Although he made a number of clocks, his most visible achievement in the horological field was the tower clock in Grant Hall Tower at the University.

Nathan F. Dupuis was born in Portland Township, Frontenac County, to the Dupuis-Wells family. His father, Joseph Wells (b. 24 May 1798 – d. 19 June 1870), chose to change his French name of Dupuis to Wells when he moved from Lower Canada to the Kingston area. Two of his children, Nathan and Thomas, preferred to use the original name of Dupuis. Nathan's mother was Eleanor Baker (b. 25 Feb. 1799 – d. 6 Nov. 1878).[8]

As a child on the farm, Nathan exhibited remarkable manual dexterity. At the age of thirteen he built a clock from materials found on his father's farm. William Smith, a watch- and clockmaker in Kingston, heard about the youth and offered him an apprenticeship. After spending four years with Mr. Smith, Nathan taught school, saving money to attend Queen's University. During this period, Nathan increased his knowledge on many subjects by reading widely.

It was not until 1863 that Nathan Dupuis entered Queen's University as a student. He graduated in 1866 with a B.A. and in 1868 with an M.A. That year he was appointed professor of chemistry and natural history at the university, at a salary of $500 per year. To help the financially troubled university he made all the laboratory apparatus, and students were stimulated by the simplicity and ingenuity of his methods. He also made and used much of the equipment at the Kingston Observatory and was an observer in his student days. His name was mentioned in an observatory report concerning the transit of Venus in 1882. There was a "very perfect mean time clock in Professor Dupuis' house with compensation pendulum and Grimthorpe gravity escapement...and electrically connected with a chronograph there and with the observatory. Both clock and chronograph are of Professor Dupuis' construction and the rate of the clock is very steady." In 1887 Dupuis made a clock with batteries to ring classroom bells for the university.

FIG. 285a

FIG. 285b

FIG. 285 *"Dominion" and "Canada" clocks are similar.*

175

FIG. 286

FIG. 287

FIG. 286 *Nathan Fellowes Dupuis.*
FIG. 287 *Grant Hall Tower.*

In the 1870s Dupuis was one of the few professors to welcome ladies at the university, and in 1892 he appointed the first woman, Netta Reid, to his staff. He also supported the teaching of practical sciences. As professor of mathematics and science from 1880 on, he refuted many of the principles of Euclid; his excellent mathematical models are still on display at the university after a period of over a hundred years. He helped to readmit the Royal College of Physicians and Surgeons to the college. He also walked 4 miles in all kinds of weather to conduct daily classes.

Nathan F. Dupuis wrote a number of books on mathematical subjects and was one of the most prolific scientific writers in the *Queen's Quarterly* founded in 1893. After his retirement in 1911 he continued to contribute to the periodical on a diversity of subjects such as mathematics, astronomy and measurement of time.

In 1894 Nathan Fellowes Dupuis became dean of the Faculty of Practical Sciences. In his honour, a portrait painted in 1901 by Robert Harris now hangs in Dupuis Hall, the chemical engineering building completed in 1966 and named after him.

Nathan Dupuis also continued to build clocks, which are now owned by Queen's University. He employed a gravity escapement similar to but less complicated than the double three-legged gravity escapement designed by Lord Grimthorpe (Edmund Beckett Denison) for the tower clock at the House of Parliament at Westminister. Because of his fascination for astronomy, most of the clocks Dupuis made are astronomical clocks.

Dupuis designed the clock that was placed in the Grant Hall Tower. He was assisted in its construction by James Connell,[9] instructor in the mechanical laboratory. The clock was completed in 1905. The parts of the clock were brought up to the top of the tower and assembled there. By 26 May, dial and hands had been attached. Although the clock was not always accurate, due to extremes of temperature as well as wind, rain, snow, and the fact that birds often perched on its hands, it fulfilled its duty for more that eighty years. The clock was retired in 1992 and removed from the tower to be placed on display.

One unique Dupuis clock has nine dials, which show mean time, sidereal time, day of the month, position of the planets in the ecliptic, position of the moon, and four positions of the sun. The planet Mercury appears to have always been missing from the orrery. The frame around the dials was made of wood from cigar boxes. The mechanism of another astronomical clock was also constructed of cigar-box wood.

On his retirement Nathan F. Dupuis received honorary degrees of L.L.D. from both McGill University and Queen's University. He continued to write and spent much time in California. Dupuis was from a Methodist family, but he considered himself to be Presbyterian, and faithfully attended Chalmers Presbyterian Church, to which he willed money. His liberal beliefs were the moving force in his fight to have Queen's University interdenominational instead of under the influence of the Presbyterian Church.

Nathan Dupuis married Amelia Ann McGinnis (b. about 1844 – d. May 1905) on 20 Aug. 1860. Their three children were James McGinnis (b. 24 May 1861 – d. 25 Sept. 1895), Helen Heloise (b. 17 Dec. 1864 – d. 9 Dec. 1941), and Eugene Leon (b. 3 Nov. 1866 – d. 15 Feb. 1933). His second marriage was to Mae Gordon Thompson (b. 1868 – d. July 1943) on 2 May 1906. Nathan died of "acute gastritis" on 20 July 1917 in Long Beach, California. He is buried in Cataraqui Cemetery, Kingston, Ontario.

FIG. 288

FIG. 289

FIG. 290

FIG. 291

FIG. 292

FIG. 293

FIG. 294

FIG. 288 *Clock by Dupuis in a Queen's University laboratory.* FIG. 289 *A regulator-type clock by Dupuis.* — *Courtesy Queen's University, Kingston, Ontario* FIG. 290 *Movement detail of clock in Fig. 289. Note gravity escapement and use of lantern pinions.* FIG. 291 *Grant Hall tower clock movement by Dupuis. The gravity escapement can be seen at centre right.* FIG. 292 *Another view of the Grant Hall tower clock movement.* FIG. 293 *Unique clock with nine dials made by Nathan F. Dupuis.* FIG. 294 *Mechanism constructed of cigar-box wood found behind the dial of an astronomical clock made by Nathan F. Dupuis.*

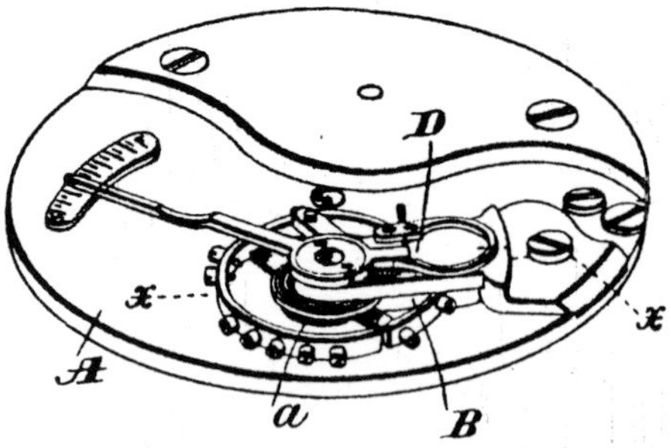

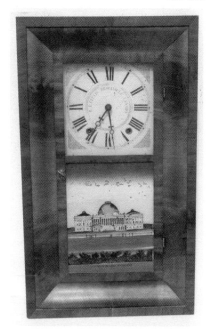

FIG. 295

FIG. 296

FIG. 297a

FIG. 295 *Drawing from the Felt patent.*
– Courtesy NAWCC
FIG. 296 *OG clock sold by R.B.Field*
FIG. 297a *Movement of clock in Fig. 296.*

HOWARD A. FELT

In 1891 Howard A. Felt (b. 2 Aug. 1860 – d. 25 Mar. 1949) of Oshawa, Ontario, applied for a United States patent, and on 4 April 1893 patent number 494919 was awarded to him. The patent was "as a means for testing and adjusting a watch balance and its connecting parts, the combination with a back movement plate, of balance staff cocks, which are each secured upon the rear face of such plate and are adapted to be turned outward so as to bring the balance outside of the edge of the movement." In addition to the fact that Howard A. Felt received a patent, a watch is known that carries the name of "Felt Bros." on the dial.

In 1886 Howard A. Felt opened a watchmaking and jewellery store on Simcoe Street South in Oshawa. By 1892 his brother Everett (b. 1859[10] – d. 1900) became his partner. When Everett died, Howard purchased his equity.

Howard's son, Neil (b. before 1900 – d. 1961), became a partner in 1935 and took over the business after Howard died. Howard's daughter, Marian (b. 12 Apr. 1900 – d. 14 June 1965), also worked in the business, which continued to be known as Felt Brothers. In 1962 Lord's Credit Jewellers of Oshawa occupied the store. There are no descendants of Howard A. Felt.

Howard A. and Everett Felt were sons of Arthur O. Felt (b. 1834 – d. before 1878), who settled in Oshawa sometime between 1864 and 1868. Although of English extraction, Albert and his wife, Ladoria (b. 1833 – d. 18 May 1913), were born in the United States, as were Howard A. and Everett.

R.B. FIELD

Rodney Burt Field (b. 25 Feb. 1809 – d. 18 Mar. 1884) came to Elizabethtown Township, Leeds County, from the New England states in 1833. Here he was married on 6 November 1833 to Louisa H. Chamberlin (b. 9 Sept. 1810 – d. 26 Jan. 1882). He and Louisa lived with John Ketchum and Mary (Louisa's sister) and family in Lyn, Elizabethtown Township. For three years Field sold weight clocks with wooden movements throughout the area. The name of a larger community, Brockville, was on the clock labels.

It is probable that Field was the first seller of clocks in Leeds County because most of the clocks that the authors have examined have wooden movements. No other wooden-movement shelf clocks with Brockville or Leeds County labels are known to exist.

Five Field wooden-movement clocks were available for study. One clock, possibly the oldest, has a standard Terry-type 30-hour movement, designated by Dr. Snowden Taylor as type 1.114. The movement was made by Terry firms, including E. Terry & Sons, Eli Terry Jr., and Henry Terry. This movement is in an OG case.

The majority of the wooden-movement clocks sold by R.B. Field were in bevel cases. The bevel case was introduced toward the end of the wooden-movement era which ended about 1840. The movements of these clocks were of the style made by Chauncey Boardman and Boardman and Wells in Connecticut. The movement is Taylor type 9.223.

Before returning to the United States, R.B. Field became associated with an unknown person or persons and thereafter "R.B. Field & Co." was printed on the labels. One bevel case clock with "R.B. Field & Co." on the label and the Boardman and Wells movement is pictured in Figs.

FIG. 297b

FIG. 298

FIG. 299

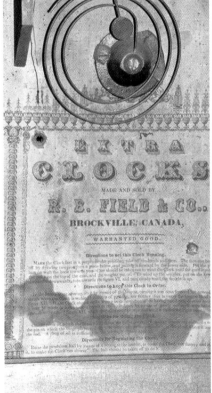

FIG. 300

FIG. 301

FIG. 302

FIG. 297b *Label of clock in Fig. 296.*   FIG. 298 *R.B. Field clock in bevel case, with original mirror.*   FIG. 299 *Movement from clock in Fig. 298, by one of the Boardman firms (Taylor type 9.223).*   FIG. 300 *Label from clock in Fig. 298.*
FIG. 301 *An R.B. Field label variant from a bevel case clock.*   FIG. 302 *A second label variant in a bevel case clock containing a wooden movement.*

FIG. 303

FIG. 304

FIG. 305

FIG. 303 *R.B. Field clock with brass movement in OG case.*
FIG. 304 *Brass movement of Field OG clock.*
FIG. 305 *Label of clock in Fig. 304.*

298, 299, and 300. The label is one of five variations of labels found on Field clocks. The identical label and movement were found in an OG case fitted with a wooden movement.

It should be noted that on the labels, the statement "made and sold by" R.B. Field (& Co.) appears. According to Brooks Palmer in his book, *The Book of American Clocks,* this statement was frequently used on the labels of clocks made by Boardman and Wells and Chauncey Boardman.

Two additional label variants are illustrated in this section.

Although R.B. Field was in Canada for two clock-selling periods, he spent most of his life in the United States. His parents lived in Vermont, and Rodney was first employed in 1824 as a clerk in a store in Brattleboro. When the business was sold, he went to Sacket Harbor, New York, where he was "engaged in the mercantile business." He sold this business in 1833 and gave his land to his father before coming to Upper Canada.

In 1836 R.B. Field and his family left Leeds County. After a short time in Ohio they moved to Michigan City, Indiana, where Rodney manufactured furniture. Between 1837 and 1839 they lived in Guilford, Vermont, where three of their children died. In 1840 they returned to Leeds County and lived with the Ketchum family, who had moved to Concession IX, Elizabethtown Township. Again, R.B. Field began to sell clocks. It is not known if he sold some of the wooden-movement clocks in Canada after 1840. It is known that he sold clocks with brass movements.

The brass-movement clock shown in Fig. 303 has an OG case, and the brass movement is type 4.511 assigned by Dr. Snowden Taylor to brass movements made by Elisha Manross. Brooks Palmer also states that Manross made wooden clock parts for Boardman.

Rodney Burt Field continued to sell clocks in Leeds County until the spring of 1851. The Field family then left Canada for Ogdensburg, New York, where their son, George Pliny, was born. R.B. Field ran a furnace business in New York, but while he was there he had a serious accident and was forced to sell his business interest. He returned to Guilford, Vermont, where he was appointed inspector of distilleries for the Second Congressional District of Vermont. He resigned in 1866. Meanwhile, he was appointed postmaster on 1 October 1865, a position he held until his death.

On 10 June 1870 he was elected member for Guilford to the Constitution of Convention of Vermont held in Montpelier. He represented the town in the legislature in the years 1870 to 1873.

R.B. Field has been mentioned in a number of books that have been written about the State of Vermont. Also, Frederick C. Pierce published a book in Chicago in 1901 called *Field Genealogy.* Much of the research appearing in this book was gathered by clockman R.B. Field. His genealogical notes have been microfilmed by the Mormon Church in Salt Lake City, Utah.

Rodney B. Field was a descendant of Hurbutus de la Field from Alsace-Lorraine. Before the 16th century, members of the family had settled in Britain, and early in that century Rodney's ancestor, John Feild [*sic*], was a famous astronomer in England. In a period when scientists whose ideas were in conflict with the teachings of the church were being burned at the stake for heresy, Feild nevertheless followed the ideas of Copernicus. He published astronomical tables in Latin, and in 1556 he produced his own almanac for 1557, written in Latin, which contained tables prepared for the latitude of London. These tables were based on the Copernican theory. The book's preface, written by famous astronomer John Dee, contained an explanation of the theory and the advantages it offered. An ancestor of R.B. Field came to the New England colonies before 1700.

Rodney B. Field died in Guilford of pneumonia on 18 March 1884. He was seventy-five years of age. He and his wife are buried in Guilford with other members of his family.

SIR SANDFORD FLEMING

Sir Sandford Fleming, "the man who synchronized time"[11] has been included in this book not only as a horologer, but also because of the very important work he did on the standardization of time measurement. He was a man of impressive stature and intelligence, who in his lifetime accomplished an astounding variety of achievements and held many important posts. From a horological viewpoint, he is best remembered for his contributions to the concept of world-wide Standard Time, with all the ensuing benefits to railways and other time-sensitive organizations. In addition his interest in time was recalled when a watch was sold at a Christie's auction in New York in October 1990. The watch, acquired by the National Museum of American History (the Smithsonian Institution) of Washington, D.C., for some $22,000 US, was designed by Fleming and used by him to demonstrate his concept of world-wide standard or "Cosmic" time. The watch was described as follows in the auction catalogue:

FIG. 306

> A unique and highly important gold open face and half hunter cased double dial watch with world time zone indications built to the design of Sir Sandford Fleming by Nicole Nielsen & Co. for E. White London. The gilt lever movement with Nicole's keyless work geared for two dials, the first labelled *Local Time* with centre seconds having black roman chapters I – XII enclosing red arabic chapters 13 – 24, the second, labelled *Cosmic Time* on a large scroll-engraved gilt hour wheel within a 24 hour annular white enamel arabic chapter ring further labelled A – Z designating Fleming's world time zones, within a gold half hunter cover similarly enameled A – Z in blue and red, blued steel hands, c *1880–51.5 mm. diam.*

FIG. 306 *Sir Sandford Fleming, May 1874. – Courtesy National Archives of Canada (PA 26439)*

Sir Sandford Fleming[12] was born at Kirkcaldy, Scotland, and even as a boy, was fascinated with the concept of time. He decided to follow the advice of Benjamin Franklin in the *Poor Richard's Almanak*, "Dost thou love life? Then do not squander time for it is the stuff life is made of."

Fleming, a surveyor, immigrated to Canada in 1845. He was a shrewd investor and used much of his own money to promote the concept of Standard Time. Although the concept was suggested by many people, including Charles F. Dowd in the United States, Sir Sandford Fleming was responsible for its world-wide acceptance. After an incident in 1876 in England when because of an error in railroad time he was seventeen hours late for an engagement, he realized the need for a complete overhaul of the time system so that it could complement the excellence of the railroad system in Europe and the United States.

Certainly the fact that North American railways crossed as many as fifty-eight time zones by the year 1880[13] was responsible for initiating a change in attitude toward time in that continent. In 1876 Fleming prepared a paper entitled "Terrestial Time" [sic], which he circulated widely. In it he reviewed the story of time measurement throughout world history and proposed a prime meridian from which time would be measured in 24 standard zones, 15 degrees apart. He also suggested a 24-hour clock instead of a 12-hour clock.

FIG. 307

FIG. 307 *Fleming's watch has two dials, for "Local" and "Cosmic" time, dividing the world into twenty-four standardized time zones. The "Cosmic" time dial is shown here and the "Local" time dial is pictured on page 1.*
– Courtesy Smithsonian Institution, Washington, D.C.

He addressed audiences on both sides of the Atlantic on the subject of "Universal Time." The Governor General for Canada used his high office to publicize the papers to many foreign governments. He found one of his first allies of the system in the government of Russia. At his own expense, Fleming, using England as a base, made many trips to European countries in the early 1880s.

After considerable documentation on the subject, the railway authorities in North America recognized the new system on 18 November 1883,[14] and it was implemented without incident but not without opposition, as indicated by some of the nation's newspapers.

Fleming continued to strive for the use of Universal Time in Europe and on New Year's Day 1885, the new system of Universal Time was adopted widely. The system comprises twenty-four standard zones, 15 degrees apart in longitude, starting with Greenwich. Within each zone the time is uniform and changes one hour in passing from one zone to the next. This system was eventually adopted almost everywhere. In some regions a compromise involving half-hour differences is used, e.g. Newfoundland. Also, during the summer months the clocks in some countries are advanced one hour, resulting in Daylight Saving Time.

The 24-hour clock proposed by Fleming was not so universally accepted, although its use is more and more common.

On arriving in Canada in 1845, Fleming wisely ignored the advice given him by a bishop in Toronto, who told him, "Go back to Scotland, my boy. There is no future here for the professional man. All the great works in this country have been completed." Sir Sandford's contributions to Canada are as varied as they are important and continued after his official retirement in 1880.

Upon entering Canada, he qualified as a land surveyor and became chief engineer of three different railways. Under his control were constructed the Intercolonial Railway and much of the Canadian Pacific Railway. He crossed Canada on foot through the Rockies in 1871 to carry out a preliminary survey for the new Canadian Pacific Railway. He recommended a northern route through the Yellowhead Pass. This was later abandoned in favour of a more southerly route, as a defense against the perceived threat of American expansion, using the Kicking Horse and Rogers passes. Fleming was eased out of office as engineer-in-chief of the Canadian Pacific Railway in February 1880 for a variety of political reasons. He went to Europe later in the year to pursue his interests in standardized time measurement.

In 1883, however, Fleming was asked to make a journey through the proposed southern route of the C.P.R. to confirm that the Rogers and Kicking Horse passes were indeed suitable. He confirmed that they were and this became the final route of the C.P.R. through the Rockies.

The involvement of Fleming and his influence on the route of the proposed railway across Canada is told in detail in *The National Dream* and *The Last Spike* by Pierre Berton. Of all the men connected with the active planning and construction of the railway, with the exception of politicians, only Fleming was present at both the beginning and the end.

Sir Sandford Fleming's main accomplishments are highlighted below:

1. He designed Canada's first postage stamp.
2. He founded the Royal Canadian Institute, a society of professional men that still exists.
3. He wrote papers not only on Universal Time, but also on many other subjects, which included "Short Daily Prayers for Households and for Travellers" and many scientific papers on railway surveys and construction. He was the author of over 150 books, pamphlets, and articles.
4. After retirement he backed the construction of a state-owned system of telegraphs throughout the British Empire and saw completion in 1902 of the laying of a Pacific cable linking Canada and Australia.
5. He was a director of the Hudson's Bay Company and president of several private companies.
6. He was chancellor of Queen's University in Kingston, Ontario, for twenty-five years, although he had never attended a university or college. He made substantial financial gifts to Queen's University.
7. He was chief engineer for the Dominion Government from 1867 to 1880.
8. He advocated confederation in the period of 1864 to 1867 and in 1891 attacked the Liberal policy of unrestricted reciprocity with U.S.A.
9. He acted as ambassador to Hawaii.

According to Pierre Berton, Fleming was an impressive man physically as well as intellectually. He wore a vast beard from the time he was in his twenties until his death. He was over 6 feet tall and was comfortable talking to persons in all walks of life.

Fleming received many honours during his career. He received the C.M.G. in 1877 and the K.C.M.G. in 1897 and he was given a number of honorary degrees, including four law degrees from St. Andrew's University in Scotland, Columbia University in New York, the University of Toronto, and Queen's University.

Sir Sandford Fleming died in Halifax on 22 July 1915.

FIG. 308

FIG. 308 *A recently discovered clock made by the Hamilton Clock Company.*

THE HAMILTON CLOCK COMPANY

The industrial manufacture of clocks in Canada is outside the scope of this book. However, because the production of clocks by the Hamilton Clock Company took place in the 19th century, a brief summary of this company's activities has been included. Details of the company and its clocks can be found in a book by Jane Varkaris and James E. Connell entitled *The Canada and Hamilton Clock Companies.*

The Hamilton Clock Company, Hamilton, Ontario, was established in 1876 by James Simpson, president, and George Lee, business manager, utilizing the equipment purchased from the defunct Canada Clock Company of Whitby, Ontario. From that organization, they also hired John Collins as mechanical superintendent. During the next four years this company manufactured a variety of clocks with both spring and weight movements. Both cases and movements were manufactured at the plant.

About 1879 John Collins left the company after a disagreement with the owners. George Lee withdrew in 1880 because of ill health. Production of clocks stopped at this point. James Simpson found other investors to start a new company, which was incorporated in 1881 as the Canada Clock Company Limited, and continued to make clocks in the same building at Cathcart and Kelly streets in Hamilton.

GEORGE HARDY

George Hardy (b. about 1783 – d. 5 Jan. 1873) was a native of Scotland and operated a watch and clock shop in Niagara, Upper Canada (now Niagara-on-the-Lake), for a period of four years between 1823 and 1827. One tall case clock bearing his name has come to the authors' attention. Three announcements appeared in the *Niagara Gleaner* concerning his activities in the four-year period:

9 August 1823 – GEORGE HARDY, Respectfully Informs the Inhabitants of Niagara, and its vicinity, that he has opened a Shop in the Town of Niagara, U.C. Where he intends to Practice Watch and Clock making in all its branches. G.H. having served a regular apprenticeship in SCOTLAND, and passed a course of instructions under the best masters in LONDON, as a Watch Finisher, etc. flatters himself he is qualified to undertake Watches & Clocks of the highest finish and most complicated constructions — Those who may please to favour him with their work may depend on the greatest attention & the strictest justice.

N.B. Mariners and Pocket Compasses made and repaired — Wedding Rings warranted Sterling Gold — Gold and Silver work neatly repaired, etc. etc. Wheels cut for the Trade.

2 July 1825 – GEORGE HARDY, Watch and Clock-Maker, Has Removed to the HOUSE of Mrs. Rogers, next door to Mr. Stocking's Hat Store.

19 May 1827 – SECOND NOTICE. Those who have WATCHES uncalled for in the hands of the Subscriber, will receive them on applying to Robert Burns Esq. Attorney. GEORGE HARDY.

These references would suggest that Hardy opened shop optimistically in 1823, moved to smaller quarters in 1825, and gave up his business in 1827.

The dial bearing Hardy's name is illustrated in Fig. 309. Unfortunately, the clock has been separated from its case in recent years. Originally the case was of cherry and was probably made locally in the Niagara area.

The case had simple lines, stood 7 feet 11½ inches tall and the hood was surmounted by "horns." The clock was known to be in the U.S.A. for a number of years. Since it did not sell quickly, the Hardy dial and movement were removed and the case was sold with another movement, as a "Pennsylvania" clock. The movement and dial have since been repatriated to Canada. The dial and movement are typically English, and it is apparent that Hardy merely added his name to an imported English clock and had it cased locally. In the arch of the dial, the figure of Britannia can be seen and also a trident and crouched lion. It is regrettable that the case has been lost. However, the incident does support the authors' belief that Niagara Peninsula tall cases were generally similar to those of Pennsylvania.

George Hardy is also listed as having been in business in Kingston, Upper Canada (Canada West), from 1829 to 1864. One tall case clock with George Hardy's name on the dial was sold in Canada West.[15] However, the location of this clock is not known.

George Hardy was established as a Kingston watchmaker by 1829. In *Canadian Silversmiths and their Marks*, John E. Langdon quoted a newspaper advertisement of 1829 in which George Hardy announced he had "moved to the house formerly occupied by Leslie & Sons, Lower end of Store Street, nigh the Post Office." In the same newspaper, he offered "Clocks, teaspoons, gold rings, silver thimbles etc."

By 1838 his residence and his shop were on Princess Street on the block between Wellington Street and Bagot Street. In 1851 a fire destroyed his shop and much of the block. He rebuilt his three-storey stone building after the fire. The 1857 and 1862–63 directories gave the number of his shop as 29 Princess Street (now 121 Princess Street). In 1861 George Hardy was training an apprentice named Mathew Gage, a seventeen-year-old youth from Ireland.

George Hardy remained in business until 1864 when he rented out the business quarters to Vincent and John Ockley, merchants.

George Hardy's wife, Isabella (b. about 1792 – d. 23 Jan. 1847), was born in England. Four children, Jesse (b. 1813 – d. 22 July 1840), George Jr. (b. 1815 – d. 9 Feb. 1840), Andrew (b. 1820 – d. 9 Dec. 1854), and William (b. 1822 – d. 29 Jan. 1892), came with their parents to Niagara. The youngest daughter, Selina (b. 1831 – d. 18 Sept. 1853), was born in Canada West and died before George Hardy made his will.

Also born in Upper Canada was a son called James. In 1871, when George wrote his will, he had had no contact for many years with James and stated in the will that James's "Conduct was a very great grief to me." The will provided for a bequest of five shillings to him and each of his lawfully begotten children.

Three other children of George Hardy were alive at the time of his death. William was in an asylum in Vermont and he was provided for. Ann was married to Mr. Fenwick and she and her son Kenneth L. were named in the will. Isabella lived with George Hardy and he appreciated her taking care of him. He provided for any of Isabella's children. Also in the will, provisions were made to evangelists to further the work of the church.

FIG. 309

FIG. 309 *Dial detail of a clock marked "Geo. Hardy Niagara U.C."*

185

George Hess (b. 1833 – d. 1891) was born in Wurtemburg, Germany, in the Black Forest region. Around 1851 he went to Switzerland, where he learned the clock- and watchmaking trade. In 1856 he set out by ship to cross the Atlantic with his wife, Christina. Unfortunately, during the voyage an epidemic broke out, taking the life of Christina. George was delayed for some time by medical authorities upon his arrival in New York, but eventually came to Canada West, to the settlement of Berne (now Blake) in Hay Township, Huron County. Here he worked as a carpenter. When he had saved enough money, he returned to Switzerland to purchase equipment to be used in his own watch- and clockmaking business.

However, ill fortune continued to follow him. His ship capsized in New York harbour and he lost most of his purchases. Nevertheless he carried on and came to the village of Zurich, Canada West (also in Hay Township). He purchased property at the corner of Zurich Road and Frederick Street, where he set up a watch and clock repair business. In 1870 he purchased property on Goshen Street, where he added photography to his business. In the 1880s he began to manufacture tower clocks. In addition, he operated an office of the Northwest Telegraph Company, making daily reports to the Meteorological Association in Toronto. He experimented to improve the telegraph system. He also became interested in battery-operated electric clocks.

George Hess married again in the 1860s to Rose Stelke (b. about 1850). He is reported to have had one son, Christian,[16] by his first wife. Christian eventually came to Canada and developed an interest in clocks. He made four clocks in the 1920s, which were decorated with pebbles from the Bayfield area.

There were three sons and a daughter by the second marriage. The eldest, Fred W. Hess (b. about 1869), carried on the business in watch and clock repairs after George's death by "blood poisoning" on 16 June 1891. Henry (b. about 1871), William (b. about 1872), and Flora (b. about 1876) were the other children of George and Rose Hess.

This brief summary of George Hess's life serves to illustrate that he was a man of considerable talent, with a wide range of interests. (He was also noted for his beard, which measured 28 inches in length.) In horological matters, he was an innovator and was successful in obtaining two Canadian patents. These were No. 30429 for improvements to tower clocks (April 1888) and No. 32485 for improvements to electric clocks (April 1889). Copies of these patents are illustrated in *Canadian Clocks and Clockmakers* by G. Edmond Burrows.

There are three known Hess tower clocks. One was installed in the tower of St. Peter's Evangelical Lutheran Church in Zurich around 1880, where it is still in service. A second was installed in 1881 at Trinity Evangelical Lutheran Church, which is on the outskirts of Tavistock, Ontario. This clock, too, continues to operate. The third clock was installed in the town hall at Exeter, Ontario, around 1888. This clock was superseded by an electric movement a few years ago, although most of the old movement was left in the tower.

Hess tower clocks are distinctive. They are all constructed with wooden frames built up from oak or other hardwoods. The arbors are of iron and run in brass bearings, which are attached to the wooden frames. The wheels are commercial cast-iron gears. The two escapements that have

FIG. 310 *George Hess and his wife, Rose Stelke, shortly after their wedding.* London Free Press.
– Courtesy the National Library of Canada

FIG. 311

FIG. 312

FIG. 313

FIG. 314

FIG. 315

FIG. 316

FIG. 311 *Saint Peter's Evangelical Lutheran Church, Zurich, Ontario, with steeple clock made by George Hess.* — *Courtesy Ken W. Yokom* FIG. 312 *The Zurich clock is of simple design on a heavy painted wooden frame.* FIG. 313 *View of time-and-strike trains in Hess clock at Zurich.* FIG. 314 *Pendulum arrangement of Zurich clock.* FIG. 315 *The Exeter movement. Pendulum suspension, escape wheel and other parts are not present.* FIG. 316 *Hess clock dial at Trinity Lutheran Church, Tavistock.*

FIG. 317

FIG. 318

FIG. 319

FIG. 317 *The Tavistock clock is in excellent condition.*
FIG. 318 *Detail of escapement in Tavistock clock. The clamps are temporary to adjust levelness of the suspension.*
FIG. 319 *Side view of the Tavistock movement showing time train on the left and two strike trains on the right.*

survived are similar, having brass escape wheels with long pointed teeth. The pallets are of hardened steel set into a brass anchor of distinctive shape. The escape wheels are of relatively small diameter and require quite long pendulums, on the order of 6 feet in length. The pendulum bobs are about 12 inches in diameter, made of cast iron. The bob at Exeter weighs about 25 pounds and the one at Tavistock may be heavier. The weights for the various clocks are quite heavy and, at Tavistock, they vary from train to train. No information is available, but the actual weights are probably between 200 and 400 pounds.

The Zurich clock is a relatively straightforward mechanism, as can be seen in Figs. 312 and 313. The frame is heavy, but simple in shape. All arbors run in the same plane, being lined up on the top of the frame in two rows. The weights are made up of boxes of scrap iron. The gears are narrow solid discs. The escape wheel has twenty-six teeth. The pendulum rod, which is wooden, is fitted with a short suspension spring. The clock has three dials, and the reduction gears that turn the hands were originally wood, although some have been replaced. The movement is located in a room immediately behind the dials. The movement has two trains, for time and hourly strike.

The Exeter clock unfortunately is in a state of disrepair. It lies in the tower, behind the dials. The frame and larger wheels are present, but important parts, such as the escapement, appear to be missing. The weights and pendulum bob still exist. It is quite different from the Zurich clock, but does resemble the example in Tavistock. As can be seen in Fig. 315, it has a finely crafted wooden frame, which supports two gear trains for time and hourly strike. A brass plate on the frame reads, "Pat. Dec. 18 1888."

The Tavistock (or Sebastopol) clock is well preserved and slightly more complex when compared with the one in Exeter. Both clocks have commercial iron gears with five or six spokes. At Tavistock there are three gear trains. The time train occupies half of the oak frame. In the other half are a quarter-hour strike train and an hour train. The quarter-hour strike is 1-2-3-4 on a small bell in the tower. The hour strike that follows the fourth quarter is struck full count on a large bell. The pallet arbor is currently running in commercial ball bearings, which are not believed to be original. The pendulum bob is suspended by hooks, which fit over pins in a yoke. This yoke, in turn, is supported on rounded knife-edge-type bearings of small diameter on which the pendulum rocks. There is no suspension spring.

It is interesting to note that the twenty-six-tooth escapement and pendulum suspension are exactly as described in Patent 30429. The patent states in part, "My invention relates to improvements in the mechanism controlling the striking of the hours; of improvements in the escapement; of a free and safe mode of hanging the pendulum, and of the lightness and simplicity of construction of the various parts."

The Tavistock clock is under the care of Mr. Carl Beynon of Tavistock, a local watch- and clockmaker, who was, interestingly, enough the last apprentice of the famous Jerry Smith, who is described elsewhere in this book. Mr. Beynon assisted in a major overhaul of the Tavistock clock in 1977. At that time the wooden pendulum rod and the escapement pallets and anchor were replaced. The new parts were fabricated to be close copies of the originals.

Another interesting feature of this clock is that its movement is located about 30 feet below the dials, which are driven by a very long shaft going vertically from the movement. The strike bells are about 10 feet above the movement.

The Hess clocks were intended to be wound once a week and both operating clocks are still wound manually. The fall of the weights at Tavistock is restricted to about 10 feet within the movement room. This is because of fears that ceilings below would not be able to withstand the fall of a weight should a cable break. There are heavy planks on the movement room floor to break such a fall. As a result, the clock must be wound twice a week or oftener. The Zurich clock has full cable extension, allowing a weekly wind, but floors there are steel reinforced. Neither clock is run in the winter. The dials of the three clocks are of painted wood and have small access doors, allowing the hands to be reached from inside. The hands themselves, where original, are of wood.

The Tavistock clock bears a brass plate containing the name of George Hess and the date 1881. On the wall of the movement room there is a penciled inscription, "F.W. Hess December 9 1881," no doubt placed there by son Fred at the time of installation. It has been said, incidentally, that the clock was brought from Zurich to Tavistock by horse and wagon, a distance of close to 50 miles.

In addition to tower clocks, a tall case clock is known which bears the inscription "Geo. Hess maker – 1888" on the large brass pendulum. According to its owner, the clock case appears to be of cherry wood. The front of the clock is mostly glass. It stands 78½ inches tall, is 16 inches wide at the waist and 18½ inches at the top of the base, which is 9½ inches deep. The dial is glass with numerals reverse painted on it. The owner points out that the hands are not original.

The clock is equipped with a time-only heavy brass movement with a wind hole at the seven o'clock position and requires winding every nineteen days. The clock keeps excellent time. The single weight is also made of heavy brass. The owner relates that "It is a pleasure to own and a joy to display." Unfortunately, it was necessary to refinish the case of the clock, which had been painted at the same time as the dining room wall by a member of the family. During the refinishing process, the dial, dial pan, and hands were misplaced. Years later, entirely by accident, the owner of the clock met the refinisher. He had located the dial and dial pan and gladly turned them over to the owner of the clock, thus making the clock complete and original except for the hands.

FIG. 320a

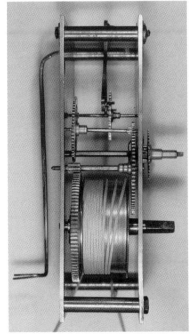

FIG. 320b

FIG. 320 *George Hess tall case clock and movement.*
– Courtesy Richard A. Kidd

189

George Hess was particularly interested in electrical clocks and his original models are still in existence. Patent number 32485 was granted to Hess in 1889. The summary of the patent states that, "My invention consists of a novel arrangement of electrical apparatus for connecting any number of clocks of all sizes and distances apart and operating the same with or without striking apparatus simultaneously from a single pendulum." The clock was powered by dry cells, which George Hess made; later, Edison gravity batteries were used. Although he was asked to relocate in a bigger city to produce electrical clocks, Hess chose to remain in Zurich.

After his father's death in 1891, Fred W. Hess took over, moving the business to Main Street, to the shop illustrated in Fig. 322. Like his father, Fred was interested in the telegraph. In 1917 he sold the Zurich business to his brother William and opened another store in Hensall, Ontario. Later, his son continued the business.

Before his father's death, William, who was in the bicycle business, began to construct telephones; it is believed that he made the first telephones in Hay Township. When his brother Fred bought the first car in the area in 1907, William took over the car dealership. After 1917 William operated the Hess jewellery and watch repair business, and his son Albert continued the business until his death in 1968. Albert's widow, Margaret Hess, was still operating the old shop in 1991, keeping the Hess name alive in clock and watch repairs in Zurich for over 120 years.

ISAAC HORNING

One clock has been reported[17] bearing the label of Isaac Horning, Simcoe. There are several references to Horning at the town of Simcoe (Ontario). His activities are mentioned in *The Cabinetmakers of Norfolk County*, first and second editions by Yeager *et al.*, where he has been credited with being a cabinetmaker, photographer, handyman, and health inspector. According to census records, Isaac Horning was born about 1819. He was a member of the large family of Hornings in Ancaster, Ontario, who, like Isaac, were listed as New Connection Methodists. It is not known when Isaac moved to Simcoe, but by January 1852 he was living there with his wife, Charlotte (b. 1827). There were four children, all born in the 1860s. Dun's Report[18] of July 1884 lists him as a photographer at Simcoe. By 1887 the only Hornings listed in Simcoe were daughters Hannah and Madilla, dressmakers.

The clock reported above is the first indication that Horning could also add "clock peddler" to his list of accomplishments. The clock itself is a bevel-cased weight clock with a 30-hour Terry-type wooden movement. Horning's label partially obscures the original maker's name, which can still be deciphered as Boardman and Wells of Bristol, Connecticut. Clocks of this style are quite late in the history of wooden movements, probably dating from the early 1840s. Horning's label does not specify whether Simcoe at the time was in Upper Canada or Canada West. This would have been a help in dating the clock more precisely, but it seems evident that the clock was sold at an early point in his career.

FIG. 321

FIG. 321 *George Hess is seen in the doorway of his photography studio and watchmaking business in Zurich, admiring the first electrical clock built in Ontario. Son William, who carried on the business, leans against the post.* – *From* London Free Press, *courtesy the National Library of Canada*
FIG. 322 *The Hess shop, Main Street, Zurich, 1991.*

FIG. 322

THE KENT BROTHERS

Kent Brothers was a watchmaker's and jeweller's business established in 1869 at 166 Yonge Street, Toronto. The owners of the enterprise were Ambrose and Benjamin Kent and it remained in the family for over eighty years.

Kent Brothers advertised that the business was "at the sign of the Indian clock." This clock attracted the attention of the public to the store for many years. According to *The Dominion Illustrated,* in a special Toronto edition published in 1892, quaint figures struck resounding bells and gongs at different hours of the day.

Two well-known horologers, H.R. Playtner and Edward Beeton, were employed in the 1880s by Kent Brothers. These men went on to become two of the most important watchmakers of Canada (see p. 156 and p. 144).

The Kent store was a three-storey structure with large plate-glass windows. It featured watches, clocks, bronzes, and leather goods. Kent Brothers manufactured a large portion of the jewellery and other goods sold in the store.

In the mid-1890s, a second Kent's store was established when Ambrose and his son William created a branch at 5–7 Richmond Street. Ambrose Kent and Son was in business until 1945. Benjamin and son Herbert B. continued the wholesale store at various addresses on Yonge Street. A third generation, S.L. Kent, worked with Herbert B. in "Kent's" until the shop went out of business in the late 1940s.

FIG. 323

FIG. 323 *The only known clock bearing Isaac Horning's label survives as a wreck.* – Courtesy Ward Francillon

FIG. 324 *Kent Brothers store on Yonge Street in 1892.* – From The Dominion Illustrated, *courtesy the National Library of Canada*

FIG. 324

JOHN N. KLINE

John Nicholas Kline (originally Klein) (b. 1791 – d. 10 Nov. 1856) was born in Hanover, Germany. He left Germany for Pennsylvania during the 1820s. However, by at least 1830 he had left Pennsylvania to settle in Vaughan Township, York County, Upper Canada, and records show that he purchased land in Concessions IV and VIII.

In the 1830s John Kline is said to have supplied wooden movements used by at least one case-maker, Reuben Burr. One Reuben Burr clock is shown in Fig. 254 and is described in the accompanying text. This is the only clock by Kline and Burr to be examined by the authors, although others are said to exist. This clock contains a typical New England 30-hour wood-movement weight-driven, with seconds bit and calendar. The dial and pull-up weights are also typical of New England clocks. Some references credit Kline with actually making movements, but on the basis of this one example, it rather appears that he was an importer.

In 1837 John Kline built a sawmill on the east half of Lot 12, Concession VIII, utilizing the water power of the big branch of the Humber River. A settlement grew up nearby and was known as Kleinburg. About 1847 Kline built a flour mill on the west branch of the Humber River, and by 1852 he owned a grist mill, a saw mill, and a cooper shop and employed eleven people. In 1852 he began to sell his property and moved to Toronto.

On 10 November 1856 John Kline died without a will and his wife, Anna Eve (b. 1796), went to live with her eldest son, Anthony (b. 1832), in Tecumseh Township, Simcoe County. The 1851 census, in addition to Anna Eve and Anthony, listed Catherine (b. 1839). Other records show that another son, John N. Kline Jr., owned Lot 24, Concession VIII in Vaughan Township, in 1851.

THE LEES FAMILY

Four generations of the Lees family have served the residents of Hamilton, Ontario, as clock- and watchmakers and jewellers, spanning the years from 1861 to 1974. Thomas Lees Sr. (b. 24 Feb. 1841 – d. June 1936) was the founder of the establishment. He was the son of George Lees (b. 30 Sept. 1801 – d. 3 Nov. 1879), a baker by trade who had arrived in Hamilton after a lengthy trip from the area of Lauder and Edinburgh, Scotland. Once settled, he sent for his wife and several children.

Thomas, born in Hamilton, was apprenticed to a watch- and clockmaker on John Street and continued to work as an employee in the shop. However, there was no money to pay him. Business was poor and the employer gave Thomas the shop to settle the debt.

After operating the John Street shop for several years, Thomas moved to 5 James Street North and remained there for forty-five years. In May 1918 he relocated his business to 17–19 King Street West. The business continued at this location until 1974, when the property was taken over by the Canadian Imperial Bank of Commerce and the building was demolished.

According to *Hamilton and Its Industries*, by E.P. Morgan and F.L. Har-

FIG. 325

FIG. 325 *Carriage clock sold by Thomas Lees.*

vey, in 1884 the store was "one of the handsomest in the city with large double plate glass windows…. The premises is roomy…and all around the store are arranged clocks of every known make – Canadian, American and French. Mr. Lees has his Waltham watches made at the factory especially and all are sold by him and are stamped with his name. No watch is sold without being first thoroughly examined by the most expert man in the trade, Mr. Lees himself!"

In addition to the many watches sold with the Lees name on the dials, clocks may be found carrying the Lees name. One such clock is pictured in Fig. 325. Although this carriage clock was purchased in the first part of this century, Lees watches exist from the 19th century. The clock was purchased by Thomas Lees from a firm called Duverdrey & Bloquel in Saint Nicolas, France, to sell to his Canadian customers.

This French company was started by A. Villon in 1867. At first, the firm was run like a co-operative with work given to families to be done at home and paid for on a piecework basis. Even the pinions and wheels were cut in private houses and workers made one item only. In those days, M. Villon even had his own brass band.

About 1900 modern tools were introduced and self-sustained factory production began. In 1887 Paul Duverdrey became director, and in 1910, when Villon died, Joseph Bloquel joined the firm. The trade mark of the factory is a lion (see Fig. 326). A great variety of carriage clocks was made by the firm and, although they were relatively inexpensive, they were well constructed.

Thomas Lees installed most of the tower clocks in the city of Hamilton, among them the city hall, post office, Canada Life Building, and the Hess Street school. The firm was also awarded the public contracts for winding, servicing, and regulating the public clocks in Hamilton and Dundas. T. Jack Lees also reports that generations of Lees men had access to the finer homes in Hamilton on their weekly round of winding and servicing the families' clocks.

FIG. 326

FIG. 326 *Back plate showing lion trademark.*

Sons of Thomas Lees, George (b. 30 Jan. 1866 – d. Nov. 1948) and Thomas Jr. (b. 29 May 1874 – d. 19 Dec. 1951), were trained by their father as watchmakers and joined him in the retail business. The business was turned over to them when Thomas Lees Sr. retired about 1920. Other children of Thomas Sr. and Agnes are William Alexander, Stephen (d. July 1937), Susan H. (b. 1868 – d. Dec. 1959), and Agnes (b. 1877 – d. June 1913).

In 1947 Thomas Jack Lees and his cousin, Ralph Wilkenson Lees, took over the business and operated as partners until 1957, when Ralph died. Jack Lees continued until 1967, at which point the business was turned over to his sons Thomas John and James Richard Lees. Richard became a real estate agent and Thomas John remained in business until 1974 when the bank took over the building. It was one of the oldest buildings in downtown Hamilton. Hand-hewn joists, pegged with wooden dowels, supported the floor; at the back of the store was an old drive shed, with a covered area to unhitch the horses. William Lyon Mackenzie had used this accommodation from time to time.

FIG. 327

FIG. 327 *Clock made by John K. Lemp.* – *Courtesy Queen's Hotel, Stratford, Ontario*

According to T. Jack Lees, "From the time the business started in 1861 until it was closed in 1974, the firm had a fine reputation for honesty and fair dealing. It enjoyed what was known as the carriage trade over many years. Thomas, always ahead of his time, had one of the first telephones in Canada in the old Baker telephone exchange, a forerunner of the Bell system."

With the exception of several years spent as a child at his uncle's farm near Oakville due to frail health, Thomas Lees Sr. lived most of his ninety-six years on the corner of Main and Queen streets in Hamilton. After church union, the family belonged to the United Church of Canada. Many of the family were Masons.

George H. Lees (b. 1860), nephew of Thomas Sr., became a watchmaker, but was more interested in manufacturing and established a factory in 1875 over the bakery in his grandfather George's building. The building was purchased by the post office about 1918 and the Hamilton Manufacturing Jewellers took over the business and established Lees Hamilton Manufacturing Jewellers at 78 Catherine Street North, where it was in 1977.

Sons of George H. Lees, Stuart Hermon (b. 28 Feb. 1884) and Wallace William (b. 17 July 1891), were in business with their father.

JOHN. K. LEMP

John K. Lemp was born about 1860 and lived in Tavistock, South Easthope Township in Perth County,[19] Ontario. His parents were both born in Germany and the family was Lutheran.

John became a carpenter and made four or five large clocks in the 1890s. One such clock is now in the Queen's Hotel in Stratford, Ontario. The clock is nearly 8 feet in height. John K. Lemp was listed in *The Mercantile Agency Reference Book* (Dun, Wiman & Co.) in 1884 and was still in business in 1937 in his machine shop on Woodstock Street North in Tavistock. He was not listed in the 1938 telephone directory.

John married Mary (b. about 1858). In 1891 there were two children, Catharine, aged nine, and John, aged seven. John Jr. became a druggist and branch manager for the Bell Telephone Company.

JOHN LESLIE

John Leslie (b. 1813 – d. 19 Nov. 1895) was a watch- and clockmaker in Bytown (Ottawa). A high-quality French carriage clock bearing his name is in the Laurier House Museum in Ottawa.

John Leslie's shop in 1848 was in "Lower Bytown" on Rideau Street. He advertised in the *Bytown Packet* of 21 June 1848 that he had received a new consignment of watches, clocks, and jewellery. In 1866 he moved to 27 Sparks Street. His home was also at this address. About 1874 he moved his shop to its final address at 62–64 Sparks Street, and the family moved to 335 Theodore Street.

John and his wife, Eliza (b. 1825 – d. after 1900), were born in Scotland. Their three children, born in Canada West, were Jane (b. 1856), James (b. 1858), and John (b. 1862). James clerked in his father's store and trained as a watchmaker with his father. In the early 1890s James managed the shop, but the business was closed by 1900.

FIG. 328

NATHAN MARKS

Nathan Marks (b. 1831 – d. 18 July 1909) was a watchmaker and jeweller who carried an inventory of many clocks in his establishment in Ottawa, Ontario. Included in the inventory were clocks purchased from the Hamilton Clock Company, a Canadian clock company that manufactured clocks from 1876 to 1880.[20] One lot of clocks was made especially for N. Marks.

Marks came to Canada in 1874 and spent a short time in Quebec. He settled in Ottawa, and by 1875 his shop was situated on Sussex Street. The following years he was located at 87 Sparks Street. He remained there until 1881, at which time he sold the business to the firm of Rosenthal and Sons and went with his family to Europe.

He returned to Ottawa and in 1884 formed a company with E. Cochenthaler. They were in the wholesale jewellery and watch business at 65 Sparks Street. Nathan Marks's son Abraham joined them around 1888. Nathan withdrew from the company in 1890 and began a short career as auctioneer at 141 Sparks Street with Charles D. Chitty.

From 1893 to 1904 Nathan was listed in the city directory as a jeweller, optician, and oculist. He retired around 1904 and in 1909 died after an illness of four months. Following a funeral in Ottawa, the interment took place in Montreal. His wife, Reba, survived Nathan and lived in Ottawa until the mid-1920s.

Available records reveal that Nathan Marks's father, Philip, was born in Germany; his mother, Rebecca Davis, came from Southampton, England.

Nathan's son Abraham, who joined him in the firm, N. Marks & Company, continued as a jeweller after Nathan left the company. He was in business on Sparks Street, and with the exception of several years when he was a civil servant he continued as a jeweller until 1900. He went to Montreal for a number of years, but returned before the war and worked for A.J. Freeman & Company.

A second son, Philip, worked in Ottawa until 1899 when he moved to Pittsburgh, U.S.A.

Anne, one of Nathan Marks's three daughters, worked in the Department of Agriculture, married James Goltman and moved to New York after the death of her husband. Anne later returned to Ottawa to live with her mother.

Two other daughters, Mary and Dorothy, married H. Buchan of Winnipeg and J.G. Preston of Ottawa respectively.

FIG. 329

FIG. 328 *Interior of Nathan Marks's establishment.* – *Courtesy the National Archives of Canada*
FIG. 329 *Hamilton Clock Company clock sold by N. Marks.*

FIG. 330

FIG. 330 *Clock by Samuel I. Moyer.*
– Photo courtesy R. Taylor and
Niagara Historical Museum

SAMUEL I. MOYER

A small number of tall case clocks was fashioned in the Niagara peninsula of Upper Canada during the early days of settlement. These were often made by cabinetmakers of considerable skill. The movements vary in origin, but most are from England, often with the local sales agent's name on the dial. In many instances the cases were made from local wild cherry or walnut lumber. Certain similarities can be seen when these cases are compared to clock cases of Pennsylvania origin and, in some instances, the cabinetmakers themselves were immigrants from Pennsylvania.

One such clock may be seen in the Niagara Historical Museum and is shown in Fig. 330. This clock case was constructed by Samuel I. Moyer, a farmer and cabinetmaker who resided near the present town of Vineland on the border of Clinton and Louth Townships. The clock movement is English, unmarked except for the designation "Wilson Birm." on the back of the moon-phase mechanism. The movement is typically English brass 8-day time-and-strike, with moon phase, calendar, and seconds display. The movement, according to museum records, was brought to Canada from England in 1830 by one John Andrews, who died on the voyage. The case is attributed to the Moyer family and probably dates from about ten years later.

The case shows fine woodworking skill. The major wood is walnut with inlays at the top of the hood. The clock stands 8 feet 1 inch high.

Clocks of similar style in other collections are attributed to either Jacob Grobb, a fellow Mennonite from Pennsylvania, or Daniel Simmerman or Henry Simmerman, both from Germany.

Moyer was born in Pennsylvania in 1795. His father, the Rev. Jacob Moyer, was instrumental in promoting the immigration of Pennsylvania Mennonites into the Niagara Peninsula. Jacob and his family came to Lincoln County in 1800, and by 1842 he and his sons John, David, Samuel, and Abraham had settled on land which they farmed in Clinton Township on Concessions VI and VII.

Samuel I. Moyer married Elizabeth (b. 1797) and there were four sons, none of whom was called Samuel I., and three daughters. Repetition of names in the families of Samuel I. Moyer and his brothers makes it difficult to identify the activities of any one person. However, the Samuel I. Moyer listed in the census record of 1871 was born in Ontario in 1818 and would be too young to be the builder of the clock.

A.S. NEWBURY, L. BEEMER AND N.F. BIRLEY

Arkle S. Newbury, alone and with several partners, imported clocks from the United States at the Port of Hamilton. In 1851 Levi Beemer and A.S. Newbury were dealers in clocks and in gold and silver watches on King Street near Hughson Street in Hamilton.

By 1857 Levi Beemer had his own shop on King Street near James, and he remained in business until 1865. In June 1860 he imported six boxes of clocks at the Port of Hamilton.

A.S. Newbury took Norris F. Birley as a partner, and in 1857 they were importers of watches at the Anglo American Hotel in Hamilton. Subsequently, they opened a shop on James Street, opposite Fountain Street. From June 1858 to January 1861 Newbury and Birley imported at least 207 boxes of clocks. The names of Newbury and Birley may be found on the dial of a clock the movement of which was made by George Darling (see p. 173).

By 1864 A.S. Newbury was in business by himself as a wholesale jeweller and importer of watches on James Street, opposite Gore Street.

Newbury, Beemer, and Birley were not mentioned in the 1865–66 *Hamilton Directory.*

THE NORTHGRAVES FAMILY

The Northgraves family can boast of more than seven members who were watch- and clockmakers in many locations in Lower and Upper Canada, Canada West, and Ontario over a period of one hundred years.

As young men, two sons of William Northgraves (b. 1764 – d. 17 June 1819) of Hull, England, came to Lower Canada about 1818. One son, William (b. about 1800 – d. 25 May 1864), settled in Quebec City at 15 Fabrique Street, a shop identified by a clock over the door. In February and March of 1819, William advertised in the *Quebec Gazette* for "an apprentice wanted immediately for watchmaking business – a youth of respectable parents who can give good references for character – who can speak English and French."

William advertised his business frequently in the newspapers. A typical advertisement in the *Quebec Gazette* in May 1822 announced that, "William Northgraves, watch- and clockmaker, silversmith and jeweller respectfully informs his friends and the public that he constantly keeps on hand a neat and general assortment of goods in his line which he will sell low – consisting of excellent 8 day and other clocks, gold and silver watches, cat gut for rackets, maps etc.… all watches and clocks repaired and cleaned."

While in Quebec City, William Northgraves[21] married Theresa Brussien, and two of their four children, William J. (b. 1820) and Frederick (b. 1822 – d. 5 June 1879), were born there. According to the *Quebec Gazette,* Mrs. Northgraves was a "subscriber to the ladies Compassionate Society" and her husband was a frequent contributor to the Quebec Fire Society.

In 1825 William Northgraves and his family moved to Montreal and commenced business at 150 St. Paul Street. By 1826 he had moved to 93 Notre Dame Street. On 18 February 1827 a daughter, Eliza Ann, was born to William Northgraves, horologer, and Theresa of Montreal. She was baptized in the Notre Dame Cathedral on 25 February 1827.

Leaving Montreal about 1829, William spent a very short time in Brockville, Upper Canada, where he worked with his brother George. He announced in the *Brockville Recorder* that he was leaving for Bytown (now Ottawa) where he was "planning to commence the watch and clockmaking business." Records show that while in Bytown, William paid for grazing of his cows on government fields. In 1829 he wrote Lieutenant Colonel By that the inhabitants of Bytown wished for their lots to be freehold and not held in lease. In 1830 his name appears with those of the other citizens in an open letter to Lord Aylmer, complaining of inequality of rents of the lots in Bytown. William was also involved in the military. He was in the 1st Carlton Militia, which was assembled on 4 June 1829 under Colonel Burk. William's son, George Richard (b. 1834 – d. 1919), who became a renowned priest, was born while the Northgraves family lived in Bytown.

From Ottawa the family moved to Kingston where George R. went to school and was present at the first mass conducted by Bishop MacDonnell of Kingston in 1839. However, from newspaper advertisements and city directories it has been established that William Northgraves moved to Belleville in 1843 and was in business there until his death.

FIG. 331

FIG. 331 *A tall case clock sold by George Northgraves in Brockville.*

In 1862 an advertisement appeared in the *Hastings Chronicle* claiming that the Northgraves establishment had been "eighteen years nearly opposite the Commercial Building in Belleville." He also claimed to be "the longest established watchmaker in Canada."

In March 1864 William sold the business to John O. Tucker, and two months later, on 25 May, he passed away at the residence of his daughter, Eliza Ann Deane of Lindsay, Canada West.

Two of William Northgraves' sons, William J. and Frederick, became watch- and clockmakers. It is probable that William J. Northgraves stayed in Kingston (when his father and family moved to Belleville) and that he formed a partnership with a man called Maden. He imported a "lot" of brass clocks from Oswego which arrived for him at the Port of Kingston in 1846 on the boat *St. Lawrence*. He moved to Belleville and advertised in the *Hastings Chronicle* in 1855 that he had for sale gold and silver watches, clocks, and lottery tickets. The store was on Front Street in the Coleman's Building.

For a brief time in 1864–1865, William J. had a shop in Napanee; his son Henry was born there in April 1864. He returned to Belleville and advertised on 2 June 1866 that he was opening a shop on Front Street. The *Province of Ontario Directory* lists his shop at 268 Front Street in 1884.

William J. Northgraves of Belleville married Alice Jane (b. about 1825). It is not known when he and his wife died or where they are buried. An A.J. Northgraves is listed in the directory for 1887, perhaps indicating that William J. had died about 1886.

However, the *Province of Ontario Directory* lists a William J. Northgraves as watchmaker in Seaforth, Ontario, in 1888–1890. It is possible that the Seaforth Northgraves is the son of William J. Northgraves of Belleville.

Another son of William Northgraves, born in Quebec, was Frederick. He worked with his father in Belleville, and by 1851 his father had opened a business, which Frederick managed, in Picton, Canada West. The shop was in business until about 1860. Frederick then went to Madoc, Canada West. He was accidentally killed in 1879 while walking on the railroad tracks.

George Northgraves, born in Hull, England (b. 23 Aug. 1803 – d. 12 Oct. 1873), was a brother of William Northgraves (of Quebec City and Belleville). George married Harriett (b. 1815 – d. May 1881) in England. George and family settled in Brockville in 1832, where he was a watch- and clockmaker in a shop on Main Street. A beautiful tall case clock, pictured in Fig. 331, has his name on the dial.

In November 1851 George decided to relocate. He announced that he was moving to Perth, Canada West, and that "all watches which had been left to be repaired might be picked up with Norman McDonald." The family arrived in Perth in 1852 and opened a shop on the corner of Gore and Harvey streets. The location is now an apartment building.

William, third son of George (b. about 1842 – d. 3 July 1908), apprenticed with his father; the business, known as "George Northgraves and Son," was located on Gore Street in the Matheson Building in Perth. In 1864 they advertised that they sold watches, clocks, plate, and jewellery. Watches and clocks were "cleaned in the best manner and on reasonable terms." By 1871 they were also dealers in gold, silver, spectacles, and musical instruments.

After his father's death in 1873, William remained in business, advertising in 1882 as watchmaker, jeweller, and seller of musical instruments and fancy goods. In 1892 William was also a telephone agent. He remained in business until May 1906 when he sold the building to Alexander Cameron.

William married Maggie (Margaret) A. Feehan (b. 1862 – d. 3 Feb. 1887), oldest daughter of P. Feehan of Trenton, Ontario, on 4 September 1883. Their only child died in 1886. William died in St. Frances de Sales hospital in Smiths Falls.

Another son of George Northgraves of Perth, George Denton (b. about 1843 – d. after 1908), was also a jeweller and watchmaker. He established his business in Almonte about 1862 and remained there until about 1871. On 1 July 1868 George D. married Miss Isabel Dodd (b. 1848) of Kitley Township. The family went west in the early 1870s and settled in Winnipeg. By 1903 they had moved to Prince Albert, Saskatchewan, and in 1907 were living in Edmonton, Alberta.

C. A. OLMSTED

Charles Albert Olmsted (b. 5 Sept. 1867 – d. 29 Sept. 1943), born in Hull, Quebec, was a watchmaker and jeweller who worked in Ottawa, Ontario, from about 1890 until 1943. While in business, he sold clocks that bear his name. One such clock is a marble mantel clock from France, which has "C.A. Olmsted, Ottawa, Ont." on the dial.

Many of the clocks that have his name on the dial, however, were purchased from the Arthur Pequegnat Clock Company, Berlin (Kitchener), Ontario. Although a number of these clocks are mantel clocks, he is best remembered for the many master school clock systems purchased for use in the Ottawa area and in Quebec schools from Gatineau to Shawville.

C.A. Olmsted apprenticed at the shop of Addison's jewellers on Sparks Street, Ottawa, where he worked until he established his own watch repair and jewellery store on 97 Sparks Street around 1890.

About 1895 Olmsted and William G. Hurdman became partners, locating their business at 67 Sparks Street. As silversmiths, they had a silver mark and during this period watches marked with their names were sold.

FIG. 332

The partnership was not a financial success, and in 1903 the store was taken over by Henry Birks and Sons with Charles Olmsted as manager until 1915.

At this time Charles ran a small watch repair shop in connection with his son Howard, who owned an optometrist business. The business was then known as "Olmsted and Son" and was located at 68 Sparks Street in the Blackburn Building. For a few years in the mid-1920s a second son, William B., was also associated with the business. Charles was president of the firm, Howard was vice-president, and William was secretary-treasurer. The name of the business changed frequently during the years of operation: it was listed in the city directory as "Olmsted and Olmsted" and, in 1936, as "Olmsted's." Although Charles A. Olmsted suffered from angina for many years, he continued to work until his death in 1943.

The name Olmsted is found with several spellings. The alternate spelling used by some of the family members in Canada and the United States is "Olmstead."

The original Olmsteds arrived in Britain from Denmark in the 11th century; the name is mentioned in the Domesday book for the County of Essex. Part of their original land is now incorporated into the Cambridge University holdings.

FIG. 333

FIG. 332 *C.A. Olmsted*
FIG. 333 *Clock sold by C.A. Olmsted.*

FIG. 334

FIG. 334 *A 21-jewel lever escapement watch with "Olmsted and Hurdman, Ottawa" on the dial.*

The first Olmsted immigrated to America around 1635. In 1783, following the American Revolution, many Olmsteds were among the United Empire Loyalists who landed in New Brunswick. Some of the families proceeded to Lower Canada, while others settled in Upper Canada, obtaining land in the early 1800s. There is much information in the form of archival material, published books, and articles concerning the ancestral families of C.A. Olmsted,[22] including the Wright family.

The Wrights and Olmsteds were important pioneers of Bytown (Ottawa), Ontario, and Wrightville (Hull), Quebec. Philemon Wright, the founder of Hull, was C.A. Olmsted's great-great-grandfather on the maternal side.

Charles married Agnes Buchanan (b. 1866 – d. 1956) around 1889. There were six children: Richard Alan (b. 26 Oct. 1890 – d. 28 May 1965), Howard R. (b. 1894 – d. 1975), William Buchanan (b. 7 Aug. 1900 – d. 16 Jan. 1973), Sarah Alice (b. 1902 – d. 1925), Janet, and Eric.

By 1889, in addition to his devotion to his work, Charles was a member of the Ottawa Board of Trade. He was an accomplished fencer and had a fondness for horses and white cats. Shortly after 1900 he moved to his father's homestead on the third concession in South Hull, near Aylmer, Quebec. From there he drove to work in Ottawa in a carriage drawn by his horse "Lady Torrington." Around 1910 he returned to Ottawa to live, residing on 5th Avenue, where he could ride in the Exhibition grounds near his home. Toward the end of his life, he moved in with his son Richard to live on Hamilton Street.

Charles Olmsted and his family were of the Presbyterian faith. He was a lifelong member of St. Andrew's Church in Ottawa, where he was an elder for seventeen years. His son, Howard, however, sang in the Anglican Cathedral choir for many years.

200

His family remembers him as a kind, gentle man. At his funeral the minister remarked that "the character of the man himself was a guarantee of fine workmanship." He is buried with members of his family in the Belleview (Bellevue) cemetery. The cemetery is on land donated by Gideon Olmsted, his great-grandfather, and was "free to all but one. No Rollins must ever be buried herein." Rollins, who owned the farm a short distance to the west of the property, took this so to heart that he sold out and returned to the United States.

ROBERT OSBORNE

Robert Osborne (b. 1815 – d. 13 Apr. 1874) was a watch- and clockmaker from Scotland. He was established in business by 1851 at 8 James Street in Hamilton, Canada West. He was also teaching the art of watchmaking to two apprentices from Scotland, David Hauters and Joseph Kerr. Although a number of the port records for Hamilton are missing, Robert Osborne did import clocks and weights in 1858. By 1861 Robert Osborne had a capital investment of $15,000 in his business.

One wall clock bearing Osborne's name on the dial is shown in Fig. 335. It is currently in the collection of the Royal Ontario Museum. The clock is large in size, with the dial measuring 16 inches in diameter at the chapter ring. Total width is 22½ inches and the height is 37½ inches. The clock has a wooden bezel, which is hinged at the top and fitted with a simple latch at the bottom. There are two decorative pieces of cast-metal scroll work at the point where the dial meets the base portion. The wood is a dark-stained hardwood.

The movement is time-only 8-day and is weight-driven. It has the general configuration of a typical banjo-type movement, with wind hole at two o'clock and the idler wheel to the left of the main arbor. The movement plates measure 2¾ by 3⅞ inches, and there are no identifying marks. The clock appears to be of U.S. origin and the style is consistent with the dates mentioned above.

There are two penciled inscriptions in the case, which give some clue to its history, namely "Bought from Canada Life Ins. Co. Feb'y 1st 1883, Colin.," and "Fitted up in office Feb 3, 1883, Colin Munro Feb 3/83, 8 days." It appears that a Mr. Colin Munro purchased the clock from the Canada Life Assurance Company of Hamilton, Ontario, and set it up in another office. The clock had evidently begun its career as an office clock at the insurance company.

Robert Osborne and his wife, Margaret (b. 1827 – d. 1 Feb. 1892), had two of their five daughters reach adulthood – Jessie (b. 1853 – d. 22 Jan. 1923) and Annie (b. 1855 – d. 1 Jan. 1919). Robert, who was a student in 1862, may have been a son.

FIG. 335

FIG. 335 *Clock sold by Robert Osborne, Hamilton, Canada West.*
– Courtesy the Royal Ontario Museum, Toronto, Canada

J.M. PATTERSON

J.M. Patterson was an importer of clocks[23] which he sold in the cities of Hamilton and Toronto, Canada West. The authors were unable to find any information about Patterson's activities either in Hamilton or Toronto. Therefore, information presented here is limited to the types of clocks found with the Patterson name on the labels.

The triple-decker clock shown in Fig. 336 is one of the clocks J.M. Patterson sold in Hamilton. The clock has a strap-brass movement and was made by Birge, Peck and Company, who were in business from 1849 to 1859. Its label was printed by the Elihu Geer Company, which was in business at 10 State Street, Hartford, Connecticut, from 1850 to 1856. Therefore, it can be assumed that the clock was made sometime between 1850 and 1856. The lower tablet is a replacement. The clock has an overpasted label stating that the clock was "Manufactured expressly for J.M. Patterson, Hamilton, Canada West, by Birge, Peck & Co."

A column-and-cornice 8-day clock was also made by Birge, Peck and Company and exhibits a label identical to that of Fig. 343, with the exception that the overpasted label shows that the clock was sold by J.M. Patterson in Hamilton.

Also while in Hamilton, Patterson sold a half-column clock bearing an overpasted label stating that the clock was made for J.M. Patterson by Seth Thomas.

Fig. 340 shows a column-and-cornice-style 8-day weight clock that was sold by J.M. Patterson in Toronto. The clock was manufactured by the Seth Thomas Clock Company, Plymouth, Connecticut. The label was printed for Patterson by the Seth Thomas Company.

R. W. PATTERSON

R.W. Patterson sold clocks in Toronto, Canada West. Little trace has been found of his other activities. However, he was quite successful in selling clocks, as Canadiana clocks bearing the R.W. Patterson label are relatively common.

The clocks sold by R.W. Patterson were obtained from three manufacturers – Seth Thomas; Birge, Peck & Co.; and Forestville Manufacturing Company (J.C. Brown). The Seth Thomas clocks were provided with custom-printed labels declaring that the clocks were "Made For and Sold By R.W. Patterson & Co. Toronto, Canada West." The Birge, Peck & Co. clocks were furnished with that maker's standard labels. However, the Birge, Peck name was concealed by an overpasted label that again states, "Made For and Sold By R.W. Patterson & Co., Tornto [sic] Canada West." The type used on the overpaste was identical to the original, so it would seem that Birge, Peck also provided clocks with ready-to-sell customized labeling. No printer's name appears on the Seth Thomas labels. Birge, Peck & Co. labels were printed by Elihu Geer, 10 State Street, Hartford, who was at that address from 1850 to 1856.

FIG. 336

FIG. 336 *Birge, Peck & Co. triple-decker clock sold by J.M. Patterson in Hamilton.*

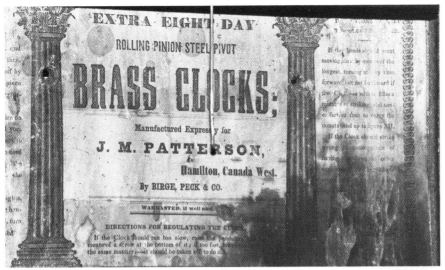

FIG. 337

FIG. 338

FIG. 339

FIG. 340b

FIG. 340a

FIG. 337 *Label from the clock in Fig. 336.* FIG. 338 *Movement from the clock in Fig. 336.* FIG. 339 *Label of another J.M. Patterson Clock sold in Hamilton.* FIG. 340 *Seth Thomas column-and-cornice clock and label sold by J.M. Patterson in Toronto.*

The majority of clocks known to the authors are of Seth Thomas origin in both 8-day and 30-hour style. The 8-day clocks are column-and-cornice style, typical of Seth Thomas production as illustrated in Fig. 341.

These clocks exist with plain or gilt half columns, but none with imitation tortoiseshell finish have been reported. These clocks are fitted with the distinctive Seth Thomas 8-day, time-and-strike, weight-driven "lyre" movement, as illustrated in Fig. 342 and used by that company for many years. This movement is type 4.23 in the Taylor system. Another clock with a similar case contains a Taylor-type 4.211 lyre movement.

The 30-hour clocks are mostly regular Seth Thomas half-column products. The illustrations show two slightly different styles. A clock similar to Fig. 344 was also made for Patterson by J. C. Brown, Forestville Manufacturing Company in Bristol, Connecticut. This clock may have been made for the English market as the original scene on the tablet is entitled "Snecler [?] Hall Training School, England."

The clocks from Birge, Peck and Co. are less common than the Seth Thomas clocks, although three examples have been reported. All are 8-day, striking, weight clocks with strap-brass, rolling pinion movements typical of Birge, Peck manufacture. There are two case styles – the earlier triple-decker case with splat and three layers of columns, and the later Empire column-and-cornice case.

Little information has come to light about R.W. Patterson. However, it seems safe to say that he was active between 1850 and 1860. The Birge, Peck firm itself spanned the period 1849 to 1859. Similarly, the Seth Thomas clocks are of this era. Chris Bailey, former curator of the American Watch and Clock Museum, commented in a 1973 reprint of the 1863 Seth Thomas catalogue that production of the half-column clocks began about 1850, and that they were still the company's main output when Seth Thomas died in 1859. This reprint also shows styles of the 30-hour half-column clocks that had been discontinued before the 1863 catalogue. These were some of the models sold by Patterson. Confirmation of the period of Patterson's clock-selling activities may be found in the fact that an R.W. Patterson was registered at the American Hotel, Toronto (corner of Front and Yonge streets), in 1856. This would suggest that Patterson had no fixed address in Toronto and that he was an itinerant peddler who bought a quantity of clocks, came to Toronto, stayed until they were sold and then departed.

This book also contains a brief reference to J.M. Patterson, who sold clocks in Hamilton and Toronto, Canada West, at approximately the same time as R.W. Patterson. There is little information on either of these men and nothing to suggest that they were related, except the similarity in their names and products. J.M. also sold clocks made by Birge, Peck and Seth Thomas with overpasted or custom labels.

FIG. 341a

FIG. 341a *Clock sold by R.W. Patterson & Co., furnished by Seth Thomas.*

FIG. 341b

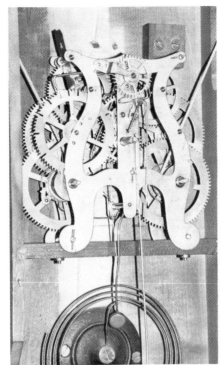

FIG. 342

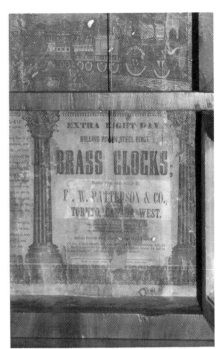

FIG. 343

FIG. 344

FIG. 345

FIG. 346

FIG. 341b *Label from the clock in Fig. 341a.*   FIG. 342 *Typical Seth Thomas 8-day "lyre" movement as found in clock in Fig. 341.*
FIG. 343 *R.W. Patterson label pasted over Birge, Peck & Co.label. Note misspelling of Toronto.*   FIG. 344 *This typical Seth Thomas*
30-hour clock sold by R.W. Patterson is one of the two variants, illustrating the shorter half columns.   FIG. 345 *This clock has*
longer half columns compared with Fig. 344, but has the same Patterson label.   FIG. 346 *The movement in the Patterson 30-hour*
clock (Fig. 344) is typical of Seth Thomas movements.

PEQUEGNAT FAMILY

Ulysse Pequegnat (b. 21 May 1826 – d. 30 Sept. 1894) and his wife, Françoise (b. 1829 – d. 1917), were the parents of a clock- and watchmaking dynasty in Canada. They left Switzerland for Canada, arriving in Berlin (now Kitchener) with fourteen children on 14 April 1874. Ulysse was a master watchmaker and a member of the watchmakers' guild in La Chaux de Fonds, Switzerland. His watchmaking skill was attested to by the fact that a watch he made was exhibited at the Great Paris Exposition of 1889.

Once settled in Berlin, all members of the family of employable age found work. They saved money to procure from Switzerland supplies and equipment to commence watchmaking in Canada.[24] Eventually, the eight sons of Ulysse Pequegnat went into the watch repairing and jewellery business at various times in Berlin, Waterloo, Ayr, Tavistock, Neustadt, Paris, New Hamburg, Brantford, Stratford, and Guelph. The Guelph jewellery store is the only such business still owned by descendants of a Pequegnat (Joseph).

Arthur U. Pequegnat (b. 22 Sept. 1851 – d. 11 Aug. 1927) was the eldest of the Pequegnat brothers. He started to learn watchmaking in Switzerland at the age of ten, and by sixteen was foreman in a Swiss watchmaking factory. In Canada he worked for over a year at the Shantz button factory in Berlin. Then, with his wife, Hortense Marchand (b. 1852 – d. 16 Oct. 1919), moved to Mildmay, Ontario, where he opened a watch repair and jewellery shop.

After several years in Mildmay, Arthur and his family returned to Berlin, where in 1880 he established a watchmaking and jewellery business in the Stubbing Block with his brother Paul

FIG. 347

FIG. 347 *The Pequegnat family, 1892.*
Standing: Georges, Léa, Paul, James,
Arthur, Marie, Leon.
Sitting: Joseph, Ulysse, Emma, Lina,
Françoise, Philemon.
Kneeling: Dina, Albert, Rachel.

(b. 8 Oct. 1853 – d. 12 Apr. 1923). The business was known as the Pequegnat Brothers. Many advertisements for the business appeared in the newspapers, one of which was the *Berliner Journal*, a newspaper printed in German. Shortly after the partnership was established, Paul and Arthur opened a second store in nearby Waterloo in the Killer Block. For a time this store was managed by a third brother, Philemon (b. 6 Dec. 1863 – d. 18 May 1945). Arthur continued to own the Waterloo shop until 1896.

Around 1890 Arthur and Paul established separate businesses. In the mid-1890s Arthur added a bicycle repair shop to the back of his watch repair and jewellery shop in Berlin. About 1898 he built a three-storey building at 53–61 Frederick Street, where he and his brother Paul manufactured bicycles for the specialist, called Racycles. In 1903 when the automobile began to replace bicycles in popularity, Arthur added the manufacturing of clocks to his business. Leon, Georges, and Philemon joined Arthur in the clock factory. The Arthur Pequegnat Clock Company was an immediate success and the company made clocks until 1942.

Clocks made by the Arthur Pequegnat Clock Company are the commonest and best-known Canadian clocks. Since they are of 20th-century manufacture, they have not been included in this book. Pequegnat clocks, however, have been extensively documented by several authors. One such reference is *The Pequegnat Story, the Family and the Clocks* by Jane and Costas Varkaris. When Arthur died in 1927 his son, Edmond (b. 1884 – d. 2 Oct. 1963), took over as manager of the factory. Arthur's son, Marcel (b. 27 April, 1886 – d. 8 Oct. 1988), was the last president of the clock company. The building was demolished in 1964.

Paul had learned watchmaking in Switzerland at the famous Longines factory. In Berlin, he established a watch repair shop on the south side of East King Street. The building was destroyed by fire in 1878.

Around 1890, when Arthur and Paul established separate businesses, Paul moved to 48 King Street East in Berlin. By 1905 Paul's shop was at 22 King Street West. In 1915 he moved his business to the front corner of the Arthur Pequegnat Clock Company factory, and he remained at this location until his death. Many watches sold by Paul bear his name. One such watch can be seen at Doon Pioneer Village, Kitchener, Ontario.

Arthur and Paul trained their younger brothers as watchmakers and also financed a business for each. Watches exist with the name of each brother on the dial.

James V. (b. 16 Nov. 1854 – d. 9 June 1922) opened a shop in Tavistock in 1881. In 1887 he moved to Stratford. His brother Joseph (b. 22 Dec. 1864 – d. 29 Nov. 1947) learned watchmaking from James and in 1892 opened a store in Guelph which descendants operate today.

Georges (b. 30 May 1882 – d. Dec. 1945) went into business in Ayr in 1886, then moved to Neustadt and remained there until 1895. In 1897 he returned to Berlin to help Arthur in the bicycle factory. Leon (b. 22 May 1858 – d. Jan. 1939) went to Mildmay with Arthur and there learned the watchmaking trade. In 1881 he opened a watch repair and jewellery business in New Hamburg. When Arthur began to make clocks in 1904, Leon left the jewellery business and continued all his active life to sell Pequegnat clocks from coast to coast. With the death of his brother Paul in 1923, Leon became president of the Arthur Pequegnat Clock Company.

The twelfth child of Ulysse, named Philemon, learned watchmaking and, with the assistance of brothers Arthur and Paul, was established in business in 1884 in Plattsville. In 1887 he moved to Paris, Ontario, but returned to the Waterloo area in 1890 to manage the store owned by Arthur in Waterloo. When Arthur converted most of his bicycle factory to a clock factory, Philemon was responsible for assembly of movements and continued to work there until retirement. Albert (b. 10 Feb. 1866 – d. 10 Jan. 1951) was the youngest son of Ulysse. He established his jewellery business in Brantford about 1893 and continued the business until his retirement around 1929.

The success of the businesses was due in part to the fact that their many shops were run in the manner of a present-day conglomerate. The Berlin store bought in quantity for all the shops owned by the Pequegnat brothers. Because of mass buying, the prices were lower than those of their competitors, and most of the businesses thrived. By 1891 more than ten stores were operated at various times by the Pequegnat brothers.

FIG. 348

FIG. 349

FIG. 348 *A lady's hunter-style watch, sold by a Pequegnat establishment.*
FIG. 349 *A watch sold by A.N. Pequegnat.*

THE PETERBOROUGH CLOCK

The Anglican Church of St. John's in Peterborough, Ontario, possesses one of the most interesting tower clocks in Canada. The St. John's church itself is a beautiful old white limestone building with fine Gothic lines. It stands on a height of land in the downtown area of the city. At the time it was built in 1836 the builders decided that a clock should be added to the tower, no doubt because of the church's dominant position overlooking the town. The tower was completed with the clock in position in 1839.

The authors are indebted to Professor Elwood Jones of Trent University, Peterborough, for details of the clock's early history. Old photographs and drawings exist that show the clock with a single dial on the front of the tower. The maker of the clock is unknown and there are several theories about its origin. It is known that the architect who designed the church and the contractor who built it both came from Kingston and that the cost of the clock was £100. Professor Jones has suggested that the clock may have been built locally or by a Kingston maker. There is, conversely, some reason to believe that it may have been imported, since no other similar wood tower clocks are known in Canada and the clock design itself is typically English and quite sophisticated.

As a public clock, unfortunately, it was a failure. Many difficulties were experienced in keeping the movement in running condition. The reasons for this are not immediately apparent, but there undoubtedly were problems associated with temperature and humidity. Perhaps the design was a little too elegant for the purpose.

Professor Jones quotes local historian Dr. T.W. Poole as stating that the clock had already been inoperable for several years by 1866, "owing to the trouble and expense

FIG. 350

FIG. 350 *The Peterborough clock, front view. The dial and hands are not original.*

attending its regulation and supervision." The clock remained in position until 1882 with its hands stopped at ten to eleven. In that year major renovations were made to the church. The movement was set aside and the tower redesigned to conceal the fact that there had ever been a clock. A bell was hung in its place and later, in 1911, a carillon. During the ensuing years the clock was shunted from place to place like an unwanted child. Fortunately, however, its quality and historic value were recognized and it survived with a minimum of damage.

From 1882 to 1911 it remained at the church, pushed to one side in the tower. In 1911, to make room for the carillon, it was taken down and stored at the new Peterborough Public Library, where it and several other "relics" were cared for. Here it remained until 1942 when it went back to the church, where it was received with mixed feelings. The rector and the church wardens decided it should be preserved and arranged to send it to the Royal Ontario Museum, Toronto, in January 1943. By this

FIG. 351

FIG. 351 *Peterborough clock, rear view. The figure of tower clock enthusiast Marvin De Boy gives a relative measurement of the movement's size.*

time the clock had lost its original hands, dial, bell, and weights. The Royal Ontario Museum stored it, but made no attempt to display it. Interestingly enough, it was featured in an early issue of the NAWCC *Bulletin*, Vol. II, February 1947.

The Royal Ontario Museum had not yet developed a Canadiana section and began to look for another recipient. Eventually, in 1949, the clock was moved to Old Fort George, a museum and National Historic Site near Niagara-on-the-Lake. Here it remained on display until 1976. The clock was again written up in the NAWCC *Bulletin*, Vol. XI April 1964. At that time it was reported that the museum staff was carrying out repairs with assistance from clock collectors in the Buffalo area. It is probable that the hands and small dial currently on the clock were added at that time.

By 1967 historically minded persons in Peterbrough were beginning to ask that the clock be returned. This finally took place after a concerted effort by Professor Jones and the rector, William Moore, in 1976 – the 140th anniversary of the church.

At the time of writing, the clock remains at St. John's church. In some respects its future is almost as obscure as its past. It sits in a storage area under a tarpaulin, awaiting a decision on how it should be restored and displayed.

From a technical standpoint the clock is intriguing, and its origin is a mystery. As noted previously, there is a feeling at Peterborough that the clock was made locally, but no concrete evidence of this has been found. During its days at the Royal Ontario Museum, it was very tentatively attributed to Eli Terry, probably because his name was so closely connected with wooden clock movements. However, in the intervening decades,

FIG. 352

FIG. 353

FIG. 354

FIG. 355

FIG. 352 *The Peterborough clock escapement.*
FIG. 353 *Wheel detail. Teeth are mounted radially in
quarter-cut oak. The metal bar in the great wheel (left) is
the maintaining power spring. Metal arbor and bearing
construction are clearly visible.* FIG. 354 *Detail of rack
and gathering pallet in Peterborough clock.*
FIG. 355 *Great wheel, time side. The inner ratchet is for
winding, the outer for maintaining power.*

210

much has been learned about clock movements, and there appears to be no connection with Eli Terry. It is true, however, that Eli's son, Henry Terry, did make a few wooden tower clocks and one is on display at the Smithsonian Institution in Washington, D.C. That movement bears little resemblance to the Peterborough clock. In the opinion of several collectors and students of horology, the most likely origin is England. There are two reasons for this assumption: First, the overall design contains features usually found in better-quality English tall case clock movements, including maintaining power, rack-and-snail strike, and a strike/silent mechanism. The second and perhaps more important reason is found in two penciled inscriptions. One, which is barely legible, includes a date of 1803. The other, which is quite clear, states, "John Robinson, Stannington Law House, Northumberland England 1814." None of this, of course, constitutes proof of origin, but the authors would suggest that the clock may well have been purchased second hand for installation at Peterborough. Investigations are continuing and someday the full story may yet be uncovered.

The clock itself is a pleasure to examine. It is large and in all respects exceptionally well made. The overall impression that one forms on examining it is that it was constructed by a master craftsman who was familiar with both clock technology and the shaping and joining of wood. There is evidence of later repairs, such as nails and screwnails, but the original wood fittings are rigid, precise, and good looking.

The plates are made up of heavy pine straps approximately 2 inches thick with widths of about 3 inches and 6 inches. The straps are held together by a combination of pins, wedges, and mortises. Most pivots run in separate wooden blocks attached to these plates. The wheel rims are made of quarter-cut oak, usually in four pieces with four oak spokes and a central hub. The teeth are made of light-coloured hardwood and are set radially into the rims, making a very strong construction. The pinions are of wood, seemingly one piece, with precisely shaped teeth. The arbors themselves are of steel.

The movement is very large with plates approximately 52 inches square and a depth between plates of 18 inches. The great wheels are about 26½ inches in diameter with ninety-six teeth. The pendulum, which is believed to be original, is relatively small for such a massive movement. It was not available for examination, but from data presented in earlier articles, it has an overall length of about 45 inches, which would yield a second beat when installed and adjusted. The rack itself is made of steel, and the gathering pallet is a metal pin. The oak snail is enormous, being over 18 inches across at its greatest width. The escapement anchor is forged steel. The escape wheel is of relatively thin metal, with teeth that have been widened at the tip by the addition of a small piece of metal brazed or soldered to the tooth. The strike/silent mechanism is a sliding wooden bar above the motion work which can be moved into position to interrupt the strike mechanism. The time side great wheel has maintaining power provided by simple linear springs within the spoke area and a ratchet on the wheel rim.

At the present time, there is some damage. Quite a number of teeth are missing and some are saved in a bag at the church. Part of the strike mechanism is broken, but has been saved. Overall, however, the damage is superficial and could readily be put right. It is hoped that a suitable haven can be found for the clock within the confines of St. John's, where it can function once again.

FIG. 356

FIG. 357

FIG. 358a

FIG. 356 *Clock sold by M. Haas*
FIG. 357 *Haas label*
FIG. 358a *Clock sold during the Pfaff
and Haas partnership.*

PFAFF AND HAAS

A surprising number of 30-hour OG clocks are known which bear labels indicating that the clocks were sold by Anthony Pfaff and Matthias Haas. It is always difficult to say whether a high survival rate is a matter of luck or of successful salesmanship. However, at least six of these clocks have survived. This is a higher number than some of the more prestigious labels in this book. Pfaff and Haas appear to have been active for only a very few years in the early 1840s and to have sold clocks only in the Vaughan Township area of York County, Canada West.

There is no evidence that either Pfaff or Haas did any actual clock-making. However, they did add distinctive labels. These are of plain brown paper with handwritten inscriptions in an elegant "copperplate" script. Some labels were pasted over the original manufacturer's label and others appear to have been applied to unlabelled clocks.

The two men sold clocks separately and in a partnership. The Port of Toronto records[25] provide some insight into these business arrangements. On 2 August 1843 Anthony Pfaff imported twenty-seven brass clocks (costing $121.75). By October of that year, Pfaff was working in partnership with Haas and the company imported eighteen brass clocks ($72). A large order of twelve boxes of clocks was received by the company on 14 December 1843 ($234). Apparently no clocks were imported by these men in 1844 or 1845. On 14 July 1846 M. Haas imported two boxes of clocks ($28.50), followed on 25 November 1846 by a case of clocks ($89.10). On 16 October Anthony Pfaff imported eleven boxes of clocks ($169.95).

Four label variants are known:
1. Anthony Pfaff Clockmaker, Lot 15 Concession 3, Vaughan Township.
2. A. Pfaff & M. Haas Clockmaker [*sic*] No. 15, 3 Con. in Vaughan 1844.
3. Pfaff, Anthony and Haas Matthias, 1844 Township of Vaughan County of York.
4. Matthias Haas Clockmaker Township of Vaughan.

The clocks are all typical 30-hour OGs, and they appear consistent with the 1840s dates. Note the early reverse painted tablet in the clock shown in Fig.356. Three clocks were reported as having Chauncey Jerome movements and one of these was dated by Snowden Taylor to be 1843 to 1845. A fourth clock had a Clarke, Gilbert movement (Taylor type 6. 111) that can be dated after 1841.

Further confirmation of this time period can be found in the land records. Pfaff purchased Lot 15, Concession III in 1842. The property faced on present day Keele Street, about one and a quarter miles south of the town of Maple.

The clock shown in Fig. 358 is still owned by the grandson of the original owner, who was a farmer living within a few miles of Lot 15, Concession III, Vaughan Township. The present owner recalls that his grandfather purchased the clock off a wagon being driven around the area by Pfaff and Haas.

Anthony Pfaff married Soloma[26] Miller on 11 April 1837. Two children were born: Josephine (Mrs. James Edward Smith) and Charlotte, a minor in 1857. In the 1840s they moved to 77 Richmond Street in Toronto where Pfaff again entered into a partnership with Matthias Haas, sharing a grocery business at 102 Yonge Street. All clocks that are known date from the years that the men lived in Vaughan Township and there is no evidence of clocks being sold at 102 Yonge Street. Anthony Pfaff died on 23 May 1857, leaving an estate of about £2,000.

Matthias Haas lived on Lot 13, Concession VIII, of Vaughan Township in the early 1840s, but a few years later he, too, moved to Toronto with his wife, Matilda, to engage in the grocery business with Pfaff. Matthias Haas died in 1850. His wife set up a new business in embroidery and fancy goods at 10 King Street West, which she continued until about 1856.

JORDAN POST

Jordan Post's name stands out in the early records of York, Upper Canada – the predecessor of present-day Toronto. In commerce he was a jeweller, silversmith, and clockmaker, as well as being an early real estate developer. Later in life he became a lumber merchant and a general storekeeper. He was active in public affairs and took part in important political meetings. He was elected to the post of town warden in 1823. He contributed generously to schools and charities and in 1820 pledged £100 toward the building of a Successionist Presbyterian Church. At home he was a member of an old and distinguished family and father of a large family. To detail all his accomplishments is beyond the scope of this book and the authors will report mainly on Post as a clockmaker.

The importance of Jordan Post (b. 1767 – d. 1845) as clockmaker derives from the fact that he was the first such artisan to arrive in Upper Canada and one of the very few in this country who actually made complete clocks by hand and also trained apprentices. His output was small, as was his market. He must be recognized, not for quantity production, but for the fact that he successfully made clocks in an early and primitive environment.

The Posts were, and are, a large and eminent family, with many well-known members. They claim descent from a German nobleman, Von Poest[27] who settled in England in 1473. A descendant, Stephen Post, came to America in 1633. Jordan Post, the clockmaker, was a direct descendant of Stephen and, by actual count, was the fourth generation to bear the name Jordan. He, in turn, had a son and grandson Jordan – all of which is genealogically impressive but adds an element of confusion for present-day researchers.

Jordan Post was born in Hebron, Connecticut, on 6 March 1767, the son of Jordan Post and Abigail Loomis. The Posts were already a well-established part of the community. However, several times in subsequent years the family moved en masse to other locations. Young Jordan Post was trained to be a clockmaker. The details of his apprenticeship are obscure. John E. Langdon in *American Silversmiths of British North America* states that he was apprenticed to Samuel Post Jr. of New London, Connecticut, although there is some doubt about this since Samuel, born in 1760, was only seven years older than Jordan and left clockmaking by 1785 to move to Philadelphia.

The first known clocks by Jordan Post bear the place name of Lanesboro, Mass. It would seem likely that they were produced in the late 1780s or early 1790s. Several members of the Post family then moved to Vergennes, Vermont, where they were active for several years. Clocks by Jordan Post at Vergennes are also known.

By the end of the century, the Posts became interested in immigrating to Upper Canada. The first family member to appear in York was Jordan's youngest brother, George Washington Post in 1799. Jordan himself arrived in 1802. They were joined by other family members, including their father, Jordan Post Sr., a baker. This was a very early period in the history of the town, when its population numbered slightly over three hundred persons.

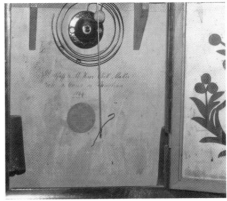

FIG. 358b

FIG. 359

FIG. 360

FIG. 358b *The label from the clock in Fig. 358a.*
FIG. 359 *Jordan Post from early daguerreotype.* *– Courtesy Thomas Fisher Rare Book Library, Toronto*
FIG. 360 *Silversmith's mark used by Jordan Post.*

Jordan Post began to advertise his presence in York in 1802. One notice of 2 September advised his customers of his temporary absence from York. On 11 December 1802 he published a general notice in the *Upper Canada Gazette:*

Jordan Post clock- and watchmaker informs the public that he carries on the above business in all its branches at the upper end of Duke St. He has a complete assortment of watch furniture – clocks and watches repaired on the shortest notice – reasonable terms etc. Gold and silver articles. NB. will also purchase old brass.

Post continued this business until about 1834, when he sold a number of his properties and moved to nearby Scarborough, where he built a sawmill on Highland Creek and shipped lumber to various Lake Ontario ports. He also ran a general store in Highland Creek village until his death on 8 May 1845. The tombstone of Jordan Post and his wife still exists in the old Highland Creek Cemetery.

In real estate matters Post was particularly astute, and much of his wealth was gained from land sales. He first applied for free land by Land Petition in 1804 and later acquired land in Hungerford Township and in Scarborough. Modern Torontonians are much impressed upon learning that he once owned the entire frontage on the south side of King Street from Yonge to Bay streets. He divided this property with two streets which he named after himself and his wife, namely Jordan and Melinda streets. These streets still exist as a reminder of the first clockmaker and the land around them is of incalculable value. He owned several other parcels of land in the centre of town. Some idea of his holdings can be gained from his advertisement in the *Weekly Register* of January 1826. Firstly, he offered the town a free plot of land off George Street, to establish a market place. Secondly, he announced the auction, for sale or lease, of four lots on King Street west of George Street, six lots on Yonge Street, twenty lots on Market (now Wellington) Street, and ten lots in front of the proposed market lot.

In 1820 he built a house and shop at 221 King Street West on the southeast corner of King and Bay streets (see Fig. 361). The large wooden clock that decorated his shop achieved notoriety in 1834 when, as a prank, some local persons attempted to steal it. A man named Craig was struck by the sign when it fell and died of his injuries. He was a distillery owner and a nephew of one of the local dignitaries, Bishop Strachan. This building, incidentally, was torn down about 1840 and replaced by a three-storey brick warehouse owned by the very well-known furniture makers, Jacques and Hay. Today the site is occupied by a bank tower.

Jordan Post, the man, was described by John Ross Robertson in *Robertson's Landmarks of Toronto* as "a tall, lean, wiry New Englander of grave address, but benevolent disposition and well liked by the community. He was a clockmaker by trade and always wore spectacles." Another author, Henry Scadding, commented that he always dressed plainly "in the Mennonite manner," although he was not of the Mennonite faith.

An interesting sidelight on Jordan Post in later life concerns the cholera epidemic of 1832. This comment appears in the diaries of James Lesslie – "Saturday 28th [July 1832]...old Jordan Post very ill of cholera.

JORDAN POST' HOUSE & SHOP S.E. COR BAY & KING STS

FIG. 361

FIG. 362

FIG. 361 *Jordan Post's shop and home. From J.R. Robertson's* Landmarks of Toronto. FIG. 362 *The dial on this Post clock is thin sheet copper screwed to an iron back plate.*

214

His physician preparing that singular yet efficacious remedy resorted to in extreme cases in Scotland – viz. injecting an artificial serum to the Blood into the veins." Jordan Post was sixty-five years old at this time and, incredibly, survived both the cholera and its rather frightening treatment.

Jordan Post married Melinda Woodruff at St. James Cathedral, York, in 1807. At this time he was about forty years of age. The Posts had two sons and four daughters[28] who survived their father. Melinda died on 21 September 1838 and is buried in the family plot in Highland Creek.

Surviving artifacts from the shop of Jordan Post are rare. A few pieces of silver are known. A pair of sugar tongs can be seen at the Royal Ontario Museum and two silver spoons have sold at auction in recent years at high prices. Each of these pieces bears his characteristic silversmith's mark, as documented by John E. Langdon in *Canadian Silversmiths 1700–1900*. An intriguing reference can be found in the diaries of David Gibson, a well-known surveyor and contemporary of Post:[29] "21 Saturday [January 1826] Left Markham at 11 o'clock, went to York, got my instrument from Mr. Post, paid $28, got receipt and returned to Alexander Milne's place pretty late." The activity of Jordan Post in surveying instruments is not entirely clear and the authors have not determined whether he made, sold, or repaired them. Post did have other dealings with Gibson. Gibson's diary speaks of having his watch repaired in 1826 and in 1832–33 of having surveyed and registered Melinda Street for Post.

The surviving clocks made by Jordan Post in York are few in number. At the time of writing, three more or less complete 8-day tall case clocks bearing his name are known. There is also a loose clock movement, found north of Toronto which exhibits the unique peculiarities of Post's work. The clock made by Post's apprentice, Charles Clinkunbroomer, has also been included in the group (see p. 166). These five movements have many common characteristics and represent all of the output of Post's shop that is known to the authors. Several earlier clocks made at Lanesboro, Massachusetts, and Vergennes, Vermont, have been reported, but not examined. The authors have been told, however, that these earlier movements differ in some respects, being less primitive in execution. The somewhat rough-and-ready aspect of the York movements suggests that Post may not have had a very well-equipped shop or that he left many details to be carried out by partially trained apprentices as he busied himself with other more lucrative activities.

CHARACTERISTICS OF POST CLOCKS
In general, movements by Jordan Post follow the proven pattern of 8-day English weight-driven, striking-brass clocks. This, in turn, was typical of the output of early New England makers of the late 1700s, when British influence was still strong. All the normal patterns for wheel count and configuration, plate layout and dimensions, use of rack striking, etc., can be found in the surviving examples from Post's shop. There are, however, a number of peculiarities that make the clocks of Post (and Clinkunbroomer) distinctive and recognizable.

1. SKELETONIZED PLATES
The movements illustrated in this book all show extensive cut-out sections, no doubt to save brass. This practice was also followed by makers in New Hampshire and other parts of New England, as Post no doubt observed in his travels. The degree of skeletonization in Post's clocks is greater than in most United States movements, particularly in the area of the strike hammer pivot.

FIG. 363

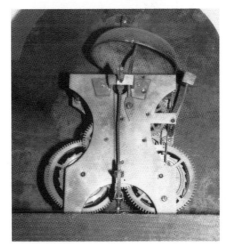

FIG. 364

FIG. 363 *This impressive clock may have been the Post family clock.*

FIG. 364 *The movement from the clock in Fig. 363 shows the skeletonizing of plates, typical of Jordan Post's York clocks.*

2. ROUGH FINISH

Close examination of the plates and wheels shows evidence of sand casting material and blowholes. The plates have been filed smooth, but all evidence of the casting material was not removed. This is possibly because of deep penetration by the sand. When the plates are viewed from the edge, they are of very irregular thickness (see information tables below). This may have been caused by an attempt to file away casting flaws or perhaps by carelessness on the part of the worker.

3. HAND-FILED TEETH

While the various gears and pinions seem normal and typical to the naked eye, under a magnifying glass many irregularities in tooth shape and thickness can be seen. Irregular notches occur at the roots of the teeth. The authors believe that Post, as mentioned previously, probably operated a fairly simple shop and may not have possessed a proper wheel-cutting engine. He appears to have been able to index the teeth reliably, perhaps cutting slots with a saw. He was then obliged to finish the tooth shapes by hand. Similar variations exist in pinions.

4. STEEL INSERTS

A novel technique was employed in two Post movements and is seen in

FIG. 365 *Post movement from clock in Fig. 362, front view.*
FIG. 366 *Rear view of clock in Fig. 362. The extensive cut-outs have saved a good deal of brass.*

FIGS. 365 & 366

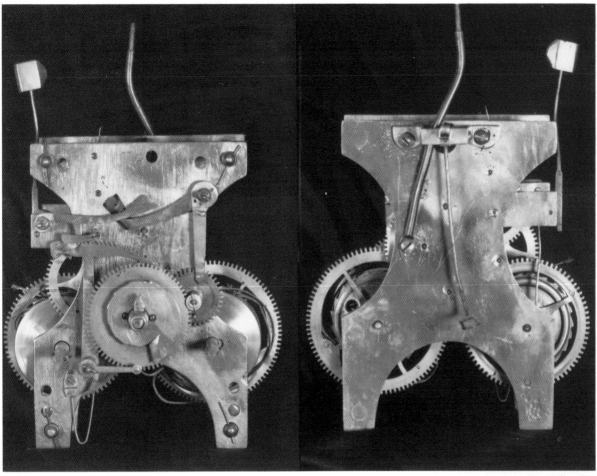

FIG. 367 & 368

the Clinkunbroomer example. As far as can be determined, this has never been seen in the work of other makers. It consists of insetting small pieces of steel into a brass escapement anchor or bell hammer at the contact points. This allows the use of brass for clock parts that would normally be made of tempered steel. Clockmakers long before had discovered that brass rubbing against brass led to rapid wear, whereas brass against steel wore much more slowly. An all-brass anchor or bell hammer would have been most unsatisfactory, but easier to make. By inserting small pieces of steel at the critical points, Post achieved a workable system.

5. VARIABLE COMPOSITION

Jordan Post advertised that he would buy old brass, which was no doubt melted down to make clock parts. It is known that there was a forge at the rear of his shop at 221 King Street West where this sort of work would have been carried out. Chemical analysis of his brass movements has not been attempted, but there are observable colour differences in some movement parts. The idler gears, i.e., hour wheel and cannon wheel, on the movement in Fig. 367, have a distinct reddish colour, suggesting high copper content. Some of the wheels and brass anchors have a golden colour more typical of bronze. These variations were probably the result of having to depend on scrap for raw material.

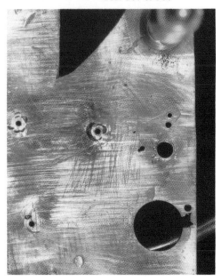

FIG. 369

FIG. 367 *Loose movement attributed to Jordan Post, front view.*
FIG. 368 *Loose Post movement, rear view.*
FIG. 369 *Inner surface of loose Post movement plate showing poor finish, rough workmanship, blowhole, etc.*

6. DIAL CONSTRUCTION

The Clinkunbroomer clock illustrated in Fig. 273 and the Post clocks in Fig. 362 and 363 all have their dials attached to posts on the front plate

FIG. 371

FIG. 370 *Great wheel from strike train of loose Post movement showing casting blowhole and irregular hand-filed teeth.*
FIG. 371 *Pinion from loose Post movement showing irregularity in tooth shape.*

of their movements, using screws that pass directly through the dial. This is a rather uncommon practice in tall case clocks. Chris Bailey[30] has commented that this is sometimes also found in clocks made by makers trained in Eastern Connecticut, where Jordan Post may well have been trained. Most of the clock dials made by Post in the United States and Canada were of brass. However, the clock illustrated in Fig. 362 has a dial made of a thin sheet of pure copper held to a steel backing plate by many screws and rivets. The Clinkunbroomer clock also has a dial made from a single sheet of heavy copper. These two clocks possess the only copper dials that have come to light. There is also a report of a solid silver dial bearing Jordan Post's name that was apparently made during his time in the United States. It was seen in New York City in recent years.

7. DIAL DECORATION
There are some interesting eccentricities in the dials. In the Clinkunbroomer section, the authors have commented on the use of Mennonite type "Fraktur Art" on that clock. The Post clock in Fig. 362 also has some uninhibited ornamentation. It is decorated with sun, moon, and stars as well as flowers and leafy ornaments. These all have red pigment rubbed into the lines of engraving. The chapter ring and numerals are black. There is also a rocking ship in the arch. This dial has another peculiarity. There are seventy-one divisions in the dial of the second bit. The chapter ring, too, has eighty-one irregular divisions, rather than the sixty minutes one would expect. The Post clock in Fig. 363 avoids these peculiarities. The rocking ship in the latter dial appears to be flying a Union Jack and is sailing between a fort and a village. The significance of this is uncertain, although it could depict the early waterfront in the town of York.

8. IDENTIFYING MARKS
The four complete clocks have the names of Jordan Post and Charles Clinkunbroomer engraved on their respective dials. In addition, the clock in Fig. 363 has "JP18" marked on the seat board of the movement, possibly indicating the eighteenth clock, or the year 1818. The clock in Fig. 362 has the name "J. Post" cast into the bell, as illustrated. This again confirms that he probably did his own casting.

9. VERBAL TRADITIONS
The clock in Fig. 363 is suggested to have been the Post family clock. This may account for its proper and conservative Roman numeral dial and the formal case. The clock in Fig. 362 may well have been the shop clock. It originally had a crude boxy case which unfortunately no longer exists. In the shop the rough case and eccentric dial would have been less important.

10. CASES
The Clinkunbroomer clock case has already been described. The only other case examined by the authors is that in Fig. 363. It appears to be original. The wood is cherry and the case probably originally had some type of feet. The styling is simple but attractive and would seem to be the work of an early local carpenter or cabinetmaker.

11. APPRENTICES
Several young men worked in the Post shop. The first indication is an advertisement in the *York Gazette* of 8 October 1807 in which Post was looking for two or three young men of good character as "apprentices to the Clock and Watch Makers, Silver Smiths and Jewellers Trades – 13 – 15 Years." There is no record of whether anyone was hired. However, a

different type of advertisement appeared in the *Gazette* of 1 November 1810 in which Post states "that he now has in his employ a professed Watch and Clock Maker (lately from Europe) who he pledges himself will repair and adjust in the best manner Watches and Clocks of every make and description entrusted to him at the shortest notice. All future favours and orders in his line will be thankfully received and executed with correctness and warranted – Jordan Post Jun'r."

The apprenticeship of Charles Clinkunbroomer is well documented, and he appears to have been the first apprentice to have completed his training. The dates are uncertain, but he was probably in training from about 1813 to 1820. He continued to work for Post until about 1833.

Other apprentices were Thomas McMurchie (around 1815), Joseph Milborne and William Watson. Milborne married Post's eldest daughter, Desdamona.

12. MECHANICAL DETAILS

TOOTH COUNT

Two Post movements, Figs. 365 and 367, and the Clinkunbroomer movement were checked. The following configuration was found with the two exceptions noted.

TIME			STRIKE		
Great wheel	96	teeth	Great wheel	78	teeth
2nd wheel/pinion	60/8	"	2nd wheel/pinion	56/8	"
3rd wheel/pinion	56/8	"	3rd wheel/pinion *	48/7	"
Escapement/pinion	30/7	"	4th wheel/pinion **	45/7	"
			Fly Pinion	7	"

Hour wheel	72	teeth
Cannon wheel	36	"
Minute wheel	36	"
Min. wheel pinion	6	"

EXCEPTIONS:
* Post Fig. 362 has 49 teeth in 3rd wheel.
** Clinkunbroomer has 8 teeth in 4th Pinion.

WEIGHTS
(a) The one surviving original Clinkunbroomer weight is 14 lbs.
(b) The original weights for the clock in Fig.363 are 11 lbs 12½ oz.

PLATE DIMENSIONS	Fig. 365	Fig. 367	Clinkunbroomer
Plate width	4⅛"	4⅛"	4⅛"
Plate length	6⅝"	6¾"	6½"
Distance between plates	2⅜"	2¼"	2⅜"
Max. thickness	0.156"	0.205"	0.180"
Min. thickness	0.094"	0.088"	0.093"

In conclusion, the evidence presented in this chapter highlights many quaint features, eccentricities, and departures from normal. There is an overall impression of rough workmanship – plates with plugged holes in the wrong places, sloppy finishing, irregular teeth, arbors not parallel, etc., etc. Certainly workmanship such as this would not have been tolerated in most shops and, indeed, does not seem to be characteristic of Post's earlier work in the United States. However, the clocks are of close enough tolerance that they run. It is perhaps unfair to dwell on these deficiencies 175 years later. Life was not easy in pioneer York, interrupted as it was by American soldiers in the War of 1812.[31] As noted previously, Post's shop was probably not well equipped and clockmaking became

FIG. 372

FIG. 373

FIG. 372 *Brass escapement anchor from loose Post movement showing steel inserts in pallet faces.*
FIG. 373 *Bell hammer from loose Post movement showing the steel insert, set into brass.*

FIG. 374

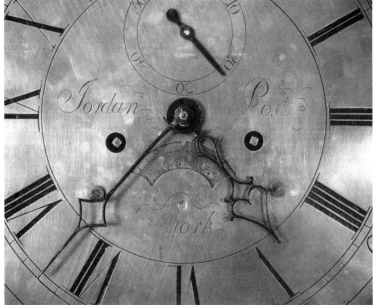

FIG. 375a

FIG. 374 *Detail of the Post clock in Fig. 363.*
FIG. 375 *Two examples of Jordan Post*
"signature." Bell is from clock in Fig. 362 and
dial is from clock in Fig. 363.

FIG. 375b

a sideline. It is also possible that some of the surviving clocks were primarily exercises for the apprentices and that very few clocks were ever made to a high standard for retail.

The problems of his shop aside, Jordan Post does stand tall in Canadian history as a pioneer maker and a busy and productive man.

ELISHA PURDY

The honour of being the first clockmaker in Upper Canada is generally given to Jordan Post, who arrived at York (now Toronto) in 1802. However, there is one earlier reference to a watchmaker who resided briefly in Niagara and York. Elisha Purdy advertised in the Niagara *Constellation* of 27 September 1799 that he "respectfully informs the public that he has taken the shop lately occupied by Mr. Whitting where he will clean and repair watches and clocks in the best manner." He also "makes and repairs all kinds of gold and silver work and will pay cash for or receive old gold and silver in payment."

Purdy arrived in York and placed an advertisement in the *Upper Canada Gazette* of 19 April 1800:

> Elisha Purdy – watchmaker begs to inform his friends and the public that he has taken a room in the house of Mr. Marther where he repairs and cleans watches of all kinds in the best manner and on the most reasonable terms. All orders left for him at said house will be duly attended to. He has a small but elegant assortment of jewellery for sale.

Samuel Marther was one of the first patentees of town lots in York. In 1800 he was given Lot 24 on the north side of Duke Street (now Richmond Street). There is no further reference to Purdy in later years. It should be noted that York boasted a total population of fewer than 325 persons in 1800 and there would have been little demand for his services.

JAMES RADFORD

James Radford (b. about 1834 – d. about 1880) was a watchmaker and jeweller in Ottawa from the early 1860s until his death. A large wall clock bears the name "Radford Ottawa" on its dial. The clock is a Number 1 extra regulator made by Seth Thomas.

Radford was born in Newfoundland. He first began business on Rideau Street, Ottawa, and for some years took William Young as a partner. By 1866 the business had moved to Sparks Street. In the mid-1870s, David Goyer replaced Young as Radford's partner. From 1875 until his death, James Radford was alone in his business, which was again on Rideau Street.

His wife, Jane (b. about 1840), was born in Quebec.

Fig. 376

Fig. 376 *Clock sold by James Radford, Ottawa.*

DAVID SAVAGE

David Savage (b. between 1803 and 1807 – d. 8 May 1857) was the son of George Savage Sr. of London, England, and Montreal (see p. 118). He came with the family to Montreal in 1818. David became a watchmaker and jeweller. Although he was in England in the 1830s, by 1841 he had established a business in Montreal at 137 Notre Dame Street. David married Ann Lawson (d. 31 Aug. 1891) of Lincolnshire, and there were nine children.

According to John Wood, David Savage was a very poor businessman. In 1847 he sold the shop, stock, and fixtures to John Wood for between £1,800 and £2,000. The business was "in such bad shape" that it was not until May 1848 that the deal was completed. David Savage declared bankruptcy in 1850.

Meanwhile, in 1848, Savage went to Guelph, Ontario, and opened a jewellery and watch repair business on Wyndham Street. In September 1851 he ordered a watch with the name "D. Savage, Guelph" on the dial from the Jackson Company, Red Lion Street, London, England. A number of clocks are also known with Savage's overpasted labels, from his shop in Guelph.

His son Benjamin (b. 27 Nov. 1844 – d. 5 Mar. 1914) married Maria Scholfield (d. 14 Mar. 1927), and continued the business after David's death in 1857. In 1885 the shop was called the Watch, Clock, Jewellery and Spectacle House. In the 1890s the business, then Savage and Company, was located on Paisley Street, Guelph. Benjamin's sons Albert, who was an optometrist, and George A. (b. 1876 – d. 13 Dec. 1935), a watchmaker, entered the business. After Albert opened his own shop, George A. continued his father's business into the 1930s. In 1939 Savage and Company was owned by Robert E. Barber and Evan D. Brill.

GEORGE SAVAGE Jr.

George Savage (d. 17 Dec. 1851) was the son of George Savage Sr. of London, England, and Montreal, Lower Canada. In London, George became a watch- and clockmaker, and in 1815 George Savage Sr. made him his partner in the firm George Savage and Company. In 1822, although working in Montreal at the time, the Savages were given a silver medal by the London Society of Arts for the development of a detached escapement for watches "that was a combination of the lever watch and the chronometer."

George Savage and Company opened a store in Toronto early in 1829 and George Jr. was put in charge of the store. The business was

FIG. 377

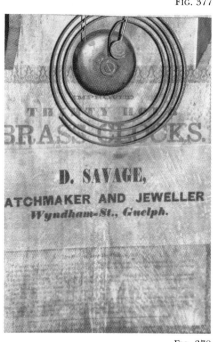

FIG. 378

FIG. 379

FIG. 377 *A 30-hour half-column clock made by Gilbert Clock Co. and sold by David Savage, Guelph, Ontario.*
FIG. 378 *Label from David Savage clock, pasted over Gilbert label.* FIG. 379 *Waterloo Buildings, King Street West, Toronto showing the Savage shop.* – *Courtesy the National Library of Canada*

located on King Street, according to an advertisement in the *York Colonial Advocate* of January 1829.

An advertisement in the *Christian Guardian* of 17 December 1834 gave notice that the Savage company had rented a store on 151 King Street "lately occupied by Mr. W.C. Ross where they have on hand an elegant and fashionable assortment of London-made watches, clocks and jewellery." They also sold plated and bronzed ware and lamps and "carefully repaired clocks and watches." In August 1834 the company thanked the people of Toronto for their patronage and took the opportunity to mention their former affiliation with London, England, and the award won by them there.

In 1845 George Savage Sr. died and from that time the Toronto business was known as "George Savage." By that time George, watchmaker, clockmaker, jeweller, and silversmith, had moved his business to the Waterloo Buildings located on the south side of King Street West. In addition to gold and silver watches and jewellery, the firm offered for sale toilet and essence bottles, ladies' and gentlemen's writing desks, cutlery, gold and silver spectacles and eye glasses, and a variety of personal brushes.

In another move, in 1850, George located the business at 54 Victoria Row, King Street East, and advertised as "importers and manufacturers of watches."

Clocks exist with the name of George Savage and Company, Toronto on the dials. George Savage married Sara Hearn and there was one child, Mary, who married Arthur R. McMaster of Toronto and moved to England.

On 17 December 1851 George died in Toronto, but was not buried until April of 1852 in the Toronto Necropolis Cemetery. No stone was erected at his grave. The plot was designed for the burial of three persons. However, he was the only person buried there and his wife sold the unused portion of the plot.

CHARLES SEWELL

Charles Sewell (d. 1849) was a watch- and clockmaker and silversmith in York (Toronto since March 1834) from 1831 until his death in 1849. In 1837 his business was at 171 King Street. In an advertisement in the *Patriot* of 24 June 1839, he gave notice of his "removal" and also that "he has been replenishing his stock since the robbery...." Therefore, it may have been in 1839 that he moved to 61 King Street East, the address given for his shop in the city directory of 1846–47.

A tall case clock bearing the name of Charles Sewell is on display in the Lynde House at Cullen Gardens and Miniature Village, Whitby, Ontario. Another tall case clock was reported in Ontario in the clock survey of 1978.

As documented in the port records at the National Archives in Ottawa, Charles Sewell imported fifty-four brass clocks with a value of $201 U.S. on 16 September 1843. This was by far the largest shipment of clocks entering the Port of Toronto in that year.

In addition to Sewell's clock and watch establishment he also made silver articles at his "manufactory." According to the *British Colonist* of 16 November 1842, it was there that Sewell, a devoted Mason of the St. Andrews Lodge number 1 of Toronto, had fashioned a large silver testimonial vase with two handles. The vase was gilded inside, decorated with Masonic insignia and engraved with an inscription. This testimonial vase was presented to Masonic Brother Thomas Gibbs Ridout in 1842.

Charles Sewell's age at his death in 1849 is unknown.

FIG. 380

FIG. 380 *Clock by Charles Sewell.*
    – Courtesy Lynde House, Cullen Gardens, Whitby, Ontario

JERRY SMITH

FIG. 381

FIG. 382

FIG. 383

FIG. 381 *Jerry Smith at fifty-four years of age.*
The Evening Telegram, *Toronto.*
FIG. 382 *Watch made by Jerry Smith.*
– *Courtesy Eugene Fuller*
FIG. 383 *Tourbillon chronometer model.*
– *Courtesy Eugene Fuller*

Jerry Smith (b. 30 Mar. 1873 – d. 8 Jan. 1953) was a watch- and clockmaker in Richmond Hill, Ontario. Jerry was one of the most successful students at the Canadian Horological Institute in Toronto and studied under H.R. Playtner (see p. 156).

While at the Institute from May 1897 to June 1899, he earned a Grade A 1 diploma. His "masterpiece" was a 16-size, 21-jewel lever watch. The watch was assigned number eighteen by H.R. Playtner.

While adjusting the watch in 1899, he built a tourbillon chronometer model as an optional part of his course. Later in Richmond Hill, where he had his shop, he made a model of the lever escapement and placed it in the window of his shop to attract attention. He also used it to explain to his customers the action of a watch.

Jerry Smith built two clocks in his spare time. In order to make the movements, he made all the necessary tools and cutters required to fashion the parts. One of the clocks was an 8-foot-tall case clock, with 8-day movement, four-quarter chime, striking Westminster chimes on gongs, and the hours on two bells. Three weights supplied the power and the pendulum beat seconds. The clock had a brass dial, steel hands, and a case of quarter-cut oak. It took fourteen months to complete.

Not happy with the accuracy of a regulator that he had purchased to regulate the time of high-grade watches, he built a clock during a twelve-year period. When tested, his "free escapement" regulator was accurate to less than three seconds a day. He checked the correct time on a device of his own making that received time signals from Arlington every noon.

In his shop in Richmond Hill, he devoted his time to repairing clocks and watches. His fame spread to England, India, and all the provinces of Canada. Jerry Smith was asked to go to the United States to set up a new watch company. However, the untimely death of the president of the proposed company put an end to the possibility of there being a "Jerry Smith" company, manufacturing clocks and watches.

Jerry Smith was born on a farm near Edgely, a small village fifteen miles northwest of Toronto. His parents were strict disciplinarians and, as Jerry said in an interview in 1927, "Children, you know, will be busy at something and if they cannot engage in foolishness,they will busy themselves with something useful." Jerry was proud of the steam engine he made as a child and of an outfit for telegraphy made from an old lever watch and other materials found around the farm. He was also adept at repairing clocks and watches for the neighbours. Although his interest in telegraphy continued, his mother, a firm believer in phrenology, took Jerry to a well-known phrenologist to decide on his future occupation. Not knowing the youth, the phrenologist decided that Jerry Smith would make a successful clockmaker. However, his continuing interest in telegraphy led him to a job as telegraph operator for the Grand Trunk Railway.

He soon became bored. He resigned on 31 May 1897 and entered the Canadian Horological Institute in Toronto. Remembering this part of his life, Jerry said, "There I found myself under the guidance of a master mechanic in the person of H.R. Playtner – the greatest horologist of all time."

After graduation in June 1899, he went to Brantford, where he opened a watchmaker's shop near Market Square. He moved to Richmond

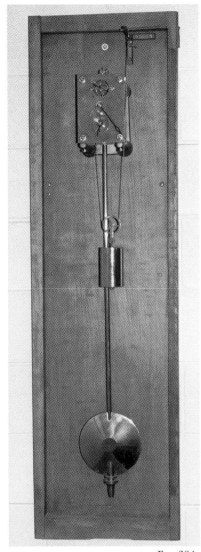

Hill in 1902 because he believed that its higher elevation was better for his health. He remained in Richmond Hill for the rest of his life.

In 1927 a writer for the *Evening Telegram* describing Jerry Smith stated, "He looked like a prophet of old, but there was youth, too, in his face and strength and beauty. Courteous dignity marked his manner."

Jerry Smith married Effie Hollingshead of Collingwood in 1900. There were ten children, none of whom carried on their father's business. A nephew, Jacob Smith, apprenticed with Jerry Smith from 1934 to 1940. Jacob made a tall case clock during this period "in the little shop in Richmond Hill." Jacob Smith then opened a shop in Stouffville, Ontario. He retired in 1975.

FIG. 384 *Jerry Smith's regulator, shown here with dial removed, had a Riefler-type escapement.*
FIG. 385 *Sign outside Jerry Smith's shop.* The Evening Telegram, *Toronto.*

FIG. 385

FIG. 386a

FIG. 386b

FIG. 387

FIG. 386 *Clock sold by*
S.J. Southworth and its label.
FIG. 387 *Label from a 30-hour*
half-column clock.

S.J. SOUTHWORTH

Stephen Jackson Southworth (b. 6 Jan. 1812 – d. 7 Feb. 1899) was one of at least five men who peddled clocks throughout the counties of Leeds and Grenville in the 19th century. A relatively large[32] number of clocks exhibiting the S.J. Southworth name can be found in private collections and museums. S.J. Southworth sold clocks made by the Seth Thomas Company of Connecticut from about 1841 to 1865 during the time that he was living in the townships of Kitley and Bastard.

Although Southworth farmed during this period, as indicated by census records, he considered himself to be a clock peddler. It is known that he also sold jewellery to his customers and perhaps offered other items for sale. Because of the length of time he distributed clocks in the area, S.J. Southworth may be considered as one of the most important clock peddlers in Canada West.

Several styles of clocks were sold by S.J. Southworth. Although several OG clocks are known, 30-hour half-column clocks and 8-day column-and-cornice clocks are found more frequently. The column-and-cornice clocks have the typical Seth Thomas 8-day lyre movements, Taylor type 4.211, and infrequently, the identical movement, but without a "Seth Thomas" stamp.

Stephen J. Southworth was born in Tinmouth, Vermont, of Irish parents. The family soon moved to Rutland, New York, where the early death of his parents forced him to fend for himself. According to his obituary, this contributed to "moulding that air of self-reliance which was one of his most distinguished characteristics." He had limited schooling, which led him in later life to place great emphasis on the advantages of education. After his parents' death, he lived with a family called Eames for a number of years, travelling on business for them to Canada. He liked Canada and decided to settle there.

After a short time in Napanee he went to Kingston and worked there at odd jobs. One winter he met a man named Phelps, who was in the clock business, and who persuaded Southworth to travel for him. Before long, Southworth went into business for himself as a clock peddler.

On 22 January 1839, while in Kingston, Stephen J. Southworth married Diantha Stoddard (b. 23 Feb. 1817 – d. 27 Jan. 1903), daughter of Arvin Stoddard of Bastard Township, Leeds County. The couple were married by the Rev. W. Smart, a Presbyterian minister. There were six children:[33] Martha (Mrs. E.A. Healy, d. 13 Feb. 1878), Simon Seaman (b. 18 July 1840 – d. 4 Jan. 1936), Donald Eames (b. about 1845 – d. 5 Dec. 1923), Arvin Stoddard (b. 20 Sept. 1850 – d. 12 Dec. 1933), Thomas (b. 16 Nov. 1855 – d. 9 Mar. 1924), and Leavitt (b. Dec. 1857 – d. Oct. 1933).

The Southworths lived in Bastard Township, where S.J. Southworth farmed on the rear of Lot 7, Concession VII, owned by his father-in-law. There were many wolves in the area, creating havoc for all farmers, and according to the *Brockville Recorder,* Southworth was one of many farmers who received a £1 certificate for each wolf pelt.

By 1851 the Southworths had built a one-storey farmhouse on the south half of Arvin Stoddard's land and in 1855 expanded their holdings to 200 acres. At the time of the census a clock peddler from the United States by the name of H.W. Cady was staying with the Southworths. It is reasonable to assume that Mr. Cady was the person who brought the clocks from Connecticut to Southworth, who then peddled the clocks in Leeds County in the 1850s. Southworth was away from home for several weeks at a time (this is substantiated by the fact that the *Brockville Recorder*

226

often reported that there were letters awaiting him at the post office). In addition to clock peddling and farming, Southworth maintained a large garden and orchard. Two men were employed in connection with his cider press, which ran by horsepower and produced almost 400 gallons of cider a year.

In November 1856 the Southworths purchased a farm near Frankville. The home, formerly an important hotel on the Old Perth Road, is pictured in Fig. 390. It was an impressive house with four fireplaces. For the next thirty years S.J. Southworth bought and sold land in Bastard and Kitley townships.

During these years Southworth continued to sell clocks and was often assisted by David Dowsley (b. 4 May 1824 – d. 3 Feb. 1912), who was a neighbour. Dowsley claimed that he could "tell some ludicrous stories of his trips and experiences selling clocks and jewellery in the Townships." Dowsley was an auctioneer and for many years president of the fair association. Therefore, it is not surprising that his friend and associate S.J. Southworth actively participated in the Leeds County fairs, as both judge and entrant, winning many prizes for his farm produce and animals.

Southworth was active in local affairs, being justice of the peace and often grand juror at the fall assizes of the counties of Leeds and Grenville. He was also a member of the Kitley Township Council in the 1860s and 1870s and, among other matters, was concerned with the condition of the jail and favoured prohibition.

In 1874 the Southworths bought a house on Keefer Street in Brockville and moved there in 1875. From this time on the sale of clocks was discontinued. However, in the early 1880s he secured the agency for the Brantford Agriculture Works and Millar's Agricultural Implements, which he sold until 1882 when he retired from business. He continued to live in Brockville, travelling to western United States from time to time with his wife to visit their sons.

In 1899 Southworth, who had been troubled with chronic bronchitis, contracted a cold and died two weeks later on 7 February 1899. His funeral was well attended, as a recognition of the high esteem that he enjoyed in the community. His wife passed away in 1903.

FIG. 388

FIG. 389

FIG. 388 *This label is believed to be a later printing.*
FIG. 389 *Tablet found on Southworth half-column clocks.*
FIG. 390 *Southworth house in Kitley Township.*

FIG. 390

CLARA STEPHENSON

Clara Stephenson (b. 1869 – d. 1941) was one of the very few known "lady watchmakers" of the last century. She became interested in this field at her father's store in Elora, Ontario, where he practised his trade as a watchmaker and jeweller. Her great interest in clocks and watches, coupled with her mechanical ability, provided her with the necessary prerequisites to learn the trade thoroughly. She apprenticed to her father and very quickly was able to repair all makes of watches and clocks.

A salesman who was fascinated by the existence of a lady watchmaker spoke of her often in his travels. Officials of the T. Eaton Company in Toronto eventually heard of Clara's occupation and offered her a job, which she accepted. A special bench in full view at the Toronto store was arranged for her work. This facilitated not only her work, but advertised her existence and drew customers to the store and she became well known. After several years she accepted an offer from P.W. Ellis, wholesale jewellers.

Due to the terminal illness of her mother, Clara left watchmaking in 1909 and returned to Elora to care for her. After her mother's death, she married and went to Iowa. In 1926 she returned to Toronto where she lived until her death.

Clara was the daughter of Samuel Stephenson, an Englishman who immigrated to Canada in 1855 and opened a store in Elora in 1865. Her mother was Kate Hall. Clara was the second oldest of eight children; four of her brothers also followed the profession of watchmaking. A fifth brother owned a large jewellery store. There was an agreement among her brothers that Clara was the best watchmaker of them all.

A picture of Clara with two sisters and a picture of her father's shop in Elora appeared in *The Canadian Jeweller* in October 1979.

FIG. 391

FIG. 391 *Sundials on the Mother House of the Sisters of Charity, Ottawa.*

SUNDIALS OF OTTAWA

Two sundials may be seen on the Mother House of the Sisters of Charity (Grey Nuns of the Cross) at the corner of Bruyere and Sussex streets, Ottawa. These dials have served to tell time for the nuns at the convent and the people in central Ottawa and are a great attraction to tourists. They tell accurate standard time.[34]

The carving of the sundials was begun in November 1850 and was completed in March of 1851. They were carved by Father Chaplain Jean François Allard, a scientist and astronomer who taught mathematics to the pupils at the school run by the nuns. The story of the sundials was told by the late Sister Paul-Emile in an English translation of excerpts from the *Annals of the Mother House*. She also told a reporter, Marian G. Rogers, of the day in 1919 when to her horror she spotted a workman painting over the Sussex Street sundial. He had been asked only to refresh the numerals and lines on the dial faces. Fortunately, the paint was of poor quality, and the outline of the numbers was still visible.

228

Years of acid rain and other pollution had almost obliterated the dials, but a restoration was completed in the late 1970s. The background is now white instead of natural grey, which blended with the stones of the building. The numbers in black can now be easily read.

J.H. THRALL

Jason H. Thrall (b. about 1842 – d. about 1906) was a watchmaker and jeweller in Almonte, Lanark County, Ontario. He established his shop on Mill Street in the late 1860s and moved to Bridge Street "opposite Almonte House" by 1884. He was listed in the *Province of Ontario Directories* until 1907.

An OG clock was sold with the name of "J.H. Thrall, watchmaker, Almonte, Ontario" on its dial. Unfortunately, the authors have not had an opportunity to examine the clock. However, from a picture, the clock may have been manufactured by the Ansonia Clock Company.

TORONTO WATCH COMPANY

Two key-wind watches marked Toronto Watch Company on both dials and plates are illustrated in Figs. 392 and 393. No record of such a company has been found. It appears that the watches were sold with customized inscriptions in the Toronto area.

This was a common practice in the closing decades of the 19th century and early years of the 20th century. Watches could be obtained with or without cases, and if ordered in reasonable numbers could be manufactured with the seller's name permanently inscribed on dials, cases, and movements. Both American and Swiss factories offered this service. The watches shown here are both of American origin.

The earlier movement (Fig. 393) has been identified as the product of a short-lived company, the New York Watch Company of Springfield, Massachusetts. It bears serial number 54958, which would place it around 1875. The other watch, which is quite similar, is numbered 80325. The New York Watch Company was reorganized in 1877 and became the Hampden Watch Company of Springfield (moving to Canton, Ohio, some years later). The second movement is believed to have been made shortly after this reorganization.

These dates appear to approximate the period when Toronto watches were sold. As noted with watches of the Canada Watch Company, these items are Canadian in name only.

FIG. 392
FIG. 393

FIG. 392 *Two watches marked "Toronto Watch Co." The earlier example is on the left.*
FIG. 393 *Movements of "Toronto Watch Co." watches.*

H. UTLEY AND COMPANY

Horace Utley owned a factory in the Niagara Falls area in Stamford Township, Upper Canada, where he made cases for clocks. The case style of the majority of clocks was pillar-and-splat (or transitional), but at least one clock has a more elaborate case, with carved columns and splat. Utley purchased wooden, weight-driven movements from Riley Whiting of Winchester, Connecticut, for his cases. Although exact dates of production are not known, it is believed that the factory made clocks from about 1824, the date on one clock, to perhaps 1835 or later. One confirmation that Utley was living in the area in 1833 is described below.

Most of the clock labels inside the cases were printed by P. Canfield, printers in Hartford, Connecticut. One clock label, however, was printed by Leavenworth, St. Catharines, and bears the coat of arms used in Canada at that time. Therefore, there are at least two label variants. Both labels contain the clear statement "Movements made by Riley Whiting, Winchester, Conn. and cased by H. Utley & Co. Niagara Falls, Upper Canada." Examination of the clocks confirms this. The movements are Terry-type, of a distinctive style made by Whiting and are easily identified by the unusual wooden bridge, which retains the escape wheel. In the Taylor numbering system, clocks examined contained movements type 1.72 and 1.73. All Utley clocks known to the authors have 30-hour movements.

In addition to the Utley clocks in private collections, clocks made by this firm may be seen at the Niagara Historical Museum, Niagara-on-the-Lake; Lundy's Lane Historical Museum, Niagara Falls; and Old Fort York and the Royal Ontario Museum in Toronto. The tablet on the Royal Ontario Museum clock was painted in 1833. A picture of this clock may be seen in *Canadian Clocks and Clockmakers* by G. Edmond Burrows. Another example is in the collection of the Henry Ford Museum, Dearborn, Michigan, and is illustrated in Fig. 394.

Little information has been uncovered about Utley himself. Mr. R. Taylor of Niagara-on-the-Lake has provided the authors with the following comments: "There is one insert in the *Niagara Gleaner* dated 21 September 1833. The postmaster, Ralph Clench, inserted a list of letters remaining at the post office in Niagara (now Niagara-on-the-Lake). The list contained the name of Horace Utley. The next list dated 5 October 1833 did not show his name. Obviously Horace picked up his mail. The mail for Niagara and the mail for Niagara Falls was always getting mixed up."

No other concrete references to Utley have been uncovered, and the location of the factory has likewise never been pinned down precisely. A good deal of early industrial development took place in Upper Canada along the Niagara River above the Horseshoe Falls. This industrial area was known as Niagara Falls Mills or Street's Mills and was variously described as being at the "head" or "foot" of Cedar Island. One reference[35] states that Utley had a clock factory near "The Gap" at the head of Cedar Island. Many other factories were in the area, including a nail factory, a woollen mill, a distillery, a tannery, etc. It is probable that Utley's factory was part of this industrial complex.

In 1887 Cedar Island, along with Dufferin Island (originally Cynthia Island), was deeded to the Crown and became part of Queen Victoria Falls Park. No sign of the original industries remains. Parkland and power developments have taken their place.

The authors have learned of one Utley clock which has an inscription written in pencil on the back of the dial plate: "This clock was first

FIG. 394a

FIG. 394b

FIG. 394 *A typical Utley 30-hour clock exhibiting label printed by P. Canfield, Hartford, CT. – Courtesy the Henry Ford Museum*

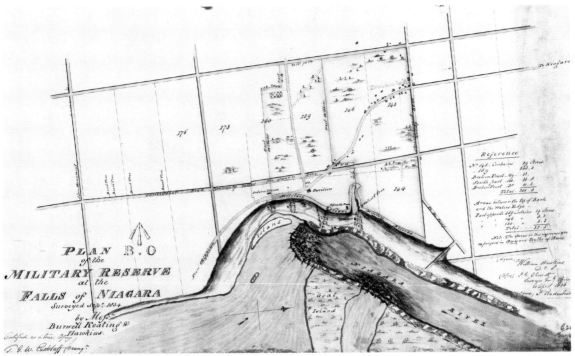

bought and owned by John Misner and Mary Misner of Silver Street, Bismark, County of York[36] [Gainsboro, Lincoln County] Ontario in the year of 1842. Passed on to their youngest son, John Nelson Misner of Bismark, Ontario. In 1931, I picked up the remains in the attic of the old homestead on Silver Street, Bismark and endeavoured to repair same after it passed into my wife's possession. [Mary Rebecca Misner was the eldest daughter of John Nelson Misner.] [signed] William F. Tritton, Fenwick, Ont., 1932."

FIG. 395 *Cedar Island (marked Island) was immediately above the Horseshoe Falls. The Utley factory was in this area. – Courtesy National Archives of Canada. Map H2/440*

One might reasonably assume that the date of 1842 represents the original date of sale. The movement of the clock is the later Whiting style (Taylor type 1.73) with a Canfield, Hartford label. This is seven years after the death of Riley Whiting, but it is quite possible that Utley continued to sell from stock for a number of years. Like many old clocks, this specimen has gone through long periods of neglect and more than one restoration, but the inscription provides an interesting insight into its history. The hamlets of Bismark and Fenwick are both within twenty-five miles of Niagara Falls, suggesting that Utley probably sold most of his clocks in the local area. The present owner obtained the clock from the Tritton estate.

The clock label bears what appears to be Utley's hand-written signature, "H. Utley and Co."

C.H. VAN NORMAN & CO.

Caleb H. and Abner E. Van Norman were watchmakers and wholesale and retail jewellers on James Street near King Street in Hamilton. The company was established between 1851 and 1857 and continued until 1865.

Port records of Hamilton indicate that the Van Norman company imported over seventy-five boxes of clocks from June 1858 to January 1861.

J. & B. VAN NORMAN

There is a small but interesting link between clockmaking in Canada and the famous Van Norman foundry at Normandale, Upper Canada. In an advertisement in a York newspaper of 1832 they listed clock weights as part of their extensive product range.

The activities of Joseph and Benjamin Van Norman have been extensively documented as part of the early industrial development of Canada. A deposit of "bog iron" was discovered in the early 1800s in Norfolk County, near the present village of Normandale on Lake Erie. Bog iron is a naturally occurring ore deposit, which results from the long-term action of specific bacteria in swampy areas where a significant amount of iron occurs in the soil.[37] In 1818 the first foundry was established on the site by John Mason. After his death it was taken over in 1822 by George Tillson, Joseph Van Norman, and Hiram Capron. Capron and Tillson pulled out in 1825, Tillson going to found the town of Tilsonburg.

Joseph Van Norman, his brother Benjamin, and other family members operated the foundry for many years, enjoying periods of considerable prosperity. Their best-known products included stoves, kettles, including tea kettles, ploughs, etc. They advertised their interest in building steam engines to order. Unfortunately, the supply of bog iron became exhausted at Normandale and the smelter was closed in 1847. Mining in other areas was financially unsuccessful, leading to bankruptcy in 1852.

Clock weights made at the foundry were probably never marked and no specific weights can be attributed to the Van Normans. However, they were active during the era of wood-movement weight clocks, and they may have offered price advantages to the clockmakers of the day.

W.H. VAN TASSEL

William Henry Van Tassel (b. 10 Nov. 1827 – d. 19 Feb. 1910) imported clocks made by the Seth Thomas Clock Company in Connecticut and sold them for the several years. He lived in Brockville, Canada West, around 1850. His activities as a clock peddler cover only a relatively short period of time, but he seems to have been quite successful. This became evident from a 1978 survey of clocks conducted by J. Varkaris. Of the dozen or so men who sold clocks in this manner, Van Tassel's name was third most common.

The clocks sold by Van Tassel consisted of typical Seth Thomas products including OGs, 30-hour half-column, and 8-day column-and-cornice styles, all with weight-driven brass movements. The lyre movement in one 8-day clock examined is Taylor type 4.211 and is similar to the movement shown in Fig. 342.

Two styles of label printing occur as illustrated in Figs. 397 and 398. The label in Fig. 398 is a later printing and the style is seen in other Seth Thomas clocks. As far as the authors can determine, this change took place around 1850, which coincides with the period of Van Tassel's activities.

W.H. Van Tassel was born in Belleville, Upper Canada, and lived there for most of his life. The family originated from the Isle of Texel, situated off the coast of Holland. Their genealogy[38] has been traced back to a man called Jan Van Texel, born about 1574. Members of the family came to America in Dutch vessels to trade with the Indians. Cornelius Jensen Van Texel, however, stayed in America and married an Indian princess on Long Island. According to a story, he was disinherited by his father, and he changed his name to Van Tassel, the name used by his descendants.

FIG. 396

FIG. 396 *W.H. Van Tassel*

232

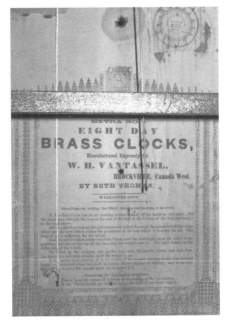

FIG. 397a
FIG. 397b
FIG. 398

FIG. 399
FIG. 400

FIG. 397 *Column-and-cornice clock and its label, sold by W.H. Van Tassel.*   FIG. 398 *Second label style found in a Van Tassel 30-hour clock. This is probably a later printing.* FIG. 399 *A 30-hour weight-driven clock. In reality a typical Seth Thomas product.*   FIG. 400 *Label from clock in Fig. 399. It is considered to be older than that shown in Fig. 398.*

Fig. 401

Fig. 401 *Tablet found frequently on "Van Tassel" clocks.*

Cornelius (b. 1625 – d. 1704) was the first Van Tassel to be born in America and one of the many members of the family to settle around Tarrytown, New York. The old Dutch church of Sleepy Hollow in Tarrytown (founded in 1684) has been immortalized by Washington Irving's story *The Legend of Sleepy Hollow*. In 1895 fifty-eight graves of Van Tassels, many with gravestones engraved in Dutch, could be found in the graveyard frequented by Ichabod Crane; and the girl Ichabod loved was named Katrina Van Tassel. In the 1970s the old Dutch church stood virtually unchanged since Washington Irving related his tale of the Van Tassels.

William H. Van Tassel's grandfather Jacob (b. 1767 – d. about 1815) was one of ten children who were the fourth generation of American Van Tassels living in New York. After Jacob's death in 1816 his widow, Lucretia, took her son, Ora (b. 3 Feb. 1800 – d. 1842), to Upper Canada, settling near Belleville. She married George Wensel and a child, George, was born on 3 April 1817.

William Van Tassel's father, Ora, married Phebe (Phoebe) Ketchison (b. 1807 – d. 1894) on 26 August 1826. The Ketchisons were one of the first pioneer families of Huntington Township, Hastings County, arriving about 1816 when the total population of the township was twenty-six souls. Ora and Phoebe left Belleville for Huntington Township in 1828 and farmed at Moira, Ontario. William Henry was the eldest child of Ora and Phoebe. Around 1849 William went to Brockville and began the first of his many occupations – clock peddling.

By January 1852, when the 1851 census was actually taken, William had left Leeds County and had returned to Hastings County, where he was listed as a carpenter. On 26 January 1854 William married Margaret Amarilla McMullen (b. 1835 – d. 25 April 1928), the daughter of Alexander McMullen and Rebecca Jones. Their children were Ann Evangeline (b. 26 Jan. 1859), who died as an infant and was buried in their Sidney Township lot, William (b. 8 Oct. 1860 – d. 16 Feb. 1911), Phoebe Jennie (Mrs. Richard Davis, b. 22 Jan. 1863), and George Warner (b. 18 Nov. 1866 – d. 2 Nov. 1951). They took up residence in 1855 at Sidney Crossing (now part of Belleville) and began farming on Lot 15, Concession 1 of Sidney Township. Van Tassel was successful here and won prizes with his produce and his cattle and sheep at the South Hastings Agricultural Fair.

Farming, however, did not completely hold Van Tassel's interest. He had learned about an invention called the "hydropult" which was sold in the United States. The hydropult, a force pump, was advertised primarily as "the most efficient fire engine in the world" and could be powered by one man and throw water at the rate of 8 gallons per minute to a height of 50 feet. It was also considered valuable "for washing windows, carriages, sprinkling plants, draining cellars, cleaning trees of insects and wetting sails." The hydropult was featured in Patterson's foundry in Belleville and sold by William H. Van Tassel at both the wholesale and retail level. The price of the hydropult was $12.50 retail.

One such pump was offered for sale at a 1979 auction in Belleville and was listed as a "Van Tassel, Canada West brass pump."

During the 1860s Van Tassel continued to raise cattle and sheep. In 1866 his wife's father gave her part of Lot 35, Concession 1 in Sidney Township. This lot, now close to the Ontario School for the Deaf, remained in the family until 1924. Once established on the new property, William bought the "milk business" from David Jones. In 1864 he advertised that he was recruiting men to go to work at Wyandott, Kansas, for the Union Pacific Railway. By 1870 Van Tassel was bored with the milk business. He established a store on Front Street in Belleville for the sale of sewing and knitting machines, and the next year advertised for cows in exchange for knitting and sewing machines. Van Tassel rented a shop next door to the "new" post office in the Bogarts building.

By 16 February 1871 he had established his "Sewing Machine Emporium," and he invited the public to come and see at least ten brands of sewing machines. A novel advertisement appeared in the newspaper of 9 March 1871 in the form of a poem. By June he had added Butterick patterns to his inventory.

Despite his continuing interest in sewing machines, Van Tassel began the last facet of his career in January 1870 by becoming an agent for the sale of agricultural implements. Farmers could use one of the "Superior Collards Iron Harrows" for several days and, if pleased with it, they could keep it and pay for it in October.

The first footbridge was constructed across the Moira River in 1873, and Van Tassel identified the location of his premises as "off Front Street near the footbridge." By 1880 he had dropped sewing machines from his inventory and was agent for the Patterson Brothers, selling agricultural implements and carriages. In addition, he continued to sell the hydropult spray pumps.

Advertising became more picturesque as Van Tassel distributed advertising cards to the public. Also, for at least a year around 1892 he published a monthly magazine called *Pointers*.

Van Tassel remained in business until 1900. His grandchildren remembered driving with him in his carriage pulled by his horse, Nelly, from the home in Sidney to the shop in Belleville.

W.H. Van Tassel was a member of the Wesleyan Methodist church, and according to his obituary, he always took a deep interest in religious matters. This is supported by the family, who recalled daily prayers and the fact that Van Tassel was frequently a lay preacher in the church. His Bible, still with a few dried flowers in its pages, remains in the family. It is inscribed, "This Bible is to the Willies if I am taken away – my Christmas gift Dec. 1909."

W.H. Van Tassel was ill for only a few weeks before he died at his home on 19 February 1910. He is buried with the members of his family in the Belleville Cemetery.

FIG. 402

FIG. 402 *Advertising cards of W.H. Van Tassel. – Courtesy Belleville Public Library*

JOHN WANLESS AND SUSAN BELL

John Wanless (b. 18 Feb. 1830 – d. 22 Feb. 1919) was born at the "School House," Longformadus, Berwickshire, Scotland. His father, William, was a well-known educator and his mother was Margaret (Graham) Wanless.

John became a watchmaker and jeweller and came to Toronto, Canada, about 1860.[39] In Toronto he managed the jewellery and watchmaking business for Susan Bell (b. 18 Nov. 1827 – d. 19 Dec. 1901). At this time the business was both wholesale and retail and sold jewellery, silverware, watches, and clocks.

A number of clocks, several made by the Seth Thomas Company, exist with the name John Wanless on the dial. One wall clock is also an advertising clock (see Fig. 403).

The business prospered under the management of John Wanless. In 1862 Wanless married Susan Bell, and after that date the business was called by his name. It was Susan Bell's second marriage. At some point, probably between 1865 and 1870, John and Susan sat for portraits painted by George T. Berthan. These portraits were inherited by their son, John Jr., and are now in the possession of the Art Gallery of Ontario.

The store was located at 172 Yonge Street. In December 1892, Mr. Wanless sold 42 feet of the property he owned immediately north of the jewellery store to Robert Simpson for $2,000 per front footage – a total of $84,000. Simpson pulled down the old buildings and put up a new structure. The Wanless establishment moved south of the Simpson store to 168 Yonge Street, the store owned previously by Ambrose Kent & Sons.

John Jr. (b. 28 Aug. 1862 – d. 15 July 1941) joined his father in the Wanless business about 1882. He had been educated at the Toronto Model School and had taken courses in banking and accounting, preparing him for a career in business. He also apprenticed as a watchmaker for his role in the John Wanless and Company enterprise. He became the firm's first gemologist and expanded his father's business, becoming a noted diamond merchant.

Other children of John Wanless and Susan Bell Wanless were Alexander (b. 1864 – d. 19 July 1872), Margaret (b. 9 June 1867 – d. 14 Nov. 1946), Clara (b. 6 Aug. 1869 – d. 22 April 1913), and Isobel May (Mrs. Arthur C. McMaster, b. 1872).

FIG. 403

FIG. 403 *Advertising clock sold by John Wanless.*

In addition to his interest in developing his business, John Wanless Sr. was an active member and elder in the Knox Presbyterian church. He was also a justice of the peace in Toronto for many years. Around 1888 the family moved to Spadina Avenue, near Bloor Street. John retired in 1908, and on 22 February 1919, he died of pneumonia, leaving an estate of $300,000. He is buried in Toronto in the Mount Pleasant Cemetery.

Susan Bell, the wife of John Wanless Sr., was a native of Otterham, Cornwall, England; she was the daughter of Daniel and Joan Kinsman. Although she and her mother were Wesleyan Methodist, some of her ancestors in England were clergy of the Church of England. There were also a number of watchmakers in the family. After her father's death, Susan came as a child to Canada with her mother. Her mother continued to live with her and her family until her death on February 17, 1882, at the age of ninety-one.

Susan Kinsman was married for the first time, around 1847, to William Bell (b. about 1801 – d. 18 Sept. 1854), a watchmaker who was originally from Jedburgh, Scotland. In Canada he established a watchmaking and jewellery business in Niagara, Upper Canada, in the 1830s. He moved to Toronto in 1840, founding a business that would continue for over one hundred years. Susan and William Bell had two children who lived to adulthood, Elizabeth (b. 25 May 1848 – d. 19 Sept. 1926) and Janet (b. 1 Oct. 1854 – d. 20 Nov. 1930). Neither daughter married. They remained close to John Wanless, their step-father, and benefited equally with his own children after his death.

William Bell's first business in Toronto was at 16 Church Street. By 1850, he had moved both his house and his business to 121 Yonge Street. Upon his death, Susan took over and for about seven years she ran her husband's business, then at 172 Yonge Street. The shop sold jewellery, clocks, watches, and "fancy goods." The Bell establishment had its own silver mark. In 1861 the value of the capital investment was $6,400 and the real estate was valued at $1,500.

FIG. 404 *John Wanless and Susan Bell Wanless. – Courtesy the Art Gallery of Ontario, Toronto*

FIG. 404a

FIG. 404b

FIG. 405

FIG. 406

FIG. 405 *This clock with the
overpasted label of T.J. Wheeler
is a typical 30-hour
New Haven OG.*
FIG. 406 *Detail of Wheeler label
from clock in Fig. 405.*

T.J. WHEELER

Thomas John Wheeler (b. 1839 – d. 1913) was born in England and worked as a watch- and clockmaker in Georgetown[40] in Halton County, Southern Ontario.

Two clocks containing Wheeler's distinctive label have been examined by the authors. The label is plain and measures about 1⅜ inches by 6¾ inches. It reads "Manufactured for and sold by T.J. Wheeler Georgetown." In each clock, the label was pasted over the original label in such a way as to completely conceal the original maker's name. Both clocks, however, were the product of the New Haven Clock Company and were in the 1880 New Haven Catalogue as the "OG" and the "Drop Octagon R.C." The OG clock in Fig. 405 has an unstamped movement of type 1.4 in the Taylor system. The octagonal clock has an unmarked 8-day spring-movement, time-only.

T.J. Wheeler came to Georgetown with his family around 1867 and remained there until his death in 1913. His wife, Mary Ann (b. 1841), born in Quebec of Irish parents, was the proprietress of a "fancy and dry goods" store in the community. A son, Thomas Alexander, was born in August 1871; by 1913 he was a commercial traveller in Toronto. A second son, Charles, was born in 1873 and became a watchmaker after being trained by his father.

Four Galbraith boys, Robert (b. 1856), Alfred (b. 1860), William H. (b. 1862), and Richard (b. 1865), from Quebec lived with the T.J. Wheeler family. William and Richard apprenticed with Thomas J. Wheeler.

T.J. Wheeler has been often confused with Thomas Wheeler who is listed as early as 1846 in the Toronto directories as a watchmaker and engraver at 6 King Street East. Thomas Wheeler was born in England about 1809 and lived on Elm Street in St. David's Ward in Toronto. His wife Eliza was born in 1823.

A.S. WHITING

Algernon Sidney Whiting (b. 1807 – d. 30 Mar. 1876) was born in New York and came to Upper Canada in the 1840s. His first place of residence was Port Hope, where he imported and distributed clocks purchased from the United States.

Whiting, a relative of the Gilberts, clock manufacturers in Connecticut, was one of the many clock importers. The customary way of bringing clocks from the United States during this period was by boat to the various Canadian ports. Whiting imported his clocks at Port Hope and Cobourg. He would then sell the clocks in the surrounding district, transporting them by a team of horses and a spring wagon. In one year alone – 1844 – he received three shipments of clocks which arrived from Rochester on the steamboats *America* and *Gore*. One shipment consisted of twelve boxes containing forty-eight clocks. The two other shipments were of thirty-eight boxes (150 clocks) and thirty-five boxes (132 clocks) respectively. As far as is known, A.S. Whiting did not add his name to any of the clocks he sold.

From Port Hope, Whiting moved to Newcastle, Bowmanville, and finally, by 1850, to the Oshawa area. For a time he represented the Winsted Manufacturing Company, an American firm that made agricultural implements. In 1852, with a capital of $75,000 and twelve stockholders, he organized the Oshawa Manufacturing Company, a business that manufactured scythes, hoes, and forks. Whiting was president of the company and the business was successful until in 1857, following the Crimean War, it was one of the many businesses that ran into financial difficulties. It failed and stockholders lost their investment.

From 1858 to 1860 Whiting rented space in the Joseph Hall "works" and began again to manufacture agricultural implements. Eventually, all his creditors were paid, and he was able to build a large plant near the waterfront, which he called Cedar Dale Scythe Works. With him in the business were E.C. Tuttle and W.J. Gilbert. At this time his brothers came to work in the business. Hiram was the bookkeeper and Edward became foreman. Homer assisted as well. Around the factory grew the village of Cedar Dale, which became part of Oshawa in 1923.

After a number of partner changes the firm changed its name to A.S. Whiting Manufacturing Company. Whiting's health failed, and after his death in 1876, his son-in-law, R.S. Hamlin, managed the business.

A.S. Whiting, a Wesleyan Methodist, married Julia (b. about 1812) while still living in the United States. Their only child, Cyrene Elizabeth, married Reuben S. Hamlin.

RILEY WHITING

Riley Whiting (b. 16 Jan. 1785 – d. 5 Aug. 1835) was a maker of wood-movement clocks in Winchester, Connecticut. Although it is unlikely that Whiting ever lived or worked in Canada, he has been included in this book because of the relatively large number of Whiting clocks which were sold here. Complete tall case 30-hour wood clocks bearing his name are frequently encountered. A few Whiting movements are also to be found in the clocks of at least one of the Twiss brothers in Montreal. Horace Utley of Niagara Falls used shelf clock movements by Whiting for all his production. Whiting movements are unique and easily identifiable. Two views of his tall case clock movement are shown in Fig. 409.

Clocks sold by Whiting had his name and place of business painted on the dials. Other manufacturers often included their name and instructions on how to care for the clock on a loose piece of paper put in with the packing. The paper was often lost or thrown away during unpacking, and in the years to come, the maker of the clock was unknown.

Riley Whiting was a native of Winchester, Connecticut. In February 1804 he married Urania Hoadley (b. 1788 – d. 1855). She was the younger sister of Luther and Samuel Hoadley, who became his partners in 1808 in a clockmaking business under the name Samuel Hoadley & Company. By 1810 they were also involved in casting clock bells and by 1812 the business expanded to include one of the earliest operations for drawing wire through hardened forms to obtain the desired diameter.

Both Hoadley brothers served in the War of 1812. Luther was killed on 8 September 1813. When Samuel returned, the name of the company was changed to Hoadley and Whiting. By 1819 Riley Whiting purchased all interests in the business and continued to manufacture clock movements under his own name. In the 1820s Riley Whiting was one of the two largest wooden clock movement makers in the country.[41] In the early 1820s most of the other makers of wooden movement tall case clocks had ceased production in favour of the shelf clock. Towards the end of the 1820s, Riley Whiting also adopted shelf clock styles. His shelf clocks were similar in style to those of other manufacturers of the period – being found with stencilled decoration or with carved columns.

During a business trip in the summer of 1835, Riley Whiting visited his daughter in Jacksonville, Illinois. During his visit he became ill and died on 5 August 1835. Chris Bailey, in his book *Two Hundred Years of*

FIG. 407

FIG. 407 *Clock made by Riley Whiting.*

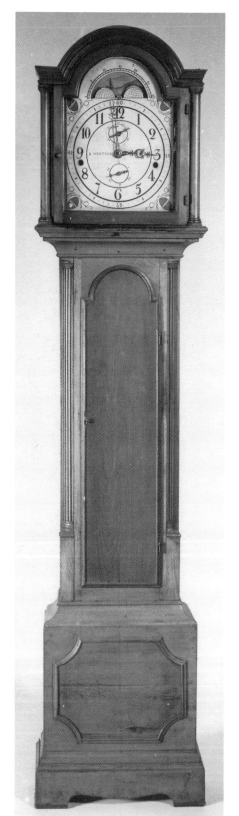

FIG. 409

FIG. 408

FIG. 410

FIG. 408 *Clock made by R. Whiting. – Courtesy Collections d'ethnographie du Québec* FIG. 409 *Two views of a Riley Whiting tall case movement.* FIG. 410 *Riley Whiting's version of the Eli Terry Patent wood movement is easily recognized by its wooden escape wheel bridge. From a shelf clock by H. Utley, Niagara Falls, Upper Canada.*

American Clocks and Watches, quotes the *Annals of Winchester* regarding Riley Whiting:

> He was a man of quiet, unassuming manner and feeble constitution who more than most men minded his own business and prosecuted it with the same perseverance in adverse, as in prosperous circumstances; and though twice compelled to assign his property, yet in both instances succeeded in paying off his debts and left a handsome estate at his death....He was a man highly esteemed. He represented the town in the legislature in the years 1818 and 1832.

Chris Bailey believes that the factory may have produced "Riley Whiting" shelf clocks for six years after his death because it was not until 1842 that his widow, then Mrs. Erasmus Darwin Calloway, sold the factory to Lucius Clarke, William Lewis Gilbert, and others.

STEPHEN WILLCOCK

Stephen Willcock (b. 28 Oct. 1845 – d. 5 Feb. 1927) was a watchmaker and jeweller in Toronto, Ontario. He was one of the few Canadians who held patents for clock innovations. His two Canadian patents are concerned with the chiming mechanism of clocks. Introducing his Canadian Patent number 36532 in 1891[42] (U.S. Patent number 451353), he claimed to have "invented a certain new useful improvement in clocks." The "aim was to design a simple mechanism by which various chimes may be introduced." His patent number 51032 in 1895, which patented the use of two separate movements and the use of a resonator, introduced further improvements. According to this patent, "The object of my invention is to devise means for increasing the volume and improving the quality of sound from the gongs or bells of a striking or chime clock."

Willcock was awarded patents in England, as well as in Canada and the United States. The English patents assigned the use of two separate movements to Thomas Sargant[43] and Reinholt Egmund Gunther. Also, the patents assigned the use of the chime mechanism to Samuel Davison and the use of the clock resonator to R.E. Gunther.

In addition, the clock resonator was patented on its own in the United States on 24 March 1896 and is U.S. Patent number 557040. Its use was assigned to Reinhold Egmund Gunther. R.E. Gunther lived in Toronto and joined the family firm of wholesale jewellers and watchmakers in 1886. This company, founded before 1859 by Frederick and Edmund Gunther, was located first at 9 King Street East and then, with Edmund and Anton Gunther as partners, at 16 Jordan Street. R.E. Gunther, the son of Edmund, rose from bookkeeper in 1889 to manager in 1893. The E. & A. Gunther Company was moved to Brantford, Ontario, in 1926 and R. Egmund Gunther continued to manage the firm until 1930. The company was the Canadian representative for the New Haven Clock Company of Connecticut.

The clocks incorporating the innovations of Willcock's patents were apparently all manufactured by the New Haven Clock Company. The clocks resemble contemporary bracket clocks, fairly large and square with pediment and finials. The depth of the clock was needed to accommodate the movement and large chime mechanism.

Two stamps mark the chime mechanism as illustrated in the Fig. 415. One stamp, in the form of a circle, identifies the 1896 Canadian patent. The other stamp was "applied for" in the U.S.A. It should be noted that the name "Wilcock" is misspelled.

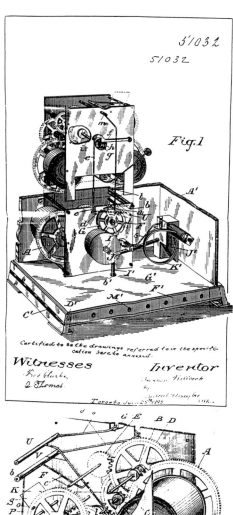

FIG. 411

FIG. 411 *Illustrations from Patent number 51032.*

Clocks using clock-chime mechanisms used two movements. The New Haven clock using the Willcock patent mechanism preceded the well-known Sonora chime clock mechanism, which was patented in 1908 and assigned to the Sonora Chime Company. (Patents were subsequently acquired and used by the Seth Thomas Company.) The clocks which have the chime mechanisms made according to the Willcock patent have the following features: the time-and-strike movement and the chime movement were placed on the same plane; both movements were mounted on an iron frame approximately a figure 8 in shape; in contrast to the usual time-and-strike movement of the New Haven Clock Company, nearly all the clocks having the chime mechanism made to the Willcock patent had a solid-brass back plate. "The centre wheel arbor and lifting wire are extended through the back plate. A four pointed star cam is mounted on the centre wheel extension to trip a lever each quarter hour and initiate the chime melody. A lever is mounted on the left wire extension. This lever is activated to strike the hour after the 16 note chime has played."[44]

Although the time-and-strike movement described above and illustrated in Fig. 413 is found most often in clocks with chime mechanisms made according to the Willcock patent, another "cut-out" movement was used in at least one model (see Fig. 420).

The Willcock chime movement, tuned to play the Westminster chimes, was spring-driven, and a resonator was used inside the case on the bot-

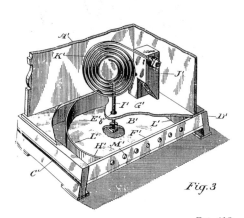

FIG. 412

FIG. 412 *Illustration of clock resonator from Patent 51032.*
FIG. 413 *Time and chime movements used in Willcock patent clocks.*

FIG. 413

FIG. 414

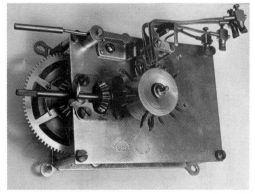

FIG. 415

tom. It was referred to in the New Haven Company catalogues as a "Gunther Resonator." An assembly of five wire gongs on a single pedestal was used for strike and chime. The larger gong was used to strike the hours. The assembly passed through a hole in the top plate of the resonator and was bolted to the bottom plate. A pinned cylinder trips the "bell" hammers to play the melody on the gongs.

The chiming mechanism itself is not self-correcting. To correct chiming, the minute hand is turned backwards to the quarter chime and then moved slowly, allowing the clock to chime and strike the correct time. An error in hour-striking can be corrected independently.

Clocks using the Willcock patent were advertised in the New Haven catalogue of 1900. Five models of these clocks have been identified, but the models differ mainly in case design. The models are numbered 3, 4, 5, 6, and 7. A sixth model may be number 2. Model 3 is 18½ inches in height. Model 5 is 22¼ inches in height and could be purchased in 1900 for $36.25.

Model 4, as assigned by Mr. Fallin,[44] has a plain case topped by three large finials. Model 6 is illustrated in *Canadian Clocks and Clockmakers* by G. Edmond Burrows. A further model of a Westminster chime clock made by the New Haven Clock Company and shown in a 1980s Price Guide by Roy Ehrhardt may also use the Willcock patent.

Model 1, an "Eight Bell Chime Clock," was shown in the New Haven catalogues. It is a bigger clock with a 7-inch dial, and it plays Canterbury chimes. No clock of this model has been located for study but, as Willcock patent chime clocks play Westminster chimes on four gongs, it has been assumed by Mr. Fallin that this model did not use Willcock chimes. Model 2 has not been identified with certainty.

Mantel clocks were available in light oak or mahogany, with oxidized

FIG. 414 *Close-up of a chime mechanism in another clock of the same model.*
FIG. 415 *Chime mechanism in clock in Fig. 421.*

FIG. 417

FIG. 416 *Model 3 and Model 5 from the New Haven Clock Co. catalogue.*
FIG. 417 *Model 3 Willcock clock with different finials and feet.*

silver or bronze trimmings. Three dial sizes were used by the New Haven Clock Company on cases with Willcock chime mechanisms. Most often found is a 6-inch square dial, with applied cast-brass spandrels in the corners and applied silver chapter ring, that was used on tentative model 2, and on models 3, 4, and 5. Two other Willcock patent clocks used an 8-inch round silvered dial with black painted numerals. This dial is found on models 6 and 7. Unnumbered models described below use a smaller dial with pie-crust bezel.

These two styles of New Haven Westminster chime clocks using the Willcock patent do not appear to have model numbers. In each of the clocks, the chime mechanism is stamped with the Willcock patent stamp (see Fig. 415). The cases of the clocks, while similar in shape to known numbered models, are distinctive in style. Fig. 420 has an ornamented, carved, curved crown. This clock has a cut-out plate on the time-and-strike movement and is not mounted on an iron frame. It also appears to be in a different plane to the chime movement, unlike the other New Haven clocks with Willcock chime movements that follow Patent No. 51032. The plaque in front of the clock identifies the date when the clock was purchased. The clock was presented to Dr. J.B. Reynolds, associated in various capacities with the Ontario Agricultural College, Guelph, Ontario,[45] by two classes of third-year students and was purchased in 1898. This clock would be one of the earlier clocks with Willcock patent chime mechanisms.

The clock in Fig. 421 has a plain curved top, but the case itself has a carved column on either side and the carving of a wreath in the area below the dial. This clock has an unusual resonator box. It is unlike all other Willcock patent clocks examined by the authors in that seven small holes on the edge of the box are missing. This clock is fitted with the New Haven movement shown in Fig. 413

Tall case models using the Willcock patent chime mechanism were made by the New Haven Clock Company between 1900 and 1920. "Wilmington," "Sherbrooke," "Annapolis," and "Minneapolis" models could be purchased in 1911 with Westminister chime movement for $72.50. The time-and-strike movement, similar to that used in mantel clocks, was modified for use of the longer pendulum. It was also weight-driven. "The ratchet wheels are installed on the great wheel arbors to accommodate chains and pull-up winding. The time and strike trains may also be wound with a crank key through holes in the dial.... The chime mechanism must be wound by pulling up the chain.... In the tall case model, the large gong for striking the hours has been removed from the assembly and is mounted to the clock hood next to the movement."[46]

Kirk Fallin has reported that "three tall case clocks with Willcock chime movements have been found and all are the Sherbrooke model. Two are in mahogany and one in antique oak. These clocks share one unusual feature – 'sandwich' pinions on the fly arbors in both the strike and chime trains. These are cut pinions, brass on the outside with fibre between, presumably to quiet the trains. A few mantel clocks have been reported using this feature, but it is rare; most mantel clocks use lantern pinions."

The key fitting the Willcock clocks is a number 6 key with an unusually long shank. "The length is needed to reach the winding square in the great wheel arbor of the chime mechanism. The chime train winds from the front on some clocks and from the side in others.... There could be front winders and side winders in the same model...a hole is drilled

FIG. 418 *Model 5 clock.*   FIG. 419 *Chime clock possibly using the Willcock patent.*   FIG. 420 *An unnumbered Willcock patent clock.*

through the case (for the side winders). A brass plug with 2 flat springs to hold it in place, is provided for the winding hole."

On one clock examined, there is a label inside the back door of the clock case giving instructions for using the clock and the model number was penciled in, on others the number was printed.

One clock enthusiast, also a distant relative of Stephen Willcock, claimed that Willcock "had some poor engineering in his movement, in particular the strike to start the chime. The clock was probably not as popular after Seth Thomas came out with a more popular chiming movement."

Stephen Willcock became a watchmaker in his youth and by 1867 had opened a watch repair and jewellery store on King Street in Toronto. By 1873 he specialized in wholesale jewellery and watch materials. He continued his Toronto business at several addresses until after the death of his father in 1893. It was during this period that Stephen patented his inventions, but by the end of the 19th century he was no longer a resident of Toronto.

The Willcock family was of Saxon origin and lived in Suffolk, England, before the Norman Conquest. Sir John Dugdale in *Visitations to the County of Suffolk* mentions over fifteen generations of the Willcock family prior to 1600, taking the family back to 1200. According to a researcher of the Willcock family, a study of the coat of arms shows that the family came from Welsh kings. Many variations of the spelling of this name exist.

Stephen's father, John (b. 13 Mar. 1807 – d. 12 June 1893), and John's brother Abel (b. about 1812 – d. 24 Aug. 1877), natives of Cornwall, England, came to Toronto in 1840 and became builders and contractors. Abel and his family, which also included a son named Stephen, continued in the lumber business in Toronto. Both Abel's and John's families were Reformers in politics and members of the Wesleyan Methodist Church. John was accompanied to Canada by his wife, Anne (b. 1806), and his eldest child, Rachel (b. 1840 – d. 7 Sept. 1930), born just before they left England. A son one year older than Stephen, a daughter, Emma, born in 1850, and Stephen were born in Canada.

Stephen Willcock and Rachel lived with their father until his death, after which Stephen apparently left Toronto. It is not known whom he married or where he made his home during the first quarter of this century. In 1927 he was living in Hespeler (now part of Cambridge), Ontario, with his daughter, Mrs. Ronald Garrett. In 1927 and 1928 the Garretts resided at 1341 King Street, Preston (also part of Cambridge). A search of town and library records in that area uncovered no additional information.

Stephen Willcock died in Hespeler on 5 February 1927 and was buried in Toronto Necropolis Cemetery with his parents and sister Rachel. His wife is not buried with him. No will was found.

FIG. 421

FIG. 422

FIG. 421 *A Willcock patent clock.*
– Courtesy Philip P. Miller
FIG. 422 *Clock in Fig. 421 showing movements.*

WILLIAM WILLOX

Another interesting clock in the collection of the Niagara Historical Museum is a primitive tall case clock by William Willox. This clock combines a movement obtained in Scotland with a simple walnut case fashioned in the Niagara Peninsula. Willox, the builder of the case, was a farmer and miller who built the case for his own use, and was neither a clockmaker nor a cabinetmaker. It is of interest to note that Willox had joined the Woodruff family of St. Davids in working the mill there. Around the same period, Richard Woodruff of St. Davids was actively selling tall case clocks in Upper Canada from his sale base at Burford.

The clock has an unmarked movement, typically British in style, 8-day brass time-and-strike, with calendar and second hands. According to the Museum records, the movement was provided by William Donald of Rhynie, Scotland. It is believed that Willox brought the movement with him to Canada when he came from Scotland.

The case, pictured in Fig. 425, is of simple styling. It is made from inch-thick walnut with a pine backboard. The arched pediment appears to have been replaced. There was probably also some sort of foot moulding, which is now missing.

The dial is made of sheet metal covered by a painted paper. The present dial is a restoration done in 1962 by a Willox family member – the grandson of the original maker. From fragments of the original dial, it has been concluded that the present dial is a copy of the original. The dial is marked "Wm. Donald Rhynie" on either side of the dial centre and "William Willox" above the Figure XII. In the arch is a folk-art rendition of Adam and Eve.

William Willox (b. 1801 – d. 17 Nov. 1870) had worked as a gardener on a large estate in Scotland. He and his wife, Mary (b. 1800 – d. 1 July 1874), and two children, James (b. 1830) and Charles (b. 1831), arrived in Canada sometime between 1831 and 1840. Another son, Edward, was born in Upper Canada. They settled on what is now Highway 8A between Queenston and St. Davids.

The Willox and Woodruff families share a burial plot in St. Davids.

FIG. 423

FIG. 423 *Tall case Willcock patent clock, "Sherbrooke" Model with replaced dial.*
FIG. 424 *Movements in tall case clock in Fig. 423.*
FIG. 425 *Clock by William Willox.*
– Courtesy R. Taylor and Niagara Historical Museum

FIG. 424

FIG. 425

R. WOODRUFF

Six tall case clocks have been reported with the name R. Woodruff on dials and cases. These clocks are basically similar and contain 30-hour, time-and-strike, "pull-up" wood movements. A distinctive feature of these clocks, where the original finish survives, is that they are usually copiously decorated with a fine gilt stencilling that was applied to both hood and body of the case. The stencilled inscription, "R. Woodruff Burford U C"

FIG. 426

FIG. 426 *This clock by R. Woodruff contains a 30-hour movement by A. Merrell, an Ohio maker.*

occurs at the top of the door in the case. The cases are of soft-wood construction and are often finished in dark brown imitation wavy wood grain. It appears that Richard Woodruff's name was added at the same time as the decoration. Other clocks are known with Woodruff's name on the case only, but with a dial bearing the name of another maker. Woodruff clocks are known with a dial by Silas Hoadley (movement not examined) and with a dial and movement by Riley Whiting. Another clock is shown in Fig. 426, where the name of Ansel Merrell, a clockmaker of Trumbull County, Ohio, appears. The movement in this clock appears also to be the work of Merrell.

The authors have been unable to determine whether Woodruff obtained the clocks from both Connecticut and Ohio or whether some of the examples cited above are the result of later "marriages." The cases shown in Figs. 426 and 428 are of a style and finish which could be typical of either area.

The existence of the Merrell/Woodruff clock and a recent publication on the clockmakers of Trumbull County, Ohio,[47] however, suggest a possible source for at least some of Woodruff's clocks. Ms. Rogers has confirmed that there was at least some commerce between Trumbull County and Canada. The Woodruff family name, too, was relatively common in Ohio. Geographically, Ohio would have been closer to Woodruff's scene of activities than Connecticut.

Burford is a township and a village that was located in Oxford County at the time of Upper Canada, but became part of Brant County after 1852. The clocks were probably sold during the period of 1825 to 1840, but no specific information about Woodruff's activities in the area has been uncovered.

The Woodruff family arrived in Upper Canada from Connecticut before 1800 and settled around St. Davids in the Niagara Peninsula. Richard (b. 10 Apr. 1784 – d. 1 June 1872) was the second of seven children of Ezekiel Woodruff. When hostilities with the United States began in the War of 1812, Richard and his brothers enlisted in the 1st Regiment of the Lincoln Militia. Richard was recommended for promotion to sergeant in the 7th Company in 1812 and subsequently received a medal for his action at Detroit.

Unfortunately, the area around St. Davids where the family settled was the scene of military operations by both the Canadians and the American forces and the area was devastated.

For his part in the war, Richard was given a military grant of land in 1820 – Lot 20, Concession IX, Zorra Township, in Oxford County. It was perhaps the reason that he went to this area and ended up in nearby Burford, where his clocks were sold. Woodruff does not appear to have remained there for long, however. In a 1913 book entitled *An Early Political and Military History of Burford,* by R. Cuthbert Muir, Woodruff's name is not mentioned. An unnamed cabinetmaker living in Burford in the 1830s could possibly have been Woodruff.

Richard Woodruff returned to St. Davids and lived there for the rest of his life. He married Ann (b. 9 Aug. 1788 – d. 30 Mar. 1873), who was born in Upper Canada, daughter of Joseph Clement of the Township of Niagara and a United Empire Loyalist. The marriage took place prior to 1811 when Ann applied for land as a daughter of a Loyalist. Richard and Ann had a large family who remained in the Niagara area in the counties of Welland and Lincoln. Sons William H. (b. about 1813) and Joseph C. (b. about 1823) were executors of Richard Woodruff's estate. When he died, R. Woodruff was a wealthy man. Ann received the homestead and furniture and all the money with the exception of bequests of $9,000. He also left to his sons Lots 97 and 98 in the Township of Niagara and two other farms (320 acres) in the Township of Stamford. He also divided among the members of his family $50,200 in stocks, mortgages, and bonds.

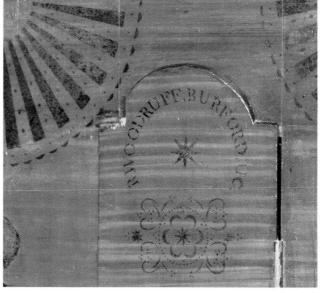

FIG. 427

FIG. 427 *Detail of a typical signature from clock in Fig. 426. "R. WOODRUFF BURFORD U.C."*

FIG. 428 *Clock and dial by R. Woodruff. Dial is marked "R. Woodruff, Burford U.C." in the circular band, under the hands. The lower left corner has been cut away, spoiling an otherwise perfect dial.*

– Courtesy Norwich District Museum, Norwich, Ontario

FIG. 428a

FIG. 428b

CHAPTER 5
PORT RECORDS

In the 19th century, most of the goods entering Canada from other countries, including the United States, came by ship to the many ports in the Maritimes and in Quebec (Canada East) and Ontario (Canada West). In 1845 there were about forty-five ports of entry in Canada West and about sixteen ports in Canada East.

At most ports of entry, detailed records were kept of each article that passed through customs. However, some ports were required to record only items on a list during certain periods of time. For example, from 1795 to 1838, the Port of Quebec and the Port of St. Johns listed only limited items which did not include clocks. The listings for some ports

FIG. 429

FIG. 429 *The* Lady of the Lake.

were aggregate listings and gave the total amount collected for customs. For instance, the duty at 12½ percent at the Port of Toronto collected for importing clocks and watches in 1854 and 1855 was more than collected for ready-made clothes and coffee and about the same as for whisky. Similarly, at the Port of Brockville, in the year ending 5 January 1845, the amount of duty collected was £17.19.1 on a value of clocks and watches worth £615.14.8.

Further listings for which there were no details: Port of Coteau du Lac, quarter ending 5 July 1834–53, clocks value £315; quarter ending Jan. 1844, 101 clocks, £43.5.0, collected at 5 percent duty & £47.11.6 at 7 percent duty. Port of Oakville, 12 clocks in quarter ending Jan. 1844, value $39.50, via schooner *A Gage* from Oswego import-

ed by P. Gage. Port of Montreal 1844, no clocks mentioned. Port of Chatham 1844, no clocks mentioned. Port of Cobourg, quarter ending July 1843, 14 clocks, value £24.8.1½, duty paid £7.6.6. Quarter ending Jan. 1844, 66 clocks, value £232.3.2, duty paid £34.11.3.

However, records for many ports have been lost. Also, for any one port, the years for which records exist are intermittent. It is therefore difficult to document either one port or one period of time.

The duty collected varied over the period. In the 1840s two separate duties were collected, one imperial and one provincial. In 1845 the duty on clocks and watches was 7 percent Imperial and 5 percent Provincial on the value of the goods. By 1856 there was one duty at the Port of Toronto – 12½ percent, which went to the province.

In the 1830s and 1840s Canada used the currency of Britain, which was pounds sterling. The United States currency was dollars. For example, in 1844 the American dollar equaled 5s 1d. The pound would have been equivalent to $4.50 to $4.75 American dollars.

The first steamer, named *Lady of the Lake,* left Ogdensburg in 1832. This ship was followed by others until 1874. On another route, it was known in 1833 that a steamer left Prescott every day for Toronto, Hamilton, and Niagara. On a third route that was often used, steamships carrying goods would sail up the Hudson River, along the Erie Canal to Oswego, cross Lake Ontario to Kingston and Belleville, then back to Kingston and Oswego. By 1845 thirty small steamships made this triangular route. This circuit would take three days if there was no delay.

The steam boat *America* made forty-four calls between July 1843 and January 1844. *Gore* made fifty-six calls and *Princess Royal* made seventy-four calls during the same period.

It is hoped that the following information gleaned from some of the port records will be of interest to readers and inspire further research in this area. Among the diverse articles imported, which included dogfish oil, porpoise oil, and goose wings, were watches and a large number of clocks.

250

Date	Carrier	Place of Origin	Imported by	Item
		PORT OF COBOURG		
14 Oct 1843	Gore	Rochester	L. Sprague	12 clocks
26 Oct 1843	Gore	Rochester	I. Fowler	1 clock
1 Nov 1843	Gore		L.C. Whiting	35 boxes clocks (132)
3 Nov 1843	Gore		L.C. Whiting	2 brass clocks
3 Nov 1843	Gore		A.S. Whiting	12 boxes clocks (48)
20 Nov 1843	America	Rochester	A.S. Whiting	38 pieces 150 brass clocks
29 Nov 1843	America	Rochester	A. Pringle	6 brass clocks
30 Nov 1843	Princess Royal	Rochester	W. Brooks	6 clocks
1 Dec 1843	Princess Royal	Rochester	E. Perry	3 brass & 5 wood clocks
Oct 43 – Jan 44	Gore	Rochester	J. Adams	6 brass clocks
Oct 43 – Jan 44	Gore	Rochester	A. Pringle	12 clocks
		PORT OF HAMILTON		
4 June 1858	Clifton		Einstein & Rudle?	6 boxes clocks
24 July 1858	Clifton		Einstein & Rudle?	13 boxes clocks
2 Nov 1858	Clifton		J. Nickerson	10 boxes clocks
2 Nov 1858	Clifton		J. Nickerson	2 boxes weights
6 Nov 1858	Clifton		Newbury and Birley	8 boxes clocks
10 Nov 1858	Clifton		Newbury and Birley	2 boxes clocks
10 Nov 1858	Clifton		Newbury and Birley	9 boxes weights
18 Nov 1858	Clifton		Einstein & M.	1 box weights
18 Nov 1858	Clifton		Einstein & M.	1 box clocks & fixtures
18 Nov 1858	Clifton		Einstein & M.	26 boxes clocks
18 Nov 1858	Clifton		Einstein & M.	5 boxes weights
18 Nov 1858	Clifton		Newbury & Birley	7 boxes clocks
23 Nov 1858	Clifton		Newbury & Birley	17 boxes clocks
24 Nov 1858	Clifton		Newbury & Birley	2 boxes clocks
27 Nov 1858	Clifton		C. H. Van Norman	7 boxes clocks
27 Nov 1858	Clifton		C. H. Van Norman	2 boxes weights
24 Dec 1858	Clifton		R. Osborne	7 boxes clocks
24 Dec 1858	Clifton		R. Osborne	2 boxes weights
28 Dec 1858	Kingston		A.W. Gage	3 boxes clocks
28 Dec 1858	Kingston		A.W. Gage	1 box weights
3 Jan 1859	Clifton		Newbury & Birley	11 boxes clocks
16 Apr 1859	Clifton		Newbury & Birley	3 boxes clocks
5 Sept 1859	Clifton		Rowes & Co.	1 box clocks
5 Oct 1859	Clifton		A. Buller	7 boxes clocks
12 Oct 1859	Clifton		Newbury & Birley	16 boxes clocks
14 Oct 1859	Clifton		C.H. Van Norman	6 boxes clocks
14 Oct 1859	Clifton		C.H. Van Norman	1 box weights
14 Oct 1859	Clifton		Newbury & Birley	25 boxes clocks
19 Dec 1859	Clifton		Newbury & Birley	1 box clocks
21 Dec 1859	Clifton		C.H. Van Norman	8 boxes clocks
21 Dec 1859	Clifton		C.H. Van Norman	2 boxes weights
27 Dec 1859	Clifton		C.H. Van Norman	3 boxes clocks
31 Dec 1859	Clifton		Newbury & Birley	20 boxes clocks
22 Feb 1860	Clifton		C.H. Van Norman	2 boxes clocks
22 Feb 1860	Clifton		C.H. Van Norman	1 box weights
13 Mar 1860	Clifton		Newbury & Birley	3 cases clocks
17 May 1860	Clifton		Newbury & Birley	5 cases clocks
6 June 1860	Clifton		Beemer & Co.	6 boxes clocks
6 June 1860	Clifton		Beemer & Co.	1 box weights
6 June 1860	Clifton		Newbury & Birley	11 boxes clocks
20 June 1860	Clifton		C.H. Van Norman	4 boxes clocks
20 June 1860	Clifton		C.H. Van Norman	2 boxes weights

Date	Carrier	Place of Origin	Imported by	Item
21 July 1860	Clifton		Newbury & B.	7 boxes clocks
24 July 1860	Clifton		Newbury & B.	7 boxes clocks
23 Aug 1860	Clifton		C.H. Van Norman	1 box clocks
9 Oct 1860	Clifton		Newbury & B.	29 boxes clocks
11 Oct 1860	Clifton		J. Van Gunter	3 boxes clocks
11 Oct 1860	Clifton		J. Van Gunter	1 box weights
11 Oct 1860	Clifton		Buchanan & Harris	6 boxes clocks
19 Oct 1860	Clifton		Newbury & B.	20 boxes clocks
22 Oct 1860	Clifton		Van Norman	8 boxes clocks
22 Oct 1860	Clifton		Van Norman	3 boxes weights
12 Nov 1860	Clifton		Meyers	4 boxes clocks
22 Nov 1860	Clifton		Newbury & B.	2 boxes clocks
23 Nov 1860	Clifton		Newbury & B.	3 boxes clocks
28 Nov 1860	Clifton		Newbury & B.	13 boxes clocks
5 Dec 1860	Clifton		Newbury & B.	7 boxes clocks
7 Dec 1860	Clifton		Beiman?	4 boxes clocks
7 Dec 1860	Clifton		Beiman?	2 boxes weights
11 Dec 1860	Clifton		Newbury and B.	3 boxes clocks
15 Dec 1860	Clifton		Newbury and B.	5 boxes clocks
19 Jan 1861	Clifton		C.H. Van Norman	6 boxes clocks
19 Jan 1861	Clifton		C.H. Van Norman	2 boxes weights
9 pages torn out of book				

PORT OF KINGSTON

Date	Carrier	Place of Origin	Imported by	Item
6 July 1843	Clinton	Oswego	J.M. Mansman	18 clocks
6 July 1843	Rochester	Oswego	C.E. Miller	65 clocks
7 July 1843	Lady of the Lake	Oswego	Rossin Bros.	24 brass clocks
12 July 1843	Telegraph	Oswego	D.H. Sines	2 mantel piece brass clocks
15 July 1843	Union	Oswego	W. Lorimer	2 wooden clocks
21 July 1843	Clinton	Rochester	D. Campbell	2 brass clocks
1 Aug 1843	sleigh ?	Turnsville?	Orvis Griggs	12 wooden clocks
4 Aug 1843	St. Lawrence	Oswego	A.A. Griggs & Co.	12 OG clocks
7 Aug 1843	Clinton	Rochester	Samuel Rose	1 wooden clock
8 Aug 1843	St. Lawrence	Oswego	J.W. Wait	12 wooden clocks
10 Aug 1843	Oneida	Oswego	Mrs. M. Madia?	1 brass clock
12 Aug 1843	Union	Oswego	W. Brophy	2 wooden clocks
14 Aug 1843	Clinton	Rochester	S. Blenheim	3 clocks
16 Aug 1843	Empress	Rochester	A. Barclay	6 brass & wooden clocks
16 Aug 1843	St. Lawrence	Oswego	J.M. Tayler	30 brass & wooden clocks
16 Aug 1843	Empress	Rochester	A.G. Buckley	2 wooden clocks
16 Aug 1843	St. Lawrence	Oswego	C. Kennedy & Co.	80 clocks
18 Aug 1843	Oneida	Rochester	J.P. Tayler	6 brass clocks
23 Aug 1843	Lady of the Lake	Oswego	C. Cropman	1 brass clock
24 Aug 1843	Lady of the Lake	Ogdensburg	P. Meydonn	1 brass clock
2 Sept 1843	St. Lawrence	Rochester	W. Borman	1 brass clock
6 Sept 1843	Clinton	Oswego	Govis Griggs	48 wooden clocks
6 Sept 1843	Oneida	Rochester	A. Fisk	1 brass clock
11 Sept 1843	Empress	Oswego	E.O. Connell	1 brass clock
23 Sept 1843	Rochester	Oswego	P. Burke	12 brass clocks
5 Oct 1843	Oneida	Oswego	T. McKinley	18 brass &wooden clocks
9 Oct 1843	Rochester	Rochester	P. Burke	66 brass clocks
12 Oct 1843	Rochester	Oswego	J. Berkley	12 brass clocks
19 Oct 1843	St. Lawrence	Rochester	C. McMillan	18 wooden bevel clocks
21 Oct 1843	Oneida	Oswego	P. Burke	42 brass & wooden clocks
26 Oct 1843	Oneida	Oswego	P. Burke	36 brass & wooden clocks
30 Oct 1843	Oneida	Rochester	C. Mc. Millan	24 wooden clocks
3 Nov 1843	Gildersleeve	Ogdensburg	A. Rowan	1 brass clock
8 Nov 1843	President	Oswego	J. Powell	30 brass & bronzed clocks

Date	Carrier	Place of Origin	Imported by	Item
8 Nov 1843	Clinton	Oswego	O. Griggs	7 boxes (42) wooden clocks
13 Nov 1843	Lady of the Lake	Oswego	A. Thrasher	6 wooden clocks
18 Nov 1843	St. Lawrence	Oswego	A. Ross	54 brass & wooden clocks
20 Nov 1843	Schooner L. Goler	Oswego	T. Murphy	2 clocks
21 Nov 1843	St. Lawrence	Oswego	J.C. Brown	1 wooden clock
15 Dec 1843	Gen. Brock	Oswego	Rossin Bros.	2 gold watches
Years missing				
7 Jan 1846	Lady of the Lake	Cape Vincent	William G. Wait	1 case clocks brass & wood (6)
12 Jan 1846	Gen. Brock	Oswego	Wait & Simmons	5 boxes clocks & weights
23 Jan 1846	Gen. Brock	Oswego	Wait & Simmons	17 boxes brass clocks & weights (102)
3 Feb 1846	sleigh	Ellisburgh	Lovell Bigelow	6 brass clocks
11 Feb 1846	sleigh	Rochester	Ben L. Davy & Co.	2 boxes (12) brass & wood clocks
26 Feb 1846	sleigh	Watertown	Calvin Gaylord	3 brass clocks
18 Mar 1846	?	Cape Vincent	Charles Heath	1 box brass clocks (6)
pages missing				
2 May 1846	Lady of the Lake	Oswego	Wm. McMillan	24 brass clocks with weights
7 May 1846	Clinton	Oswego	Wm. McMillan	26 boxes brass clocks &weights
18 May 1846	Clinton	Oswego	J. Carruthers	2 cases clocks
18 May 1846	Clinton	Oswego	J. Carruthers	4 boxes clocks
23 May 1846	Wheeler	Oswego	J. Paterson	27 boxes clocks
26 May 1846	Lady of the Lake	Oswego	G. Spandenburg	2 clocks
26 May 1846	Lady of the Lake	Oswego	W. Hurdle	9 boxes clocks
27 May 1846	Lady of the Lake	Oswego	W. Hurdle	1 case clocks
2 June 1846	Rochester	Oswego	G. Spandenburg	4 boxes clocks
2 June 1846	Rochester	Oswego	C.W. Jenkins	6 boxes clocks
3 June 1846	Niagara	Oswego	J. Pul...	1 box clocks
3 June 1846	Niagara	Oswego	J. Fraser	12 boxes clocks
3 June 1846	Winnebago	Oswego	McBean & Strong	1 box clocks
9 June 1846	St. Lawrence	Oswego	J. Rankin	5 boxes clocks
25 June 1846	Clinton	Oswego	W & B Wait	4 boxes clocks
25 June 1846	Pulaski	Oswego	Blair Wait	2 boxes clocks
26 June 1846	Clinton	Oswego	Murphy	4 boxes clocks
27 June 1846	Lady of the Lake	Oswego	G. Spandenburg	4 boxes clocks
27 June 1846	Gen. Brock	Oswego	Wait & Simmons	10 boxes brass clocks
29 June 1846	Pulaski	Oswego	Blair Wait	3 boxes clocks
8 July 1846	Clinton	Oswego	J.B. Wait	30 brass clocks
11 July 1846	St. Lawrence	Rochester	C. Covert	10 brass clocks
17 July 1846	Pulaski	Oswego	Blair & Wait	6 brass clocks
17 July 1846	Clinton	Oswego	J.B. Wait	24 brass clocks
20 July 1846	Rochester	Oswego	J. Patterson	262 clocks
22 July 1846	Rochester	Rochester	S. Perkins	1 brass clock
27 July 1846	Clinton	Oswego	J.B. Wait	30 brass clocks
28 July 1846	Lady of the Lake	Rochester	J.B. Wait	1 clock
29 July 1846	Lady of the Lake	Oswego	W. Hurdle	2 boxes brass clocks (5)
5 Aug 1846	Niagara	Oswego	J. Murphy	7 boxes clocks (42)
6 Aug 1846	Lady of the Lake	Oswego	W.G. & J.B. Wait	30 brass clocks
8 Aug 1846	Niagara	Oswego	A. Hall(Hales)?	1 brass clock
12 Aug 1846	St. Lawrence	Oswego	Wm. G. Northgraves	6 brass clocks
27 Aug 1846	Lady of the Lake	Oswego	J. Patterson	brass clocks
29 Aug 1846	Lady of the Lake	Oswego	W.G. & J.B. Wait	24 brass clocks
3 Sept 1846	Niagara	Oswego	James Linton	60 brass clocks
8 Sept 1846	Lady of the Lake	Oswego	W.G. & J.B. Wait	12 brass clocks
10 Sept 1846	J. Miller		S. Lucas	1 wooden clock
14 Sept 1846	Lady of the Lake	Oswego	W.G. & J.B. Wait	30 brass clocks
28 Sept 1846	Pulaski	Oswego	Blair Wait	42 brass clocks with wts.

DATE	CARRIER	PLACE OF ORIGIN	IMPORTED BY	ITEM

PORT OF PHILLIPSBURG

DATE	CARRIER	PLACE OF ORIGIN	IMPORTED BY	ITEM
18 Oct 1842	waggon [*sic*]	Vermont	Allen	18 clocks
11 Nov 1842	waggon	Vermont	Wm H. Stearns	12 brass clocks
15 Nov 1842	waggon	Vermont	S. Sweet	1 clock
8 Dec 1842	sleigh	Vermont	L.P. Rexford	1 wooden clock
22 Dec 1842	sleigh	Vermont	R Arms?	20 unfinished clock movements
28 Dec 1842	sleigh	Vermont	Cal...	1 brass clock
6 July 1843	waggon	Vermont	T.W. Pease	21 clocks
30 Aug 1843	waggon	Vermont	R. Whiting	3 brass clocks
21 Sept 1843	waggon	Vermont	A. Everett	6 brass clocks
30 Sept 1843	waggon	Vermont	S. Root	5 clocks
30 Sept 1843	waggon	Vermont	T. Brown	3 clocks
5 Oct 1843	waggon	Vermont	Root	5 clocks
10 Oct 1843	waggon	Vermont	Root	3 clocks
13 Nov 1843	waggon	Vermont	A. Everett	18 clocks
12 Dec 1843	sleigh	Vermont	G. Simmons	12 clocks

PORT OF SARNIA

DATE	CARRIER	PLACE OF ORIGIN	IMPORTED BY	ITEM
Sept 1840 to June 1843 – no clocks				
30 June 1843	Red Jacket	Michigan	J. House	1 clock
5 Aug 1843	Red Jacket	Michigan	J.S. House	4 clocks
24 Aug 1843	Red Jacket	Michigan	Wm Dowling	1 clock
19 Oct 1843	Sans Berette?	Michigan	A. Leys	3 wooden clocks
27 Oct 1843	Red Jacket	Michigan	J.S. House	6 clocks
6 Nov 1843	Red Jacket	Michigan	A. Young	1 clock
16 Nov 1843	Red Jacket	Michigan	T. McGlaskan?	1 wooden clock
20 May 1844	Red Jacket	Michigan	Thos. Fisher	1 clock
24 May 1844	Red Jacket	Michigan	Geo. Durand	1 clock
20 June 1844	Huron	Michigan	Thos. Johnson	1 clock
8 Oct 1844	Charlotte	Michigan	Alex Leys	3 clocks
23 Nov 1844	Huron	New York	Wm. Catheart	1 clock
13 Oct 1845	Huron	Michigan	W. Morrison	1 clock
14 Oct 1845	Scow	Michigan	W. Watson	6 wooden clocks

PORT OF TORONTO

DATE	CARRIER	PLACE OF ORIGIN	IMPORTED BY	ITEM
26 May 1838			Johnson	1 clock
26 May 1838			H. Stafford	clocks
21 July 1838			Michael Harroghy	1 clock
1 Oct. 1838			Thos. Rigney	1 clock
1 Jan 1839			Thos. Baker	4 pr. clock hands
				6 cards watch hands
				watch brushes
				1 doz. watch oil
				1 doz watch hoods
				13 gross watch glasses
				7 watch guards
1 Jan 1839			R. Bailey	134 clocks
29 May 1839			T. Rigney	2 gilt & marble clocks
14 June 1839			Thompson & Lawson	4 cases
14 June 1839			T. Rigney	2 clocks
2 Sept 1839			Thompson & Lawson	12 clocks
14 Nov 1839			Lejolie & Brother	4 German clocks
18 July 1840			T. McMurray	1 case clocks
30 Jan 1841			Jas. Simon	6 clocks
13 Apr 1843			A. Cosgrove	1 wooden clock
15 May 1843			Mrs. Frazer	1 wooden clock
10 June 1843			Mrs. Jacke	1 brass clock
27 June 1843			Thompson Smith	1 wooden clock

DATE	CARRIER	PLACE OF ORIGIN	IMPORTED BY	ITEM
4 July 1843			John Gunn	2 brass clocks
				3 wooden clocks
6 July 1843			P. O. Neill	6 brass clocks
				6 wood clocks
				24 bevel clocks
11 July 1843			Patrick Burke	6 brass clocks
				1 brass clock
20 July 1843			Wm. Allan	3 brass clocks
				1 OG brass clock
21 July 1843			Hiram Piper	4 (8day) brass clocks
				6 (30 hr) brass clocks
				6 (30hr) wood clocks
				12 wood clocks
27 July 1843			Wm. Allan	2 clocks
2 Aug 1843			Anthony Pfaff	27 brass clocks
3 Aug 1843			R.H. Brett	1 brass clock
3 Aug 1843			Kirby Marsh	6 brass clocks
7 Aug 1843			I. Gales	1 brass clock
12 Aug 1843			A. Keough	7 brass clocks
				6 bevel clocks
16 Aug 1843			A. Mc Bride	6 brass clocks
19 Aug 1843			P. O'Neill & Co	28 clocks
				14 brass clocks
21 Aug 1843			P. Patterson	12 brass clocks
22 Aug 1843			K. Marsh	12 brass clocks
23 Aug 1843			A. Kehoe	6 brass clocks
25 Aug 1843			Richard Marshall	24 wooden clocks
31 Aug 1843			Thos. Camey	1 clock
2 Sept 1843			P. Burke	6 brass clocks
5 Sept 1843			D. Maitland	6 brass clocks
6 Sept 1843			James Austin	6 brass & 6 wooden clocks
7 Sept 1843			Mathew Nemay	3 brass clocks
				6 wooden clocks
9 Sept 1843			Croizer	6 wooden clocks
				9 brass clocks
12 Sept 1843			Will Wakefield	12 brass clocks
16 Sept 1843			Charles Sewell	54 brass clocks
23 Sept 1843			Stephen Fox	1 brass clock
23 Sept 1843			T. Rigney	19 wooden clocks
				12 wooden clocks
				6 brass clocks
23 Sept 1843			Mathew Nemay	6 brass clocks
				6 wooden clocks
25 Sept 1843			John Birch	5 brass clocks
5 Oct 1843			M. Nemay	6 brass & 6 wooden clocks
6 Oct 1843	America	Rochester	M. Lawrence	2 clocks
6 Oct 1843	Union	Rochester	Pfaff & Co.	18 brass clocks
9 Oct 1843	Admiral	Rochester	I. Kellogg	6 clocks
16 Oct 1843	Union	Rochester	M. Nemay	24 wood & 6 brass clocks
23 Oct 1843	Admiral	Rochester	I. Kellogg	12 clocks
25 Oct 1843	Gore	Rochester	Wm. Malcolm	12 wooden clocks
6 Nov 1843	Gore	Rochester	F. Watkins	2 boxes clocks
9 Nov 1843	America	Rochester	J. Snell	1 clock
13 Nov 1843	Gore	Rochester	I. Kellogg	1 box clocks
14 Nov 1843	Gore	Rochester	J.S. Foster	6 brass & 12 wooden clocks
20 Nov 1843	America	Rochester	J. Terry	4 boxes clocks
20 Nov 1843	America	Rochester	A. Buckley	2 boxes clocks
23 Nov 1843	Gore	Rochester	P. Burke	6 wooden clocks
24 Nov 1843	Union	Rochester	Hugh Monroe	1 box clocks
1 Dec 1843	Transit	Lewiston	Chas. Harte	5 boxes clocks
8 Dec 1843	Richmond	Rochester	Thos. Rignet	10 boxes clocks
14 Dec 1843	Union	Rochester	Pfaff & Co	12 boxes clocks

Date	Carrier	Place of Origin	Imported by	Item
14 Dec 1843	Transit	Lewiston	J. Snell	2 boxes clocks
4 Jan 1844	Gore	Rochester	Metcalfe & Co.	2 boxes clocks
Pages missing				
20 May 1844			Thos. Fisher	1 clock
24 May 1844			Geo. Durand	1 clock
20 June 1844			T. Johnson	1 clock
8 Oct 1844			Alex Leys	3 clocks
23 Nov 1844			Wm. Cathcart	1 clock
8 Jan 1845	Transit	Lewiston	F. Watkins	1 case clocks
14 Jan 1845	Transit	Lewiston	Thos. Dick	8 boxes clocks
1 Feb 1845	Admiral	Oswego	Metcalfe & Co.	36 boxes clocks
3 Feb 1845	Admiral	Oswego	G. Barnes	1 case clocks
20 Mar 1845	American	Rochester	Murphy	1 case clocks
21 Mar 1845	Transit	Lewiston	E. Alport	1 clock
28 Mar 1845	Chief	Lewiston	J. Sibley	17 cases clocks
29 Mar 1845	Gore	Rochester	P. Sinclair	1 wood clock
31 Mar 1845	America	Rochester	Wm. Clarke	4 cases clocks
1 Apr 1845	Chief	Rochester	Wm. Knox	1 clock
pages missing				
8 July 1845	America	Rochester	Wm. Clarke	3 cases clocks
9 July 1845	Chief	Lewiston	G. Hughes	2 boxes clocks
24 July 1845	America	?	C. Robertson	4 boxes clocks
26 July 1845	Transit	Lewiston	Wm. Morrison	1 clock
6 Aug 1845	Chief	Lewiston	Jackson	7 boxes clocks
6 Aug 1845	Chief	Lewiston	McDowall	2 clocks
6 Aug 1845	Chief	Lewiston	O'Neill Bros.	11 cases clocks
6 Aug 1845	Chief	Lewiston		1 box
6 Aug 1845	Chief	Lewiston	D. Rose	2 cases clocks
16 Aug 1845	Chief	Lewiston	J. Joseph	2 cases clocks
28 Aug 1845	America	Rochester	T. Platt	1 case clocks
30 Aug 1845	Chief	Lewiston	Foy & Austin	clocks
30 Aug 1845	Chief	Lewiston	O'Neill & Co.	2 boxes clocks
30 Aug 1845	Chief	Lewiston	B. Tinamid	5 boxes clocks
4 Sept 1845	America	Rochester	Metcalfe & Co.	6 cases clocks
6 Sept 1845	Chief	Lewiston	Rigney & Co.	10 boxes clocks
6 Sept 1845	Chief	Lewiston	J. Young	16 cases clocks
11 Sept 1845	Transit	Lewiston	A. Brown	2 boxes clocks
16 Sept 1845	America	Rochester	R. Killenn?	1 box clocks
22 Sept 1845	Transit	Lewiston	B. Thomas & Co.	15 cases clocks
25 Sept 1845	Transit	Lewiston	O'Neill & Co.	6 boxes clocks
25 Sept 1845	Transit	Lewiston	Romaine & Co.	5 boxes clocks
10 Oct 1845	Chief	Lewiston	F. Perkins	6 cases clocks
13 Oct 1845			W. Morrison	1 clock
14 Oct 1845			W. Watson	6 wooden clocks
2 Jan 1846	Admiral	Lewiston	Rossin Bros	2 silver watches
7 Jan 1846	Admiral	Rochester	M. Newton	6 pkgs clocks
9 Jan 1846	Admiral	Rochester	E.B. Clarke	3 pkgs clocks
9 Jan 1846	Admiral	Rochester	Tho. Spratt	3 pkgs clocks
9 Jan 1846	Admiral	Lewiston	M. Newton	6 boxes clocks
17 Jan 1846	Transit	Lewiston	J.B. Clarke	2 cases clocks
20 Jan 1846	Admiral	Lewiston	J. Cummow	3 cases clocks
20 Apr 1846	Queen	Lewiston	A. Ogilvie & Co.	2 brass clocks
21 Apr 1846	American	Rochester	T. Rigney & Co.	clocks
23 Apr 1846	American	Rochester	J. Lemmon	4 boxes clocks
23 Apr 1846	Chief	Lewiston	J.S. Clarke	3 cases clocks
29 Apr 1846	Transit	Lewiston	Wm McMaster	clocks
30 Apr 1846	Chief	Lewiston	Foy & Austin	clocks
7 May 1846	Transit	Lewiston	ONeill Bros.	6 cases clocks
13 May 1846	Telegraph	Oswego	Romaine & Co.	playing cards & clocks
15 May 1846	Transit	Lewiston	J. Froman?	2 boxes clocks
18 May 1846	Weeks	Oswego	T. Rogaul	7 boxes clocks

DATE	CARRIER	PLACE OF ORIGIN	IMPORTED BY	ITEM
19 May 1846	Weeks	Rochester	Wm. Floon?	1 clock
22 May 1846	Manhatton	Oswego	B. Thonne & Co.	25 boxes clocks
22 May 1846	American	Rochester	Romaine Bros.	clocks
22 May 1846	American	Rochester	Rossin Bros.	watch materials
23 May 1846	American	Rochester	A.&S. Waisland?	3 boxes clocks
26 May 1846	American	Rochester	J.C. Joseph	8 boxes clocks
27 May 1846	Transit	Lewiston	Wm. Bell	5 boxes clocks
27 May 1846	Transit	Lewiston	Moffat, Murray & Co.	5 cases clocks+ other goods
28 May 1846	American	Rochester	J. Harre...& Co.	clocks
29 May 1846	American	Rochester	O'Neill Bros.	clocks
30 May 1846	Admiral	Lewiston	Wm. Sutherland	clocks
1 June 1846	Transit	Lewiston	Rossin Bros.	4 cases clocks
2 June 1846	Transit	Lewiston	B. Torrence	clocks
4 June 1846	Transit	Lewiston	Rossin Bros.	2 cases clocks
6 June 1846	Transit	Lewiston	H. Jackson	4 boxes clocks
9 June 1846	Admiral	Lewiston	E. Morphy	2 boxes clocks
10 June 1846	American	Rochester	G. Glenson	1 brass clock
15 June 1846	Transit	Lewiston	J.A. Clarke	8 boxes clocks
15 June 1846	American	Rochester	Thos. Rigney & Co.	3 cases clocks
16 June 1846	American	Rochester	Wm. Spense	1 clock
23 June 1846		Lewiston	Rossin Bros.	3 boxes watchmaking tools
24 June 1846	Transit	Rochester	James Young	1 case clocks
25 June 1846	American	Rochester	Wm. M. Newton	7 boxes clocks
29 June 1846	American	Rochester	Rossin Bros.	4 cases clocks
3 July 1846	Transit	Lewiston	R. Machel	clocks
6 July 1846	Transit	Rochester	Lewis Bradt...	13 cases clocks
8 July 1846	Transit	Lewiston	S. Miller	1 brass clock
11 July 1846	Queen	Lewiston	Urquhart?	1 clock
14 July 1846	American	Rochester	M. Haas	2 boxes clocks
25 July 1846	Queen	Lewiston	O'Neill Bros.	74 brass clocks
27 July 1846	Queen	Lewiston	A. Clarke	12 clocks
4 Aug 1846	Transit	Lewiston	Romaine Bros.	14 boxes clocks
15 Aug 1846	Admiral	Lewiston	T. Ryr... Co.	22 cases clocks
15 Aug 1846	Admiral	Lewiston	Charles Lynus?	7 cases clocks
28 Aug 1846	Admiral	Lewiston	William Graham	1 case clocks
29 Aug 1846	Admiral	Lewiston	M. Balfe	1 brass clock
10 Sept 1846	American	Rochester	G.M. Chenny & Co.	13 cases clocks
1 Oct 1846	Admiral	Lint...	Romaine Bros.	clocks
6 Oct 1846	Admiral	Lint.	T.G. Ridout	clocks
10 Oct 1846	Admiral	Lewiston	Rossin Bros	41 boxes clocks
16 Oct 1846	Transit	Lewiston	A. Pfaff	11 boxes of clocks
17 Oct 1846	Transit	Lewiston	Collins	1 clock
20 Oct 1846	Transit	Lewiston	Clarke	1 clock
4 Nov 1846	America	Rochester	J. McCooy	2 clocks
6 Nov 1846	Chief	Lewiston	Pay...	17 boxes clocks
11 Nov 1846	Chief	Lewiston	Ellis	6 cases clocks
24 Nov 1846	America	Lewiston	J. Montgomery	1 box clocks
25 Nov 1846	Chief	Lewiston	M. Haas	1 case clocks
1 Dec 1846	Beagle	Rochester	J. Murphy	1 box clocks
3 Dec 1846	Chief	Lewiston	Wm. Gamble	2 cases clocks
11 Dec 1846	Chief	Lewiston	Clarke	1 case clocks
20 Dec 1846	Quebec	Oswego	Thompson	1 pkg clocks
Years missing				
17 July 1849				8 pkgs clocks & weights
27 July 1849				1 pkg clocks
6 Aug 1849				6 pkgs clocks
22 Aug 1849				3 pkg clocks
4 Sept 1849				1 pkg clocks
6 Sept 1849				1 pkg clocks
22 Sept 1849				3 pkgs clocks
24 Sept 1849				1 pkg clocks

ADDENDUM

As noted at the outset, research is seldom if ever, complete. After this manuscript was completed, and before it was published, interesting and relevant pieces of information continued to turn up. Rather than lose the chance to include this material, the authors have provided this small addendum.

ASAHEL BARNES

The name of Asahel Barnes (also spelled Barns) has been seen on a number of wood-movement shelf clocks sold in the area of Vergennes, Vermont. Recently, one tall case clock has been reported, bearing the name of Asahel Barnes of L'Acadie and Montreal, with a label printed entirely in French.

Asahel Barnes Sr. (b. 21 Sept. 1777 – d. June 17 1859), born in Bristol, Connecticut, and his wife, Katura Ives, were the parents of nine children. Both were said to be related to prominent clockmaking families in Bristol. They moved to New Haven, Vermont, in 1814. In 1821, they moved to L'Acadie, Lower Canada, where Barnes was reportedly in the foundry business. About two years later they returned to Addison, Vermont. For the next few years Barnes followed a number of occupations, including postmaster, hotel and ferry service operator, and, in 1825, proprietor of a cabinet and clock shop. He moved to Burlington, Vermont, in 1841.

During his two-year period in L'Acadie, Barnes obviously engaged in some clockmaking activity. The tall case clock that bears his name is illustrated in Fig. 430 and is fitted with what appears to be a typical 30-hour New England wood movement, with seconds bit and pull-up weights. The movement has not been examined. The case is made of pine, finished in a clear varnish. The finish is old and appears to be original. The case is of simple design, having two "horns" or scrolls at the top of the hood and three gilded wood finials. The all-French label states, "Horologes faites par Asahel Barns a son fabrique a Montreal et L'Acadie." (Clocks made by Asahel Barns at his factory at Montreal and L'Acadie.)

The authors were unable to find any reference to Barnes's activities in Canada and are indebted to Mr. Fred Ringer Jr. of Vermont for most of the above information, from American archives.

GEORGE CAIN JR.

One clock bearing the label of George Cain, Jr. is shown in illustration Fig. 431. It is a fairly typical 30-hour half-column weight clock, with the exception of an inlaid veneer design, which can be seen on the front of the case below the door. The movement is made by Seth Thomas (Taylor type 1.241) and bears the Seth Thomas stamp.

The label itself has the ornamentation typical of Seth Thomas labels. As noted elsewhere in this text, the Thomas company offered customized labels for several Canadian vendors. In this example, the printed instructions have been rearranged and a large British coat of arms has been inserted in the centre. No mention is made of Seth Thomas as maker, but the clock is described as having been "Expressly Manufactured for GEORGE CAIN JR., Lindsay, Victoria Co. C.W."

Cain's name is recorded in the directory and census information for 1871, where he is recorded as a cabinetmaker in Lindsay.

LUKE MADDOCK

Old magazines sometimes contain interesting tidbits of information. As an example, the excellent but short-lived publication, *Canadian Antiques and Art Review,* in the December/January issue of 1979/80 (Vol. 1 No. 4) illustrates a

FIG. 430

FIG. 431a

FIG. 430 *Tall case clock by Asahel Barnes, Montreal and L'Acadie.*
FIG. 431a *Clock by George Cain Jr.*

FIG. 431b

FIG. 432a

FIG. 432b

late 18th-century gold watch by Luke Maddock of St. John's, Newfoundland, and describes it as follows: "The Newfoundland Museum, St. John's Newfoundland, purchased from a dealer in Dorset, England, in March 1979, a watch made by Luke Maddock in 1772. He operated as a watchmaker in St. John's, Newfoundland during the 1790s and was the only watchmaker listed in the 1794–95 Newfoundland census.

"The Maddocks were a family of watchmakers who plied their trade in eighteenth century Waterford, Ireland. The watch case is decorated with naval emblems and it may be assumed that the timepiece was prepared for an officer of the Royal Navy.

"The works are marked in script, 'L. Maddock, St. John's N.F.Land.' The number 32 is placed beside this inscription. The initials 'P.M.L.' appear in script on the back of the case. Hallmarks from the London Assay Office are leopard's head, crowned; lion passant; date letter and maker's mark."

This is the earliest reference the authors have found to watchmakers in Canada.

RUSSELL TWISS

Throughout the history of the Twiss family in the Montreal area, as described earlier in this book, the name of Russell Twiss comes up regularly as a participant in the family business. Clocks bearing his initials are found in partnership with his brother Joseph and his father Hiram. He was also the last member of the family to remain in business in the area, at Ligouri (between 1837 and 1851). Until recently it was believed that no clocks were sold bearing the name of R. Twiss only. However, one has turned up in time to be included here. This clock is shown in Figs. 432 and 433.

As can be seen in Fig. 432, the clock closely resembles other Twiss tall case clocks in style and finish. The chief point of interest is the movement, shown in Fig. 433. It is clearly the product of Riley Whiting. As reported previously, Whiting died in 1835. It has always been believed that his widow continued for several years after his death to sell material from Whiting's shop, and this movement would seem to confirm it.

FIG. 433

FIG. 431b *Label from the clock in Fig. 431a.*
FIG. 432 *Clock and dial by Russell Twiss.*
FIG. 433 *Movement by Riley Whiting in the clock in Fig. 432.*

APPENDIX 1

CLINKUNBROOMER

Children of Nicholas and Sarah Klingenbrumer:

1. Charles (b. 1799 – d. 12 Jan. 1881), clockmaker m. Hannah Anderson (b. 1811 – d. 30 Oct. 1872)

    Children: Emily (b. about 1835)

        Mary (b. 1836 – d. 14 Mar. 1918) in Stouffville

        Charles E. (b. about 1839) m. Martha Campbell

            Children: Minnie (b. about 1863)

                        Edward C. (b. about 1865)

                        Emma (b. about 1869)

                        Clara (b. 1870 – d. 1885)

                        John (b. 1873 – d. 1874)

                        Lotty (b. about 1878)

                        Frank (b. 1880 – d. 1881)

        Hannah Maria (b. about 1842) m. Daniel Valley

        Nathan (b. about 1844)

        Eliza (b. about 1846) m. John Alexander

        Sarah Agnes (b. about 1849)

        Clara Jane (b. 1852 – d. 24 Sept. 1869) of TB

        Thomas A. (b. about 1856) florist

            m. 1. Sarah Wright (d. 19 Mar, 1892)

                child: Lillie

            m. 2. Fanny Wright

                Children: Frederick

                Wood

2. Joseph (b. 21 Mar. 1801 – d. 20 May 1884), tailor

    m. 1. Theresa Hale

    m. 2. Nancy Flock, called Ann

        Children: Susan m. Alfred Driffet

            Louisa or Sarah – no record

            Martha Ann (b. about 1835 – d. before 1884) m. Peter Gilchrist

            William Henry (b. 1834 – d. 19 May 1813) m. Elizabeth Gilchrist (b. 1832 – d. 1904)

                Children: Charles (b. 1863 – d. Mar. 1888)

                      Joseph D.

                      Peter H.

                      Henry (b. after 1867)

                      Jessie Louise (d. 5 Aug. 1957) m. Alton Isaac Taylor (d. 1935)

                          Children: Mable

                                    Ross

                                    Roy

            Margaret (b. about 1845 – d. 25 Mar. 1909) m. Joseph H. Driffet

            James H. (b. about 1847) m. Mary Anna Sweetman

            Jane (b. about 1849)

            Also buried in same plot: Harold W. (d. 27 Dec. 1890) and Fred G. (d. 16 Dec. 1919)

3. Xavier or Exaveras, mason

    Child: Lava m. George Carroll in 1860

APPENDIX 2

THE NORTHGRAVES FAMILY

William (b. 1764 – d. 17 June 1819) from Hull, England

William (b. 1800 – d. 25 May 1864) m. Theresa Brussien

Children: William J. (b. 1820) m. Alice Jane (b. 1825)

Children: Caroline Louise (1845) musician; m. Bernard Doyle, lawyer on 29 Nov. 1871.

Theresa (b. about 1847)

Alice Jane (b. about 1852)

William (b. about 1858)

Anna Maria (b. about 1863) in Belleville

Henry (b. 11 Apr. 1864) in Napanee

Frederick (b. 1822 – d. 5 June 1879)

Eliza Ann (b. 18 Feb. 1827) m. Michael Deane, surveyor in Lindsay on 6 Sept. 1859

George Richard (b. 1834 – d. 1919), died from injuries sustained in a fall. Obituary in the *Globe* – "oldest and probably the most noted priest in America."

George (b. 23 Aug. 1803 – d. 12 Oct. 1873) m. Harriett Mary (b. 1815 – d. 2 May 1881)

Children: Ann Elizabeth (b. about 1838 – d. 21 Mar. 1903)

Margaret (b. about 1839 – d. after 1907)

Richard James (b. about 1840 – d. after 1907) m. Julia Courtney of Albany, New York, on 25 Apr. 1871

Child: Anna

William (b. about 1842 – d. 7 July 1908) m. Maggie Feehan (b. 1862 – d. 3 Feb. 1887) on 4 Sept. 1883

George Denton (b. about 1843 – d. after 1907) m. Bella Dodd (b. 1848) on 1 July 1868

Children: Harriett (b. 1869)

Alma Bella (b. 14 Sept. 1877 – d. 24 July 1878)

Harriett Winnifred (b. 1844 – d. 23 Apr. 1868)

Hannah M. (b. about 1846 – d. after 1907) m. John S. Feehan on 14 July 1881, lived in Omaha, Nebraska in 1908

Child: Ita Catherine

Belleville newspapers record that Mary Northgraves married William Walsh of Montreal in June 1874. No other record of this person has been found.

APPENDIX 3

POST FAMILY

Jordan Post Jr. (b. 1767 – d. 1845) m. Melinda in 1807
  Children: Desdamona m. Joseph Milborne
        Children: Jordan Milborne
              William Milborne
      Saphronia m. Archibald McMillan
      Clarissa Maria did not marry and lived with her father.
      Jordan III m. Almira
        Children: Jordan IV
              Woodruff
      Woodruff who inherited his father's "verge alarm antiquarian clock."
      Malinda (or Mallinda) (d. before 1844) m. 30 Jan. 1834 Charles Cornell of Scarborough
        Children: Jordan Cornell
              Desdamona Cornell
              Maria Cornell
              Malinda Cornell
              Amy Cornell

All children and grandchildren received land on the death of Jordan Post Jr. Almira, wife of his son Jordan, was excluded from inheriting any property belonging to Jordan and was allowed one penny a day for seven years following his death.

In 1875–1880 Woodruff Post was a gilder in Toronto. By 1884 his son Sylvester Post inherited the house.

Other Posts mentioned in documents:

Ezekial Post, brother of Jordan Post Jr., applied for land in Scarborough Township in 1805. He was accompanied to Upper Canada by a wife and son under sixteen years of age. Another child was born in 1806.

George Washington Post (b. 1770 – d. 1828) was a younger brother of Jordan Post. He kept a tavern in Scarborough and later moved to Pickering Township where his tavern was about three miles west of Whitby.

Hiram Post – Trafalgar Township, Halton County, obtained a marriage bond on 15 March 1830.

Ephraim Post Jr., of Trafalgar – 13 October 1847 married Jane Mills of Halton County.

Christina Post of Trafalgar – 21 September 1853 married Austin Abby of Toronto.

APPENDIX 4

SAVAGE FAMILY

After the death of his first wife, Louisa Farr, George Savage Sr. married Maria Watters (b. 1771 – d. 22 Oct. 1866), who came with him to Canada. Because of the lack of birth dates for some of the children, the exact order of birth cannot be ascertained.

1. John – died unmarried in England.

2. Joseph (b. about 1899 – d. 6 Feb. 1859) was a jeweller and watchmaker in Montreal. His wife, Abigail (b. 1807 – d. mid-1860s), was the sister of Theodore Lyman, who eventually became a partner in the firm Savage and Lyman. Joseph was a partner with his father in the firm G. Savage & Son. In 1849, he signed an annexation manifesto advocating the annexation of Canada to the United States. After Joseph's death the family lived at 4 Beaver Hall Square.

3. Sarah (b. 1800 – d. 1832).

4. George (d. 17 Dec. 1851) was a watchmaker and jeweller in Montreal and Toronto (see p. 222). He married Sarah Hearn. There was one child.

5. David (b. between 1803 and 1807 – d. 8 May 1857) was a jeweller and watchmaker. He was in business in Montreal and Guelph (see p. 222). He married Ann Lawson (d. 31 Aug. 1891). There were nine children.

6. Mary (d. between 1834 and 1840) m. Joseph Farr. They had one child.

7. Thomas (b. 1808 – d. 3 Oct. 1872) Thomas was in New Orleans in 1834 and served in the Civil War in the U.S.A. He was in Montreal in 1841. He married Eliza Ann Davenport.

8. Joel (b. 1811 – d. 10 Dec. 1825)

9. Alfred (b. 1814 – d. 9 Jan. 1889) was a druggist in Montreal. He married Jane Donaldson (b. 1815 – d. 15 Sept. 1891) in 1870. There were six children.

APPENDIX 5

TRANSLATION OF AUSTIN TWISS LABEL FROM FRENCH

CLOCKS
MADE BY
AUSTIN TWISS
IN HIS CARDING MILL FACTORY AT COTE DES NEIGES, AT ONE LEAGUE
FROM MONTREAL.
WHERE ARE MADE ALL THE PARTS THAT CONSTITUTE
THESE KINDS OF CLOCKS.
Guaranteed correct, if put in a case and properly maintained.
DIRECTIONS FOR WINDING AND REGULATING THESE CLOCKS.

THE movement of the CLOCK in terms of the time it indicates, depends on the length of the pendulum; if the pendulum is too long, the clock runs too slowly; if it is too short the clock goes too fast; however if one turns by ⅛ inch the screw which is at the bottom, one can obtain a difference of 5 minutes per hour. If the hands are not correct, you can turn longest one ahead, but never backwards and this when the clock has only 15 minutes to go before it strikes: and you will never at any time turn backwards, and never go beyond XII number. If the clock does not strike the correct time, get it to ring all the hours until the clock is at the right time: and this is what we do in lowering by applying pressure on the little thread [wire?] on the side of the clock where the clock strikes.

Having given all these instructions, we must warn you that there is no clock that can last for many years with precision unless it is put in a case; as much as possible, a clock must be protected from dust. As these clocks are made of wood, this material is so good and movements so regularly made that they are absolutely immune to change of seasons; this is an advantage not found in copper clocks to which so much oil must be applied that they are likely to stop when exposed to the cold. Wooden clocks do not require oil, or so little: a small amount of olive oil or almond oil to be applied occasionally to the copper wheel.

The public is warned not to buy clocks on which the certificate does not appear.

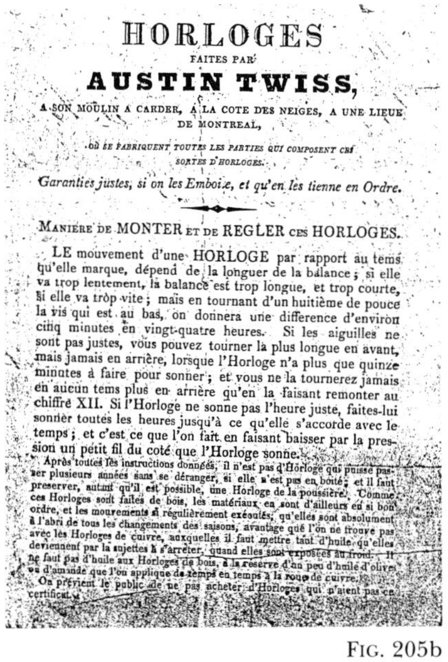

HORLOGES
FAITES PAR
AUSTIN TWISS,
A SON MOULIN A CARDER, A LA COTE DES NEIGES, A UNE LIEUE DE MONTREAL,
OÙ SE FABRIQUENT TOUTES LES PARTIES QUI COMPOSENT CES SORTES D'HORLOGES.
Garanties justes, si on les Emboi.e, et qu'en les tienne en Ordre.

MANIÈRE DE MONTER ET DE REGLER CES HORLOGES.

LE mouvement d'une HORLOGE par rapport au tems qu'elle marque, dépend de la longuer de la balance; si elle va trop lentement, la balance est trop longue, et trop courte, si elle va tròp vite; mais en tournant d'un huitième de pouce la vis qui est au bas, on donnera une difference d'environ cinq minutes en vingt-quatre heures. Si les aiguilles ne sont pas justes, vous pouvez tourner la plus longue en avant, mais jamais en arrière, lorsque l'Horloge n'a plus que quinze minutes à faire pour sonner; et vous ne la tournerez jamais en aucun tems plus en arrière qu'en la faisant remonter au chiffre XII. Si l'Horloge ne sonne pas l'heure juste, faites-lui sonner toutes les heures jusqu'à ce qu'elle s'accorde avec le temps; et c'est ce que l'on fait en faisant baisser par la pression un petit fil du coté que l'Horloge sonne.

Après toutes les instructions données, il n'est pas d'Horloge qui puisse passer plusieurs années sans se déranger, si elle n'est pas en boîte; et il faut preserver, autant qu'il est possible, une Horloge de la poussière. Comme ces Horloges sont faites de bois, les matériaux en sont d'ailleurs en si bon ordre, et les mouvements si régulièrement exécutés, qu'elles sont absolument à l'abri de tous les changements des saisons, avantage que l'on ne trouve pas avec les Horloges de cuivre, auxquelles il faut mettre tant d'huile qu'elles deviennent par la sujettes à s'arrêter, quand elles sont exposées au froid. Il ne faut pas d'huile aux Horloges de bois, à la réserve d'un peu d'huile d'olives ou d'amande que l'on applique de temps en temps à la roue de cuivre.

On prévient le public de ne pas acheter d'Horloges qui n'aient pas ce certificat.

FIG. 205b

FIG. 205b *Label from an Austin Twiss clock.*
– Courtesy the Vaudreuil Museum, Quebec

BILL OF SALE

28 December of the year 1830
Clocks sold to
Jerome[?] Lacoure[?] for
8 cords of elm and 8 pieces
of silver paid in the
course of a year
for Ira Twiss
on account.
.....sain Filion

CHAPTER ENDNOTES

CHAPTER 1

1. For more detail see *A Statutory Chronology of Ontario* by Thomas A. Hillman, published in Gananoque, Ontario, in 1988.

2. NAWCC *Bulletin,* Vol. XXVII, August 1985 "Unique Quebec Longcase Styling" by James E. Connell.

CHAPTER 2

1. Spoons exist bearing Agnew silver marks – "J.A." in cartouche with pseudo marks of Justin Spahnn and "J.A." between two "Lions Passant."

2. Dr. Taylor also pointed out that a later repairer of the movement has substituted a lady's hairpin for a count hook spring!

3. *Eli Terry and the Connecticut Shelf Clock* by Kenneth D. Roberts.

4. Ref. G.S.C. Ensko in *American Silversmiths and their Marks.*

5. *Eli Terry and the Connecticut Shelf Clock* by Kenneth D. Roberts.

6. *Nova Scotia Review,* Vol. 6, No. 2.

7. *A Seaport Legacy* by Paul O'Neill.

8. A ring and brooch made by Heber Earle in 1888 are in the possession of a grandson of Earle's daughter, Rose (Mrs. Spry).

9. Also, a third son, Franklin Germanus, came to Halifax with the family.

10. In 1788 he applied for and received 1,000 acres of land on the Musquodoboit River. It is not known to what purpose he put this land.

11. The complete inventory is shown later in this section.

12. See "The Town Clock of Halifax – 1803" by Kenneth Fram, NAWCC *Bulletin,* Vol. 28, No. 3 – June 1986.

13. Courtesy Environment Canada – Parks, Halifax Citadel National Historic Park.

14. Other children: James L. (b. 1811 – d. 14 Jan. 1834), Lisa (b. 1812 – d. Aug. 1831), George Jr. (b. 1818 – d. 6 July 1891).

15. For detail about the time ball and its function, see *The Beginning of the Long Dash* by Malcolm Thompson.

16. Most of the information used in this chapter is based on an article by W.R. Topham in the NAWCC *Bulletin,* Vol. 29 No. 249. August 1987.

17. See "Vox Temporis" in NAWCC *Bulletin,* Vol. 30, whole number 256 – October 1988.

18. Little & Elmer – "Chronometer makers of Bridgeton, New Jersey." See NAWCC *Bulletin,* Vol. XXIV, whole number 230 – June 1984.

19. See *Falmouth, a New England Township in Nova Scotia* by John V. Duncanson.

20. Mary (b. 2 May 1763) and Ann (b. 1765) were born after their parents' arrival in Nova Scotia.

21. *Silversmiths and Related Craftsmen of the Atlantic Provinces.*

22. Newfoundland joined Canada in 1949.

23. Hair of a deceased loved one was worked into a wreath or picture for posterity.

24. The Fairbanks shop in Saint John was two doors North of King Street.

25. The name Charlottetown was separated and hyphenated for many years.

CHAPTER 3

1. The concept of mass producing wood movements was not perfected until many years later by Eli Terry.

2. NAWCC *Bulletin,* Vol. XXIV, June 1982.

3. "Charly" was Prince Charles Edward Stuart or "Bonnie Prince Charlie," the last Stuart pretender to the British throne, a romantic figure in British history.

4. NAWCC *Bulletin,* Vol. XVIII, April 1976.

5. See NAWCC *Bulletin,* Vol. 34, Aug. 1992, p. 455.

6. *Revue d'ethnologie du Québec / 9.*

7. William Learmont was in business in Montreal from 1841 to 1868.

8. Quebec Census 1744.

9. According to the records of the Provincial Museum of Quebec.

10. Quoted in *Revue d'ethnologie du Québec / 9.*

11. See *The Canada and Hamilton Clock Companies* by Jane Varkaris and James E. Connell.

12. As noted below, major revisions of the St. Sulpice escapement were carried out during the second repair.

13. *Le Seminaire de St. Sulpice de Montreal* by J. Michaud and B. Harel.

14. See Appendix 4 for family.

15. For Savage silver marks see *Canadian Silversmiths 1700–1900* by John E. Langdon.

16. There were slight changes to the name of the company through the years.

17. For an English translation of the "Austin Twiss" label in Fig. 205b and bill of sale see Appendix 5.

18. This area of Côte des Neiges was annexed to Montreal around 1909.

19. NAWCC *Bulletin*, Vol. XIV, April 1970.

20. Renumbered 177 Notre Dame Street in 1854.

CHAPTER 4

1. See *Families* (Ontario Genealogical Society Publication), Vol.22, No. 1, 1983 for family.

2. As determined by a survey conducted in 1978 by J. Varkaris.

3. The first jewellery store was owned by David Butchen, in business from 1827 to 1831.

4. See Burrows *Canadian Clocks and Clockmakers*, p. 96.

5. See NAWCC *Bulletin*, Vol. 29, Whole No. 248, June 1987.

6. Now Toronto.

7. See Appendix 1.

8. More details about the life and family of Nathan F. Dupuis may be found in *Nathan Fellowes Dupuis, Professor and Clockmaker of Queen's University* by Jane and Costas Varkaris.

9. No relation to the author.

10. Cemetery records conflict with date carved on stone.

11. See "The Man who Synchronized Time" by W.R. Topham, NAWCC *Bulletin*, Vol. XXV, Oct. 1983.

12. See "The Man who Synchronized Time" by W.R. Topham, NAWCC *Bulletin*, Vol. XXV, Oct. 1983.

13. In the State of Michigan there were twenty-seven different times, most of them established by local jewellers.

14. Universal Time was not passed into law in the U.S.A. until 1918.

15. From survey conducted by J. Varkaris in 1978.

16. Not to be confused with Christian Hess the potter, who was born about 1835.

17. See NAWCC *Bulletin*, Vol. 33 Whole No. 1, Feb. 1991.

18. *The Mercantile Agency Reference Book* etc., Dun, Wiman & Co. (Canada, July 1884).

19. By 1891 Easthope South Township was in Oxford County.

20. For details about this company see *The Canada and Hamilton Clock Companies* by Jane Varkaris and James E. Connell.

21. For known Northgraves family members see Appendix 2.

22. Information about the Olmsted and Wright families is available from J. Varkaris.

23. Records of the Port of Kingston reveal that a "J. Patterson" imported several hundred clocks between May and August 1846.

24. For detailed account of the family and their businesses see *The Pequegnat Story, the Family and the Clocks* by Jane and Costas Varkaris.

25. Records for 1842, 1847, and 1848 are missing.

26. Records also show the name spelled Solomina.

27. Reference – *Directory of the Ancestral Heads of New England Families 1620–1700*, Compiled by Frank R. Holmes.

28. See Appendix 3 for family.

29. Courtesy Gibson House Museum, Willowdale, North York.

30. Former curator, American Watch and Clock Museum, Bristol, CT, U.S.A.

31. Jordan Post was compensated £75 for losses he suffered when the enemy plundered his property.

32. As established by a 1978 survey of clocks in Canada conducted by J.Varkaris.

33. Further information about the sons may be found in the *Leeds and Grenville News* – the branch newsletter of the Ontario Genealogical Society, Toronto, Ontario, 1989.

34. A sundial, erected about 1827 by Colonel John By on Parliament Hill overlooking the canal entrance although older, is not accurate.

35. *Ontario Historical Society Papers & Records,* Vol. XXIII, 1926.

36. The reference to County of York is confusing. Bismark is definitely located in Gainsboro Township, Lincoln County, and Fenwick is not far away. York County boundaries never extended into the Niagara peninsula.

37. See "The Long Point Furnace" by W.J. Patterson, *Ontario Historical Society Papers and Records* Vol. 36.

38. See *The Van Tassels and Allied Lines* by Mary Pazurik.

39. A brother, Alexander, went to London, Ontario.

40. Now part of Halton Hills.

41. The other maker was Silas Hoadley of Plymouth, Connecticut.

42. Reprints of patents are available from the Dept. of Consumer and Corporate Affairs, Canada.

43. Thomas Sargant was a general agent and appraiser in Toronto from 1889 to 1893.

44. See "The Willcock Clock" by Kirk Fallin, *NAWCC Bulletin,* Vol. 29, Whole No. 248, June 1987.

45. Now University of Guelph.

46. Kirk Fallin.

47. See Monograph, *Trumbull County Clock Industry, 1812–1835* by Rebecca M. Rogers, presented at the 1991 Seminar of the National Association of Watch and Clock Collectors.

SOURCES OF INFORMATION

Algonquin College, Lanark Campus, Perth, Ontario
American Clock & Watch Museum, Bristol, Connecticut.
Art Gallery of Ontario, Toronto, Ontario
Belleville Public Library
Bennington Museum, Vermont
Bytown Museum, Ottawa, Ontario
Canadian Museum of Civilization, Hull, Quebec
Charlotte County Historical Society, Prince Edward Island.
Chateau du Ramezay, Montreal, Quebec
Connecticut Historical Society
Cullen Gardens, Whitby, Ontario
Detroit Public Library
Dundas Historical Society Museum
Dundurn Castle, Hamilton, Ontario
Eva Brook Donly Museum, Simcoe, Ontario
Georgetown Public Library
Gibson House Museum, North York, Ontario
Haskell Free Library, Derby Line, Vermont
Henry Ford Museum, Dearborn, Michigan
Illinois State Historical Society & Library
Laurier House Museum, Ottawa, Ontario
Lundy's Lane Historical Museum, Niagara Falls, Ontario
Macdonald House Museum, Kingston, Ontario
Musée de la Civilisation, Quebec City
National Archives of Canada, Ottawa, Ontario
National Archives of Quebec, Quebec City
National Library of Canada, Ottawa, Ontario
National Museum of American History (Smithsonian Institution), Washington, D.C.

National Association of Watch and Clock Collectors – Library and NAWCC Bulletins, Columbia, Pennsylvania
Niagara Historical Museum, Niagara-on-the-Lake, Ontario
Newfoundland Historical Society, St. John's
Nova Scotia Museum, Halifax, Nova Scotia
Perth Public Library
Point Ellis House Museum, Victoria, British Columbia
Prince Edward Island Museum and Heritage Foundation, Charlottetown
Provincial Archives of Nova Scotia, Halifax
Province of New Brunswick Museum, Saint John
Province of Ontario Archives, Toronto, Ontario
Province of Quebec Museum, Quebec City
Public Archives of New Brunswick, Fredericton
Queen's University, Kingston, Ontario
Royal Ontario Museum, Toronto, Ontario
Seminaire de Saint Sulpice, Montreal, Quebec
Simcoe Public Library
Stanstead Historical Society
Toronto Trust Cemeteries
Tweedsmuir Histories
University of Toronto, Thomas Fisher Rare Book Library
Upper Thames River Conservation Authority, London, Ontario
Vermont Historical Society
Wolfville Historical Society, Nova Scotia
York-Sunbury Museum, Fredericton, New Brunswick

BIBLIOGRAPHY

Badey, T.M. *Dictionary of Hamilton Biography*. Hamilton, Ontario, 1981.

Bailey, Chris. *Two Hundred Years of American Clocks and Watches*. Englewood Cliffs, New Jersey, 1975.

Baillie, G.H. *Watchmakers and Clockmakers of the World*. London, England, 1969 Edition.

Beers J.H. *Commemorative Biographical Record of the County of York*. 1907.

Bonis, Robert R. *A History of Scarborough*. Scarborough, Ontario, 1968.

Bosworth, Newton. *Hochelaga Depicta*. Montreal, Quebec, 1839.

Broad Brook Grange No. 151. *Official History of Guilford, Vermont*. Guilford, Vermont, 1961.

Brown, Lewis. *History of Simcoe:1829–1929*. Simcoe, Ontario, 1929.

Burrows, G. Edmond. *Canadian Clocks and Clockmakers*. Oshawa, Ontario, 1973.

Chadwick, Edward M. *Ontarian Families, Vols. 1 & 2*. Toronto, Ontario, 1898.

Cochrane, Rev. Wm. *The Canadian Album*. Brantford, Ontario, 1891- 1896.

Consolidated Illustrating Co. *Toronto Illustrated*. Toronto, Ontario, 1893.

Curtis, Corinne R. *Rooms*. New York City, 1979.

de Volpi, Charles P. *Toronto, a Pictorial Record*. Montreal, Quebec, 1963.

_____ *Montreal, a Pictorial Record*. Montreal, Quebec, 1965.

Drepperd, Carl W. *American Clocks and Clock Makers*. Garden City, New York, 1947.

Drouin, Gabriel. *Dictionnaire National des Canadiens Français*. Montreal, Québec, 1979.

Duncanson, John V. *Falmouth a New England Township in Nova Scotia*. Windsor, Nova Scotia, 1965.

_____ *Township of Falmouth, Nova Scotia*. Belleville, Ontario, 1983.

Ela, Chipman P. *The Banjo Timepiece*. Lexington, Massachusetts, 1978.

Ensko, G.S.C. *American Silversmiths and their Marks*. New York City, 1983.

Evans, Patrick M.O. *The Wrights*. Ottawa, Ontario, 1978.

Firth, Edith. *The Town of York, 1793–1815*. Toronto, Ontario, 1962.

Fisher, Sidney T. *Merchant Millers of the Humber Valley*. Toronto, Ontario, 1985.

Foss, Charles H. & Vroom Richard. *Cabinetmakers of the Eastern Seaboard*. Toronto, Ontario, 1977.

Gilroy, Marion. *Loyalists and Land Settlement in Nova Scotia*. Halifax, Nova Scotia, 1937.

Gubby, Aline. *The Mountain and the River*. Montreal, Quebec, 1981.

Guillet, Edwin C. *Toronto: from Trading Post to Great City*. Toronto, Ontario, 1934.

Haliburton, Thomas C. *A Historical and Statistical Account of Nova Scotia*. Halifax, Nova Scotia, 1829.

_____ *History of Nova Scotia Vols. I and II*. Belleville, Ont. 1973.

_____ *The Clockmaker or the Sayings and Doings of Samuel Slick of Slickville*. Toronto, 1836.

Hannay, J. & Reynolds, W.K. *St. John and its Business 1875*. Saint John, New Brunswick, 1875.

Hart, Patricia W. *Pioneering in North York*. Toronto, Ontario, 1968.

Hathaway, E.J. *Jesse Ketchum and his Times*. Toronto, Ontario, 1929.

Hector Central Trust. *Nineteenth Century Pictou County Furniture*. Halifax, Nova Scotia, 1977.

Hemenway, Abby Maria. *Vermont Historical Gazetteer, Vol. V*. Brandon, Vermont, 1891.

Hillman, Thomas A. *A Statutory Chronology of Ontario*. Gananoque, Ontario, 1985.

Holmes, Frank R. *Directory of the Ancestral Heads of New England Families 1620–1700*. New York City, 1923.

Hoopes, Penrose R. *Connecticut Clockmakers of the Eighteenth Century*. Rutland, Vermont, reissued 1975.

Langdon, John E. *Canadian Silversmiths and their Marks 1667–1867*. Lunenburgh, Vermont, 1960.

_____ *American Silversmiths of British North America*. Toronto, Ontario, 1970.

_____ *Canadian Silversmiths 1700–1900*. Toronto, Ontario, 1966.

_____ *Clock and Watchmakers in Canada 1700–1900*. Toronto, 1976.

_____ *Guide to Marks On Early Canadian Silver*. Toronto, Ontario, 1968.

Lawrence, J.W. *Foot Prints*. Saint John, New Brunswick, 1883.

Leavitt, Thaddeus W.H. *History of Leeds and Grenville*. Brockville, Ontario, 1897.

_____ *Illustrated Historical Atlas of Leeds and Grenville*. Kingston, Ontario, 1862.

Lockwood, Glenn. *Kitley, 1795–1975*. Prescott, Ontario, 1974.

Loomis, B. *Watchmakers and Clockmakers of the World, Vol. 2*. London, England, 1976.

Mackay, Donald C. *Silversmiths and Related Craftsmen of the Atlantic Provinces*. Halifax, Nova Scotia, 1973.

MacLaren, George. *Antique Furniture by Nova Scotia Craftsmen*. Toronto, Ontario, 1961.

_____*Nova Scotia Furniture*. Halifax, Nova Scotia, 1969.

MacLeod, K.O. *The First Century* (Birks). Montreal, Quebec, 1979.

MacPhie, J.P. *Pictonians at Home and Abroad.* Boston, Massachusetts, 1914.

McKenzie, Donald. *Death Notices from the Christian Guardian.* Lambertville, New Jersey, 1984.

_____ *Obituaries from Ontario's Christian Guardian.* Lambertville, New Jersey, 1988.

McKenzie, Ruth. *Leeds and Grenville: their first Two Hundred Years.* Toronto and Montreal, 1967.

Mellick, Andrew D. *The Story of an Old Farm.* Somerville, New Jersey, 1889.

Michaud, Josette & Harel, Bruno. *Le Séminaire de Saint-Sulpice de Montréal.*, Québec, 1990.

Middleton, Jesse E. *Municipality of Toronto – A History.* Toronto, Ontario, 1923.

Miller, R.S. *Misi Gete.* Australia, *1975.*

Morgan, E.P. & Harvey, F.L. *Hamilton and Its Industries.* Hamilton, Ontario, 1884.

Mosser, Christine (Ed). *York Upper Canada Minutes of Town Meetings and Lists of Inhabitants.* Toronto, Ontario, 1984.

O'Neill, Paul. *A Seaport Legacy.* Erin, Ontario. 1976.

Osborne, S. & Swainson, D. *Kingston, Building on the Past.* Westport, Ontario, 1988.

Pain, Howard. *The Heritage of Upper Canadian Furniture.* Toronto, Ontario, 1978.

Palardy, Jean. *The Early Furniture of French Canada.* Toronto, Ontario, 1963.

Palmer, Brooks. *The Book of American Clocks.* New York City, 1967.

Patterson, George. *Missionary Life Among the Cannibals.* Toronto, Ontario, 1882.

Pazurik, Mary M. *The Van Tassels and Allied Lines – 1574–1974.* Gibsons, British Columbia, 1974.

Pierce, Frederick C. *Field Genealogy Vol 2.* Chicago, Illinois, 1901.

Piers, Harry & Mackay, Donald. *Master Goldsmiths and Silversmiths of Nova Scotia and their Marks.* Halifax, Nova Scotia, 1948.

Post, James. *Post Family History.* Unpublished.

Reaman, G.Elmore. *A History of Vaughan Township.* Toronto, Ontario, 1971.

Reid, W.D. *Marriage Notices of Ontario.* Lambertville, New Jersey, 1980.

Roberts, Kenneth D. *Eli Terry and the Connecticut Shelf Clock.* Bristol, Connecticut, 1973.

_____ *The Contributions of Joseph Ives to Connecticut Clock Technology 1810–1862.* Bristol, Connecticut, 1970 & 1988.

Robertson, J.R. *Robertson's Landmarks of Toronto.* Toronto, Ontario, 1894–1914.

Robinson, C.B. *History of Toronto and County of York.* Toronto, Ontario, 1885.

Rogers, Joseph S. *Rogers' Photographic Advertising Album.* Halifax, Nova Scotia, 1871.

Ryder, Huia G. *Antique Furniture by New Brunswick Craftsmen.* Toronto, Ontario, 1965.

Scadding, Henry. *Toronto of Old.* Toronto, Ontario, 1873.

Schmeisser, Barbara *Town Clock 1803–1860.* Halifax, Nova Scotia, 1984.

Séguin, Robert L. *Revue d'ethnologie du Québec / 9.* Ottawa, Ontario, 1979.

Shackleton, Philip. *The Furniture of Old Ontario.* Toronto, Ontario, 1973.

Stevens, Gerald. *In a Canadian Attic.* Toronto, Ontario, 1955.

Stevens, John Grier. *The Descendants of John Grier with Histories of Allied Families.* Baltimore, Maryland, 1964.

Tanguay, L'Abbe Cyprien. *Dictionnaire Genealogique des Familles Canadiennes.* New York City, 1871–1890.

Thompson, Malcolm M. *The Beginning of the Long Dash.* Toronto, Ontario, 1978.

Thompson, Robert. *A Brief Sketch of the Early History of Guelph.* Guelph, Ontario, 1977.

Throop, Herbert. *Throop Genealogy.* Ottawa, Ontario, 1931.

Throop, W.F. & E.B. *The Throop Tree.* La Mirada, California, 1971.

Traquair, Ramsay. *The Old Silver of Quebec.* Toronto, Ontario, 1940.

Varkaris, J. & C. *Nathan Fellowes Dupuis.* Toronto, Ontario, 1980.

_____ *The Pequegnat Story, the Family and the Clocks.* Dubuque, Iowa, 1982.

Varkaris, J. & Connell, James E. *The Canada and Hamilton Clock Companies.* Erin, Ontario, 1986.

Webster, Donald Blake. *English-Canadian Furniture of the Georgian Period.* Toronto, Ontario, 1979.

Webster, Donald Blake, ed. *The Book of Canadian Antiques.* Toronto, Ontario, 1974.

West, Bruce. *Toronto.* New York City, 1967.

White, G.A. *Halifax and its Business.* Halifax, Nova Scotia, 1876.

Williams, David M. *Montreal, 1850–1870.* Toronto, Ontario, 1971.

Wilson, Thomas B. *Ontario Marriage Notices.* Lambertville, New Jersey, 1982.

Woodhouse, T. Roy. *The History of Dundas.* Dundas, Ontario, 1947.

_____ *A Short History of Dundas.* Dundas, Ontario, 1947.

Yeager, William. *The Cabinetmakers of Norfolk County.* Simcoe, Ontario, 1975 (and revised edition, 1976).

(No Author) *Dictionary of Canadian Biography,* 1966–1990.